Fast and Effective Embedded Systems Design

Please visit the Companion Website to

Fast and Effective Embedded Systems Design

www.embeddedacademic.com

The companion website contains instructor support, program examples, errata, updates on parts and answers to frequently asked questions.

Fast and Effective Embedded Systems Design

Applying the ARM mbed

Rob Toulson
Tim Wilmshurst

AMSTERDAM • BOSTON • HEIDELBERG • LONDON • NEW YORK • OXFORD
PARIS • SAN DIEGO • SAN FRANCISCO • SINGAPORE • SYDNEY • TOKYO
Newnes is an imprint of Elsevier

Newnes is an imprint of Elsevier
The Boulevard, Langford Lane, Kidlington, Oxford, OX5 1GB
225 Wyman Street, Waltham, MA 02451, USA

First published 2012

British Library Cataloging-in-Publication Data
A catalogue record for this book is available from the British Library

Library of Congress Cataloging-in-Publication Data
A catalog record for this book is available from the Library of Congress

ISBN: 978-0-08-097768-3

For information on all Newnes publications
visit our website at **store.elsevier.com**

Transferred to Digital Printing in 2015

Contents

Introduction

Microprocessors are everywhere, providing 'intelligence' in cars, mobile phones, household and office equipment, televisions and entertainment systems, medical products, aircraft – the list seems endless. Those everyday products, where a microprocessor is hidden inside to add intelligence, are called *embedded systems*.

Not so long ago, designers of embedded systems had to be electronics experts, software experts, or both. Now, with user-friendly and sophisticated building blocks available for our use, both the specialist and the beginner can quickly engage in successful embedded system design. One such building block is the *mbed*, recently launched by the renowned computer giant ARM. The mbed is the central theme of this book; through it all the main topics of embedded system design are introduced. The aim of the book is to teach the key elements of embedded system design through the use of the mbed.

This book falls into two parts. Chapters 1–10 provide a wide-ranging introduction to embedded systems, using the mbed and demonstrating how it can be applied to rapidly produce successful embedded designs. These chapters aim to give full support to the reader, moving you through a series of carefully constructed concepts and exercises. They start from basic principles and simple projects, and move towards more advanced system design. Chapters 11–15 build on this foundation, but enter a number of more specialist fields. The pace here may be a little faster, and you may find that you need to contribute more background research.

All this book asks of you at the very beginning is a basic grasp of electrical/electronic theory. The book adopts a 'learning through doing' approach. To make best use of it you will need an mbed, an Internet-connected computer and the various additional electronic components identified through the book. You will not need every single one of these if you choose not to do a certain experiment or book section. You will also need a digital voltmeter and ideally access to an oscilloscope.

Each chapter is based around a major topic in embedded systems. Each has some theoretical introduction, and may have more theory within the chapter. The chapter then proceeds as a series of practical experiments. Have your mbed ready to hook up the next circuit, and download and compile the next example program. Run the program, and aim to understand what is going on. As your mbed confidence grows, so will your creativity and originality; start to turn your own ideas into working projects.

You will find that this book rapidly helps you:

- to understand and apply the key aspects of embedded systems
- to understand and apply the key aspects of the ARM mbed
- to learn from scratch, or develop your skills in embedded C/C++ programming
- to develop your understanding of electronic components and configurations
- to understand how the mbed can be applied in some of the most exciting and innovative intelligent products emerging today
- to produce designs and innovations you never thought you were capable of!

If you get stuck, or have questions, support is available for all readers through the book website and the mbed website, or by email discussion with the authors.

If you are a university or college instructor, then this book offers a 'complete solution' to your embedded systems course. Both authors are experienced university lecturers and had your students in mind when writing this book. The book contains a structured sequence of practical and theoretical learning activity. Ideally, you should equip every student or student pair with an mbed, a prototyping breadboard and a component kit. These are highly portable, so development work is not confined to the college or university laboratory. Later in the course, students will start networking their mbeds together. Complete Microsoft PowerPoint presentations for each chapter are available to instructors via the book website, as well as answers to Quiz questions, and example solution code for the exercises and mini-projects.

Part 1 (Chapters 1–10) of this book can be used to provide a complete introductory course in embedded system design, with practical examples using the mbed. Part 2 (Chapters 11–15) can be used to enhance the introductory course, as the basis of further reading or for a more advanced course.

This book is appropriate for any course or module that wants to introduce the concepts of embedded systems, in a hands-on and interesting way. Because the need for electronic theory is limited, the book is accessible to those disciplines that would not normally aim to take embedded systems. The book is meant to be accessible to Year 1 undergraduates, although we expect it will more often be used in the years that follow. Students are likely to be studying one of the branches of engineering, physics or computer science. The book will also be of interest to the practicing professional and the hobbyist.

This book is written by Rob Toulson, Research Fellow at Anglia Ruskin University in Cambridge, and Tim Wilmshurst, Head of Electronics at the University of Derby. After completing his PhD, Rob spent a number of years in industry, where he worked on digital signal processing and control systems engineering projects, predominantly in audio and automotive fields. He then moved to an academic career, where his main focus is now in developing collaborative research between the technical and creative industries. Tim led the Electronics Development Group in the Engineering Department of Cambridge University for a number of years, before moving to Derby. His design career has spanned much of the history of microcontrollers and embedded systems. Aside from our shared interest in embedded systems, we share an interest in music, and music technology. The book brings together our wide range of experiences. Having planned the general layout of the book, and after some initial work, we divided the chapters between us. Tim took responsibility for most of the early ones, and issues relating to electronic and computing hardware. Rob worked mainly on the later chapters, and the more advanced mbed applications. This division of labor was mainly for convenience, and at publication we both take responsibility for all the chapters. Because Tim has written several books on embedded systems in the past, a few background sections and diagrams have been taken from these and adapted for inclusion in this book. There seemed no point in 'reinventing the wheel' where background explanations were needed.

Acknowledgments

Throughout the writing of this book we have been grateful for the support of the ARM staff who are creators of the mbed project, notably Simon Ford and Chris Styles. Supporting experimental work and proof-reading have been conducted by Phillip Richardson at ARU and Bheki Nyathi at Derby, to whom we are most grateful. A number of smart and innovative people have contributed libraries of code to the mbed online cookbook; we have referred to some of these in the book. Sincere thanks to those people. We also express our thanks to those who contribute to the discussion strands on the mbed site. We hope this book responds to some of your comments and questions. Among these we thank Martin Smith for his account of prototyping an mbed-based project, mentioned in Chapter 15.

CHAPTER 1

Embedded Systems, Microcontrollers and ARM

Chapter Outline

1.1 Introducing Embedded Systems

1.1.1 What is an Embedded System?

We are all familiar with the idea of a desktop or laptop computer, and the amazing processing that they can do. These computers are general purpose; we can get them to do different things at different times, depending on the application or program we run on them. At the very heart of such computers is a *microprocessor*, a tiny and fantastically complicated electronic circuit which contains the core features of a computer. All of this is fabricated on a single slice of silicon, called an *integrated circuit* (IC). Some people, particularly those who are not engineers themselves, call these circuits *microchips*, or just *chips*.

Fast and Effective Embedded Systems Design. DOI: 10.1016/B978-0-08-097768-3.00001-5

What is less familiar to many people is the idea that instead of putting a microprocessor in a general-purpose computer, it can also be placed inside a product which has nothing to do with computing, like a washing machine, toaster or camera. The microprocessor is then customized to control that product. The computer is there, inside the product; but it cannot be seen, and the user probably does not even know it is there. Moreover, those add-ons that are normally associated with a computer, like keyboard, screen or mouse, are nowhere to be seen. Such products are called *embedded systems*, because the computer that controls them is embedded right inside. Because they tend to focus on control, in many cases the microprocessors used in embedded systems develop different characteristics from the ones used in more general-purpose machines. They are called *microcontrollers.* Although much less visible than their microprocessor cousins, microcontrollers sell in far greater volume and their impact has been enormous. To the electronic and system designer they offer huge opportunities.

Embedded systems come in many forms and guises. They are extremely common in the home, the motor vehicle and the workplace. Most modern domestic appliances, like washing machines, dishwashers, ovens, central heating and burglar alarms, are embedded systems. The motor car is full of them, in engine management, security (for example, locking and anti-theft devices), air-conditioning, brakes, radio, and so on. They are found across industry and commerce, in machine control, factory automation, robotics, electronic commerce and office equipment. The list has almost no end, and it continues to grow.

Figure 1.1 expresses the embedded system as a simple block diagram. There is a set of inputs from the controlled system. The embedded computer, usually a microcontroller, runs

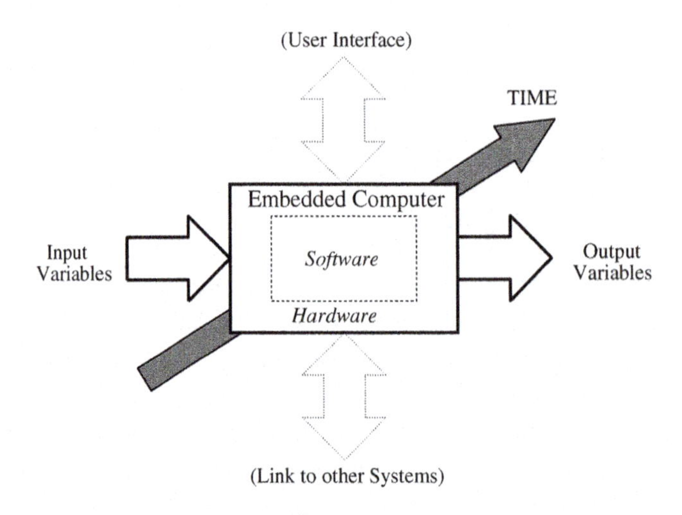

Figure 1.1:
The embedded system

a program dedicated to this application, permanently stored in its memory. Unlike the general-purpose desktop computer, which runs many programs, this is the only program it ever runs. Based on information supplied from the inputs, the microcontroller computes certain outputs, which are connected to things like actuators within the system. The actual electronic circuit, along with any electromechanical components, is often called the *hardware*; the program running on it is often called the *software*. Aside from all of this, there may also be interaction with a user, for example via keypad and display, and there may be interaction with other subsystems elsewhere, although neither of these is essential to the general concept. One other variable will affect all that we do in embedded systems, and this is time, represented as a dominating arrow which cuts across the figure. We will need to be able to measure time, make things happen at precisely predetermined times, generate data streams or other signals with a strong time dependence, and respond to unexpected things in a timely fashion.

This chapter introduces or reviews many concepts relating to computers, microprocessors, microcontrollers and embedded systems. It does this in overview form, to give a platform for further learning. We return to most concepts in later chapters, building on them and adding detail. More details can also be found in Reference 1.1.

1.1.2 An Example Embedded System

A snack vending machine is a good example of an embedded system. One is represented in Figure 1.2, in block diagram form. At its heart is a single microcontroller. As the diagram shows, this accepts a number of input signals, from the user keypad, the coin-counting module and the dispensing mechanism itself. It generates output signals dependent on those inputs.

A hungry customer may approach the machine and start jabbing at the buttons or feeding in coins. In the first case, the keypad sends signals back to the microcontroller, so that it can recognize individual keys pressed and then hook these key values together to decipher a more complex message. The coin-counting module will also send information on the amount of money paid. The microcontroller will attempt to make deductions from the information it receives and will output status information on the liquid crystal display. Has a valid product been selected? Has enough money been paid? If yes, then it will energize an actuator to dispense the product. If no, it will display a message asking for more money or a re-entry of the product code. A good machine will be ready to give change. If it does dispense, there will be sensors on the mechanism to ensure that the product is available, and finally that the action is complete. These are shown in the diagram as gate position sensors. A bad machine (and haven't we all met these?) will give annoying or useless messages on its display, demand more money when you know you've given the right amount, or leave that chocolate bar you desperately want teetering on a knife-edge, not dropping down into the dispensing tray.

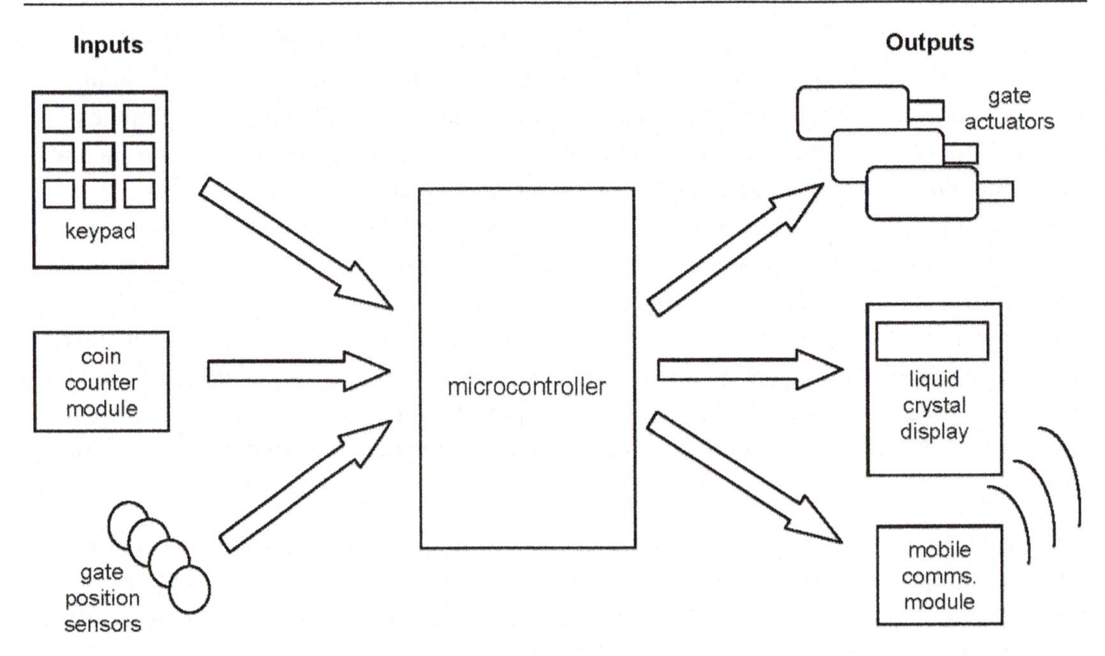

Figure 1.2:
Vending machine embedded system

What has been described so far is a fairly conventional machine. We can, however, take it further. In modern vending systems there might be a mobile communications feature which allows the vending machine to report directly to the maintenance team if a fault is diagnosed; they will then come and fix the fault. Similarly, the machine might report stock levels through mobile or Internet communications to allow the service team to visit and replenish whenever supplies run low.

This simple example reflects exactly the diagram of Figure 1.1. The microcontroller accepts input variables, makes calculations and decisions with this, and generates outputs in response. It does this in a timely manner, and correct use of time is implicit in these actions. There is in this case a user interface, and in a modern machine there is a network interface. While the above paragraphs seem to be describing the physical hardware of the system, in fact it is all controlled by the software that the designer has written. This runs in the microcontroller and determines what the system actually does.

1.2 Microprocessors and Microcontrollers

Let us go back to the microcontroller, which sits at the heart of any embedded system. As the microcontroller is in essence a type of computer, it will be useful for us to get a grasp on basic computer details. We do this in complete overview here, but return to some of these features in later chapters.

1.2.1 Some Computer Essentials

Figure 1.3 shows the essential elements of any computer system. As its very purpose for existence, a computer can perform arithmetic or logical calculations. It does this in a digital electronic circuit called the arithmetic logic unit (ALU). The ALU is placed within a larger circuit, called the central processing unit (CPU), which provides some of the supporting features that it needs. The ALU can undertake a number of simple arithmetic and logic calculations. Which one it does depends on a code which is fed to it, called an instruction. Now we are getting closer to what a computer is all about. If we can keep the ALU busy, by feeding it a sensible sequence of instructions, and also pass it the data it needs to work on, then we have the makings of a very useful machine indeed.

The ability to keep feeding the ALU with instructions and data is provided by the control circuit which sits around it. It is worth noting that any one of these instructions performs a very simple function. However, because the typical computer runs so incredibly fast, the overall effect is one of very great computational power. The series of instructions is called a program, normally held in an area of memory called program memory. This memory needs to be permanent. If it is, then the program is retained indefinitely, whether power is applied or not, and it is ready to run as soon as power is applied. Memory like this, which keeps its contents when power is removed, is called non-volatile memory. The old-fashioned name for this is read only memory (ROM). This latter terminology is still sometimes used, even though it is no longer accurate for new memory technology. The control circuit needs to keep accessing the program memory, to find out what the next instruction is. The data that the ALU works on may be drawn from the data memory, with the result placed there after the calculation is complete. Usually this is temporary data. This memory type therefore need not be permanent, although there is no harm if it is. Memory that loses its contents when power is removed is called volatile memory. The old-fashioned name for this type of memory is

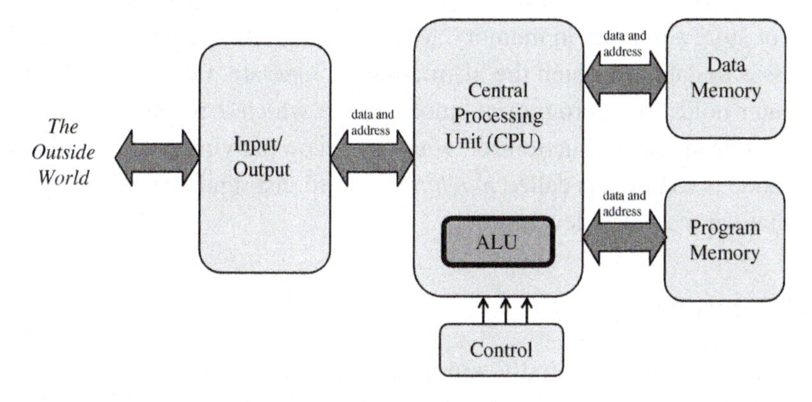

Figure 1.3:
Essentials of a computer

random access memory (RAM). This terminology is still used, although it conveys little useful information.

To be of any use, the computer must be able to communicate with the outside world, and it does this through its input/output (I/O). On a personal computer this implies human interaction, through things like the keyboard, visual display unit (VDU) and printer. In an embedded system, at least a simple one, the communication is likely to be primarily with the physical world around it, through sensors and actuators. Data coming in from the outside world might be quickly transferred to the ALU for processing, or it might be stored in data memory. Data being sent out to the outside world is likely to be the result of a recent calculation in the ALU.

Finally, there must be data paths between each of these main blocks, as shown by the block arrows in the diagram. These are collections of wires, which carry digital information in either direction. One set of wires carries the data itself, for example from program memory to the CPU; this is called the *data bus*. The other set of wires carries address information and is called the *address bus*. The address is a digital number that indicates which place in memory the data should be stored in, or retrieved from. The wires in each of the data and address buses could be tracks on a printed circuit board or interconnections within an IC.

One of the defining features of any computer is the size of its ALU. Simple, old processors were 8 bit, and some of that size still have useful roles to play. This means that, with their 8 bits, they can represent a number between 0 and 255. (Check Appendix A if you are unfamiliar with binary numbers.) More recent machines are 32 or 64 bit. This gives them far greater processing power, but adds to their complexity. Given an ALU size, it generally follows that many other features take the same size, for example memory locations, data bus, and so on.

As already suggested, the CPU has an *instruction set*, which is a set of binary codes that it can recognize and respond to. For example, certain instructions will require it to add or subtract two numbers, or store a number in memory. Many instructions must also be accompanied by data, or addresses to data, on which the instruction can operate. Fundamentally, the program that the computer holds in its program memory and to which it responds is a list of instructions taken from the instruction set, with any accompanying data or addresses that are needed. Such code is sometimes called *machine code*, to distinguish it from other versions of the program that we may also develop.

1.2.2 The Microcontroller

A microcontroller takes the essential features of a computer as just described, and adds to these the features that are needed for it to perform its control functions. It is useful to think of it as being made up of three parts – core, memory and peripherals – as shown in the block

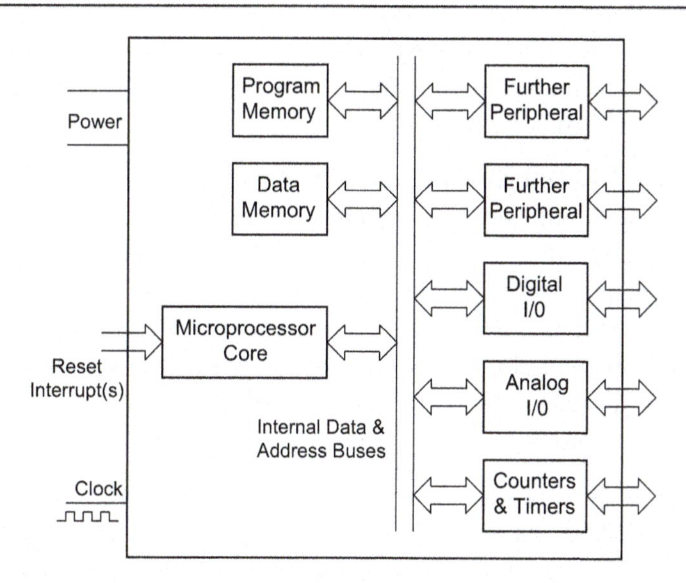

Figure 1.4:
Features of an example microcontroller: core + memory + peripherals

diagram of Figure 1.4. The core is the CPU and its control circuitry. Alongside this go the program and data memory. Finally, there are the peripherals. These are the elements that distinguish a microcontroller from a microprocessor, for they are the elements that allow the wide-ranging interaction with the outside world that the microcontroller needs. Peripherals can include digital or analog input/output, serial ports, timers, counters and many other useful subsystems.

It almost goes without saying, but must not be forgotten, that the microcontroller needs power, in the form of a stable direct current (DC) supply. A further requirement, as with any personal computer (PC), is to have a *clock* signal. This is the signal that endlessly steps the microcontroller circuit through the sequence of actions that its program dictates. The clock is often derived from a quartz oscillator, such as you may have in your wristwatch. This gives a stable and reliable clock frequency. The clock often has a very important secondary function, of providing essential timing information for the activities of the microcontroller, for example in timing the events it initiates, or in controlling the timing of serial data.

1.3 Development Processes in Embedded Systems

1.3.1 Programming Languages: What is so Special about C/C++?

The CPU instruction set was mentioned in Section 1.2.1. As programmers, it is our ultimate goal to produce a program that makes the microcontroller do what we want it to do; that program must be a listing of instructions, in binary, drawn from the instruction set. The program in this raw binary form is what we have already called machine code. It is extremely tedious, to the point of being near impossible, for us as humans to work with the binary numbers that form the instruction set.

A first step towards sanity in programming is called Assembler. In Assembler each instruction gets its own *mnemonic*, a little word which a human can remember and work with. The instruction set is represented by a set of mnemonics. The program is then written in these mnemonics, and a computer program called a cross-assembler converts all those mnemonics into the actual binary code that is loaded into program memory. Assembler has its uses, for one it allows us to work very closely with the CPU capabilities; Assembler programs can therefore be very fast and efficient. Yet it is easy to make mistakes in Assembler and hard to find them, and the programming process is time consuming.

A further step to programming sanity is to use a *high-level language* (HLL). In this case we use a language like C, Java or Fortran to write the program, following the rules of that language. Then a computer program called a compiler reads the program we have written, and converts (compiles) that into a listing of instructions from the instruction set. This assumes we have made no mistakes in writing the program! This list of instructions becomes the binary code that we can download to the program memory. Our needs in the embedded world, however, are not like those of other programmers. We want to be able to control the hardware we are working with, and write programs that execute quickly, in predictable times. Not all HLLs are equally good at meeting these requirements. People get very excited, and debate endlessly, about what is the best language in the embedded environment. However, most agree that, for many applications, C has clear advantages. This is because it is simple and has features that allow us to get at the hardware when required. A step up from C is C++, which is also widely used for more advanced embedded applications.

1.3.2 The Development Cycle

This book is about developing embedded systems, and developing them quickly and reliably, so it is worth getting an early picture of what that development process is all about.

In the early days of embedded systems, the microcontrollers were very simple, without on-chip memory and with few peripherals. It took a lot of effort just to design and build the hardware. Moreover, memory was very limited, so programs had to be short. The main development effort was spent on hardware design, and programming consisted of writing rather simple programs in Assembler. Over the years, the microcontrollers became more and more sophisticated, and memory much more plentiful. By now, many suppliers were selling predesigned circuit boards containing the microcontroller and all associated circuitry. Where these were used there was very little development effort needed for the hardware. Now attention could be turned to writing complex and sophisticated programs, using all the

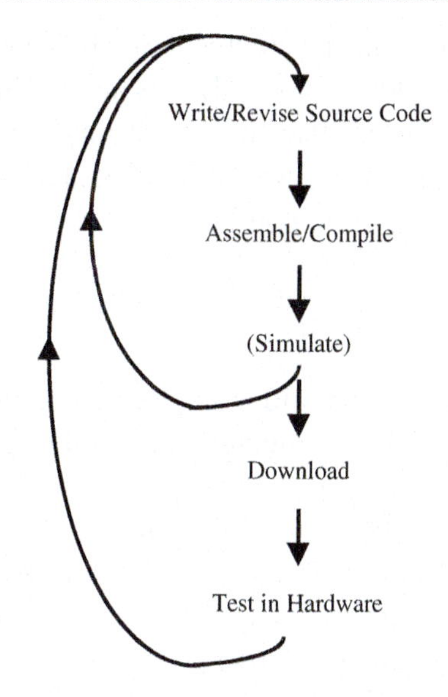

Figure 1.5:
The embedded program development cycle

memory that had become available. This tends to be the situation we find ourselves in nowadays.

Despite these changes, the program development cycle is still based around the simple loop shown in Figure 1.5. In this book, we will mainly write source code using C. The diagram of Figure 1.5 implies some of the equipment we are likely to need. The source code will need to be written on a computer – let's call it the host computer – for example a PC, using some sort of text editor. The source code will need to be converted into the binary machine code which is downloaded to the microcontroller, placed within the circuit or system that it will control (let's call this the *target system*). That conversion is done by the compiler, running on the host computer. Program download to the target system requires temporary connection to the host computer. There is considerable cleverness in how the program data is actually written into the program memory, but that need not concern us here. Before the program is downloaded, it is often possible to simulate it on the host computer; this allows a program to be developed to a good level before going to the trouble of downloading. However, the true test of the program is to see it running correctly in the target system. When program errors are found at this stage, as they inevitably are, the program can be rewritten, and compiled and downloaded once more.

1.4 The World of ARM

The development of computers and microprocessors has at different times been driven forward by giant corporations or by tiny start-ups; but always it has been driven forward by very talented individuals or teams. The development of ARM has seen a combination of all of these. The full history of the development of ARM is an absolutely fascinating one, like so many other hi-tech start-ups in the past 30 or so years. A tiny summary is given below; do read it in a fuller version in one of the several websites devoted to this topic. Then watch that history continue to unfold in the years to come!

1.4.1 A Little History

In 1981 the British Broadcasting Corporation launched a computer education project and asked for companies to bid to supply a computer that could be used for this. The winner was the Acorn computer. This became an extremely popular machine, and was very widely used in schools and universities in the UK. The Acorn used a 6502 microprocessor, made by a company called MOS Technology. This was a little 8-bit processor, which would not be taken very seriously these days, but was respected in its time. Responding to the growing interest in personal or desk-top computers, IBM in 1981 produced its very first PC, based on a more powerful Intel 16-bit microprocessor, the 8088. There were many companies producing similar computers at the time, including of course Apple. These early machines were pretty much incompatible with each other, and it was quite unclear whether one would finally dominate. Throughout the 1980s, however, the influence of the IBM PC grew, and its smaller competitors began to fade. Despite Acorn's UK success it did not export well, and its future no longer looked bright.

It was around this time that those clever designers at Acorn made three intellectual leaps. They wanted to launch a new computer. This would inevitably mean moving on from the 6502, but they just could not find a suitable processor for the sort of upgrade they needed. Their first leap was the realization that they had the capability to design the microprocessor itself and did not need to buy it in from elsewhere. Being a small team and experiencing intense commercial pressure, they designed a small processor, but one with real sophistication. The computer they built with this, the Archimedes, was a very advanced machine, but struggled against the commercial might of IBM. The company found themselves looking at computer sales which just were not sufficient, but holding the design of an extremely clever microprocessor. They realized – their second leap – that their future may not lie in selling the completed computer itself. Therefore, in 1990, Acorn computers cofounded another Cambridge company, called Advanced RISC Machines Ltd, ARM for short. They also began to realize – their third leap – that you do not need to manufacture silicon to be a successful designer, what mattered was the ideas inside the design. These can be sold as intellectual property (IP).

The ARM concept continued to prosper, with a sequence of smart microprocessor designs being sold as IP to an increasing number of major manufacturers around the world. The company has enjoyed huge success, and is currently called ARM Holdings. Those who buy the ARM designs incorporate them into their own products. For example, we will soon see that the mbed – the subject of this book – uses the ARM Cortex core. However, this is to be found in the LPC1768 microcontroller which sits in the mbed. This microcontroller is *not* made by ARM, but by NXP Semiconductors. ARM have sold NXP a license to include the Cortex core in their LPC1768 microcontroller, which ARM then buy back to put in their mbed. Got that?

1.4.2 Some Technical Detail: What does this RISC Word Mean?

Because ARM chose originally to place the RISC concept in its very name, and because it remains a central feature of ARM designs, it is worth checking out what RISC stands for. We have seen that any microcontroller executes a program which is drawn from its instruction set, which is defined by the CPU hardware itself. In the earlier days of microprocessor development, designers were trying to make the instructions set as advanced and sophisticated as possible. The price they were paying was that this was also making the computer hardware more complex, expensive and slower. Such a microprocessor is called a *complex instruction set computer* (CISC). Both the 6502 and 8088 mentioned above belong to the era when the CISC approach was dominant, and are CISC machines. One characteristic of the CISC approach is that instructions have different levels of complexity. Simple ones can be expressed in a short instruction code, say one byte of data, and execute quickly. Complex ones may need several bytes of code to define them, and take a long time to execute.

As compilers improved and high-level computer languages developed, it became less useful to focus on the capabilities of the raw instruction set itself. After all, if you are programming in a high-level language, the compiler should solve most of your programming problems with little difficulty.

Another approach to CPU design is therefore to insist on keeping things simple, and have a limited instruction set. This leads to the RISC approach – the *reduced instruction set computer*. The RISC approach looks like a 'back to basics' move. A simple RISC CPU can execute code rapidly, but it may need to execute more instructions to complete a given task, compared to its CISC cousin. With memory becoming ever cheaper and of higher density, and more efficient compilers for program code generation, this disadvantage is diminishing. One characteristic of the RISC approach is that each instruction is contained within a single binary word. That word must hold all information necessary, including the instruction code, as well as any address or data information needed. A further characteristic, an outcome of the simplicity of the approach, is that every instruction normally takes the same amount of time to execute. This allows other useful computer design features to be implemented. A good example is pipelining, an approach by which as one instruction is being executed, the next is already being fetched from memory. It is easy to do this with a RISC architecture, where all (or most) instructions take the same amount of time to complete.

An interesting subplot to the RISC concept is the fact that, owing to their simplicity, RISC designs tend to lead to low power consumption. This is hugely important for anything that is battery powered, and helps to explain why ARM products find their way into so many mobile phones.

1.4.3 The Cortex Core

The Cortex microprocessor core is a 32-bit device, and follows a long line of distinguished ARM processors. A very simplified block diagram of one version, the M3, is shown in Figure 1.6. This diagram allows us to see again some computer features we have already identified, and add some new ideas. Somewhere in the middle you can see the ALU, already described as the calculating heart of the computer. The instruction codes are fed into this through the Instruction fetch mechanism, drawing instructions in turn from the program memory. Pipelining is applied to instruction fetch, so that as one instruction is being executed, the next is being decoded, and the one after is being fetched from memory. As it executes each instruction, the ALU simultaneously receives data from memory and/or transfers it back to memory. This happens through the interface blocks seen. The memory itself is not part of the Cortex core. The ALU also has a block of registers associated with it. These act as a tiny chunk of local memory, which can be accessed quickly and used to hold temporary data as

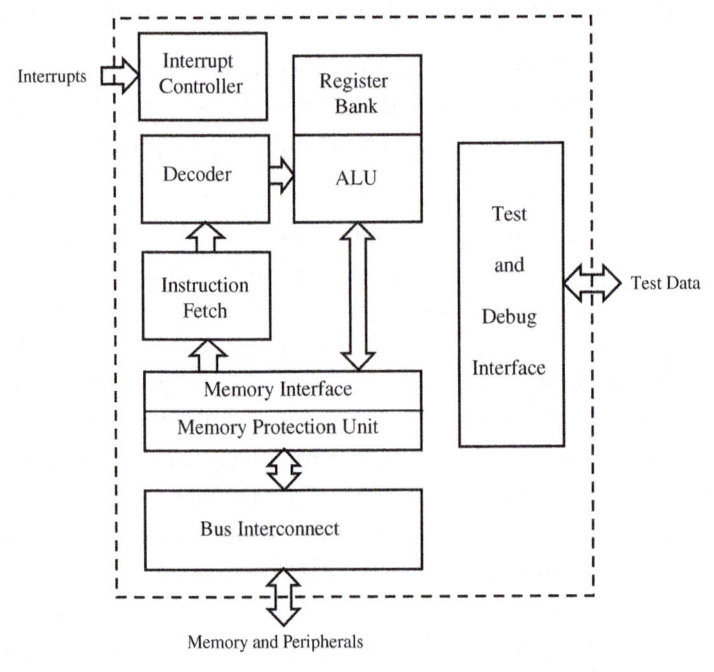

Figure 1.6:
The Cortex-M3 core, simplified diagram

a calculation is undertaken. The Cortex core also includes an interrupt interface. Interrupts are an important feature of any computer structure. They are external inputs that can be used to force the CPU to divert from the program section it is currently executing and jump to some other section of code. The interrupt controller manages the various interrupt inputs. It should not be too difficult to imagine this microprocessor core being dropped into the microcontroller diagram of Figure 1.4.

There are several versions of the Cortex. The Cortex-M4 is the 'smartest' of the set, with digital signal processing capability. The Cortex-M3, the one we will be using, is targeted at embedded applications, including automotive and industrial. The Cortex-M1 is a small processor intended to be embedded in a field programmable gate array (FPGA) – a digital electronic circuit on a chip, which is configured every time power is applied, and can even be reconfigured during operation if needed. The Cortex-M0 is the simplest of the group, with minimum size and power consumption.

We will be meeting the Cortex-M3 core, as it is the one used in the mbed. A very detailed guide to this core is given in Reference 1.2. Do not, however, try reading this book unless you are really keen to get into the fine detail; otherwise, leave it alone – it is very complex!

Chapter Review

- An embedded system contains one or more tiny computers, which control it and give it a sense of intelligence.
- The embedded computer usually takes the form of a microcontroller, which combines microprocessor core, memory and peripherals.
- Embedded system design combines hardware (electronic, electrical and electromechanical) and software (program) design.
- The embedded microcontroller has an instruction set. It is the ultimate goal of the programmer to develop code which is made up of instructions from this instruction set.
- Most programming is done in a high-level language, with a compiler being used to convert that program into the binary code, drawn from the instruction set and recognized by the microcontroller.
- ARM has developed a range of effective microprocessor and microcontroller designs, widely applied in embedded systems.

Quiz

1. Explain the following abbreviations: IC, I/O, ALU, CPU.
2. Describe an embedded system in less than 100 words.
3. What are the differences between a microprocessor and a microcontroller?

4. What range of numbers can be represented by a 16-bit ALU?
5. What is a 'bus' in the context of embedded systems? Describe two types of bus that might be found in an embedded system.
6. Describe the term 'instruction set' and explain how use of the instruction set differs for high- and low-level programming.
7. What are the main steps in the embedded program development cycle?
8. Explain the terms RISC and CISC and give advantages and disadvantages for each.
9. What is pipelining?
10. What did the acronym and company name ARM stand for?

References

1.1. Wilmshurst, T. (2001). An Introduction to the Design of Small-Scale Embedded Systems. Palgrave.
1.2. Yiu, J. (2010). The Definitive Guide to the ARM Cortex-M3. 2nd edition. 2010 Newnes.

CHAPTER 2

Introducing the mbed

Chapter Outline

2.1 Introducing the mbed

Chapter 1 reviewed some of the core features of computers, microprocessors and microcontrollers. Now we are going to apply that knowledge and enter the main material of this book, a study of the ARM *mbed.*

In very broad terms, the mbed takes a microcontroller, such as we saw in Figure 1.4, and surrounds it with some very useful support circuitry. It places this on a conveniently sized little printed circuit board (PCB) and supports it with an online compiler, program library and handbook. This gives a complete embedded system development environment, allowing users to develop and prototype embedded systems simply, efficiently and rapidly. Fast prototyping is one of the key features of the mbed approach.

The mbed takes the form of a 2 inch by 1 inch (53 mm by 26 mm) PCB, with 40 pins arranged in two rows of 20, with 0.1 inch spacing between the pins. This spacing is a standard in many electronic components. Figure 2.1 shows different mbed views. Looking at the main features, labeled in Figure 2.1b, we see that the mbed is based around the LPC1768 microcontroller. This is made by a company called NXP semiconductors, and contains an ARM Cortex-M3

Fast and Effective Embedded Systems Design. DOI: 10.1016/B978-0-08-097768-3.00002-7

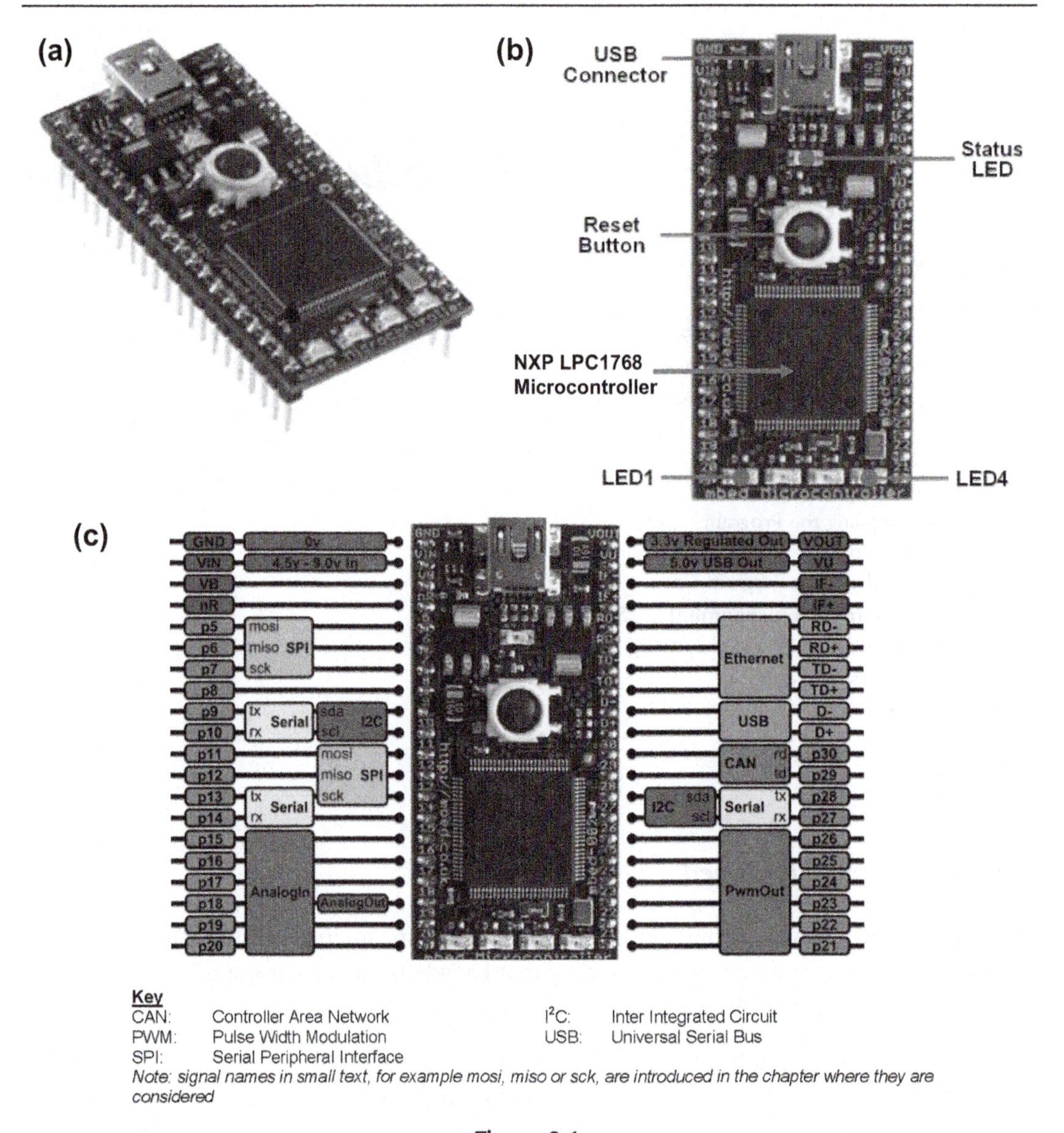

Figure 2.1:
The ARM mbed. *(Image reproduced with permission of ARM Holdings)*

core. Program download to the mbed is achieved through a universal serial bus (USB) connector; this can also power the mbed. Usefully, there are five light-emitting diodes (LEDs) on the board, one for status and four that are connected to four microcontroller digital outputs. These allow a minimum system to be tested with no external component connections needed. A reset switch is included, to force restart of the current program.

The mbed pins are clearly identified in Figure 2.1c, providing a summary of what each pin does. In many instances the pins are shared between several features to allow a number of design options. Top left we can see the ground and power supply pins. The actual internal circuit runs from 3.3 V. However, the board accepts any supply voltage within the range 4.5 to 9.0 V, while an onboard voltage regulator drops this to the required voltage. A regulated 3.3 V output voltage is available on the top right pin, with a 5 V output on the next pin down. The remainder of the pins connect to the mbed peripherals. These are almost all the subject of later chapters; we will quickly overview them here, though they may have limited meaning to you now. There are no fewer than five serial interface types on the mbed: I^2C, SPI, CAN, USB and Ethernet. Then there is a set of analog inputs, essential for reading sensor values, and a set of PWM outputs useful for control of external power devices, for example DC motors. While not immediately evident from the figure, pins 5 to 30 can also be configured for general digital input/output.

The mbed is constructed to allow easy prototyping, which is of course its very purpose. While the PCB itself is very high density, interconnection is achieved through the very robust and traditional dual-in-line pin layout.

Background information for the mbed and its support tools can be found at the mbed home page (Reference 2.1). While this book is intended to give you all information that you need to start work with the mbed, it is inevitable that you will want to keep a close eye on this site, with its cookbook, handbook, blog and forum. Above all else, it provides the entry point to the mbed compiler, through which you will develop all your programs.

2.1.1 The mbed Architecture

A block diagram representation of the mbed architecture is shown in Figure 2.2. It is possible, and useful, to relate the blocks shown here to the actual mbed. At the heart of the mbed is the LPC1768 microcontroller, clearly seen in both Figures 2.1 and 2.2. The signal pins of the mbed, as seen in Figure 2.1c, connect directly to the microcontroller. Thus, when in the coming chapters we use an mbed digital input or output, or the analog input, or any other of the peripherals, we will be connecting directly to the microcontroller within the mbed, and relying on its features. An interesting aside to this, however, is that the LPC1768 has 100 pins, but the mbed has only 40. Thus, when we get deeper into understanding the LPC1768, we will find that there are some features that are simply inaccessible to us as mbed users. This is, however, unlikely to be a limiting factor.

There is a second microcontroller on the mbed, which interfaces with the USB. This is called the interface microcontroller in Figure 2.2, and is the largest integrated circuit (IC) on the underside of the mbed PCB. The cleverness of the mbed hardware design is the way which this device manages the USB link and acts as a USB terminal to the host computer. In most common use it receives program code files through the USB, and transfers those programs to a 16 Mbit memory, which acts as the 'USB disk'. When a program 'binary' is downloaded

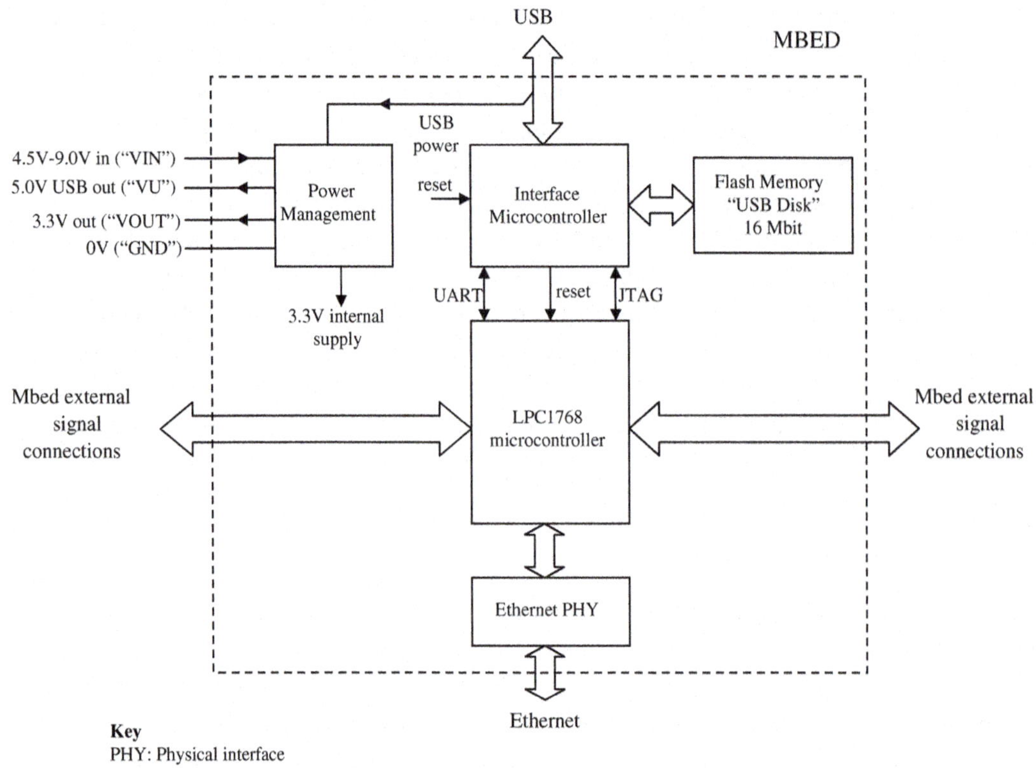

Figure 2.2:
Block diagram of mbed architecture.

to the mbed, it is placed in the USB disk. When the reset button is pressed, the program with the latest timestamp is transferred to the flash memory of the LPC1768, and program execution commences. Data transfer between interface microcontroller and the LPC1768 goes as serial data through the UART (which stands for universal asynchronous receiver/ transmitter – a serial data link, let's not get into the detail now) port of the LPC1768.

The 'Power Management' unit is made up of two voltage regulators, which lie either side of the status LED. There is also a current-limiting IC, which lies at the top left of the mbed. The mbed can be powered from the USB; this is a common way to use it, particularly for simple applications. For more power-hungry applications, or those that require a higher voltage, it can also be powered from an external 4.5 to 9.0 V input, supplied to pin 2 (labeled VIN). Power can also be sourced from mbed pins 39 and 40 (labeled VU and VOUT, respectively). The VU connection supplies 5 V, taken almost directly from the USB link; hence it is only available if the USB is connected. The VOUT pin supplies a regulated 3.3 V, which is derived either from the USB or from the VIN input.

For those who are inclined, the mbed circuit diagrams are available on the mbed website (Reference 2.2).

2.1.2 The LPC1768 Microcontroller

A block diagram of the LPC1768 microcontroller is shown in Figure 2.3. This looks complicated, and we do not want to go into all the details of what is a hugely sophisticated digital circuit. However, the figure is in a way the agenda for this book, as it contains all the capability of the mbed, so it is worth getting a feel for the main features. If you want to

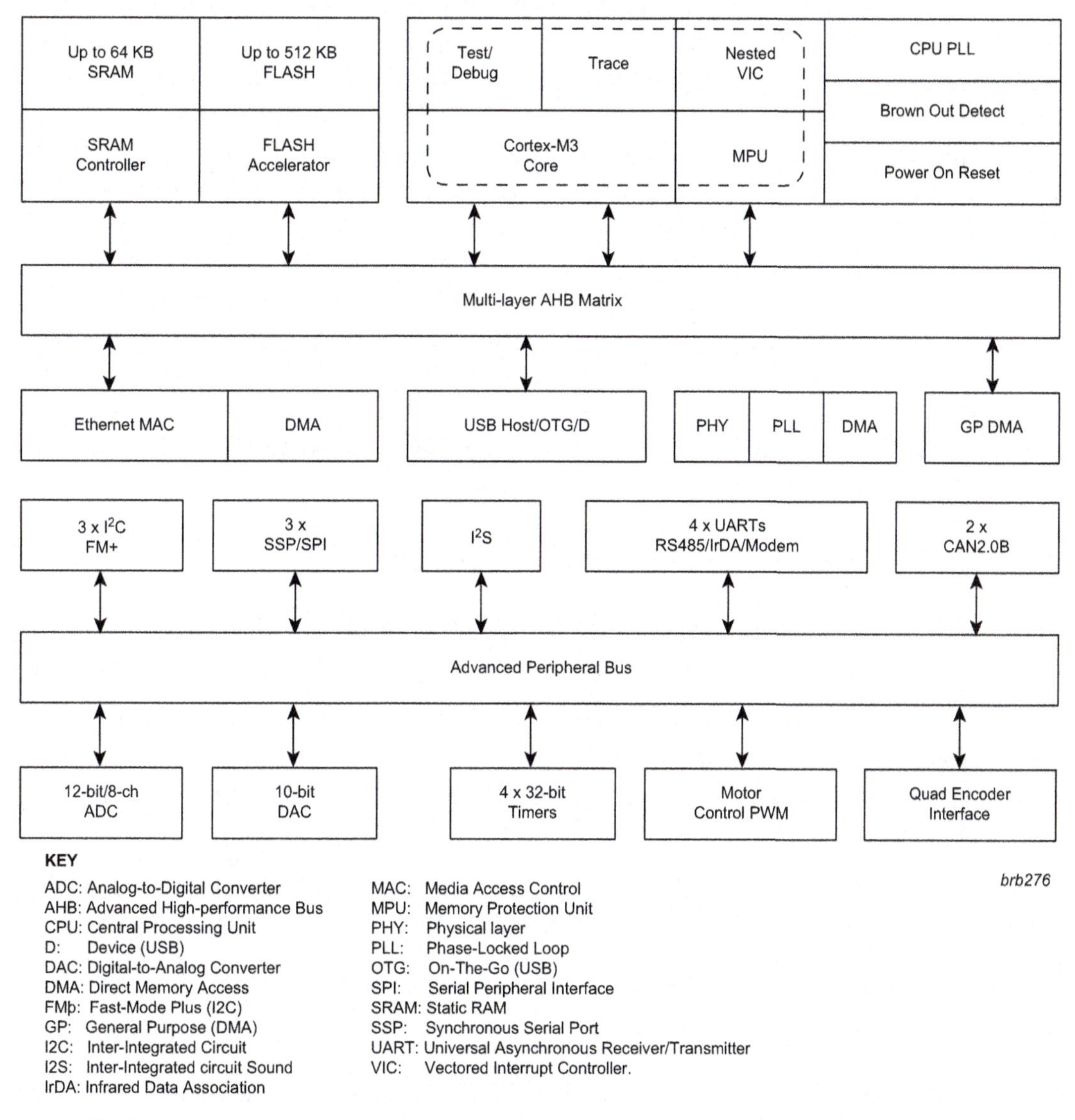

Note: The LPC1768 has 64 KB of SRAM and 512 KB of flash.
The Cortex core is made up of those items enclosed within the dashed line.
See also key to Figure 2.1 (Reproduced from mbed NXP LPC1768 prototyping board. 2009. NXP BV. Document no. 9397 750 16802)

Figure 2.3:
The LPC1768 block diagram.

get complete detail of this microcontroller, then consult one or more of References 1.2, 2.3 and 2.4. Although these references are mentioned from time to time in the book, consulting them is not necessary for a complete reading of the book.

Remember that a microcontroller is made up of microprocessor core *plus* memory *plus* peripherals, as shown in Figure 1.4. Let's look for these. Top center in Figure 2.3, contained within the dotted line, is the core of this microcontroller, the ARM Cortex-M3. This is a compressed version of Figure 1.6, the M3 outline that was considered in Chapter 1. To the left of the core are the memories: the program memory, made with Flash technology, is used for program storage; to the left of that is the static RAM (random access memory), used for holding temporary data. That leaves most of the rest of the diagram to show the peripherals, which give the microcontroller its embedded capability. These lie in the center and lower half of the diagram, and reflect almost exactly what the mbed can do. It is interesting to compare the peripherals seen here with the mbed inputs and outputs seen in Figure 2.1c. Finally, all these things need to be connected together, a task done by the address and data buses. Clever though they are, we have almost no interest in this side of the microcontroller design, at least not for this book. It is sufficient to note that the peripherals connect through something called the advanced peripheral bus. This in turns connects back through a bus interconnect called the advanced high-performance bus matrix, and from there to the central processing unit (CPU). This interconnection is not completely shown in this diagram, and we have neither need nor wish to think about it further.

2.2 Getting Started with the mbed: A Tutorial

Now comes the big moment when you connect the mbed for the first time, and run a first program. We will follow the procedure given on the mbed website and use the introductory program on the compiler, a simple flashing LED example. You will need:

- an mbed microcontroller with its USB lead
- a computer running Windows (XP, Vista or 7), Mac OS X or GNU/Linux
- a web browser, for example Internet Explorer or Firefox.

Now follow the sequence of instructions below. The purpose of this tutorial is to explain the main steps in getting a program running on the mbed. We will look into the detail of the program in the next chapter.

Step 1. Connecting the mbed to the PC

Connect the mbed to the PC using the USB lead. The Status light will come on, indicating that the mbed has power. After a few seconds of activity, the PC will recognize the mbed as a standard removable drive, and it will appear on the devices linked to the computer, as seen in Figure 2.4.

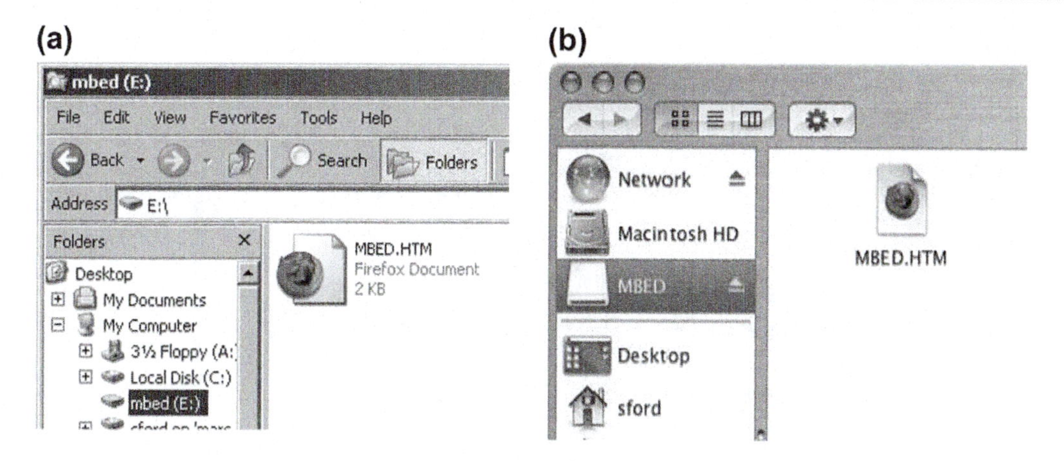

Figure 2.4:
Locating the mbed: (a) Windows XP example; (b) Mac OS X example

Step 2. Creating an mbed Account

Open the *MBED.HTM* file found on the mbed in your web browser and click the *Create a new mbed Account* link. Follow the instructions to create an mbed Account. This will lead you to the website, as seen in Figure 2.5. From here you can link to the compiler, libraries and documentation.

Step 3. Running a Program

Open the compiler using the link in the site menu, i.e. at the right of Figure 2.5. By doing this you enter your allocated personal program workspace. The compiler will open in a new tab or window. Follow these steps to create a new program:

- As seen in Figure 2.6a, right-click (Mac users, Ctrl-click) on 'My Programs' and select 'New Program ...'.
- Choose and enter a name for the new program (for example Prog_Ex_2_1) and click 'OK'. Do not leave spaces in the program name.
- Your new program folder will be created under 'My Programs'.

Click on the 'main.cpp' file in your new program to open it in the file editor window, as seen in Figure 2.6b. This is the main source code file in your program. Whenever you create

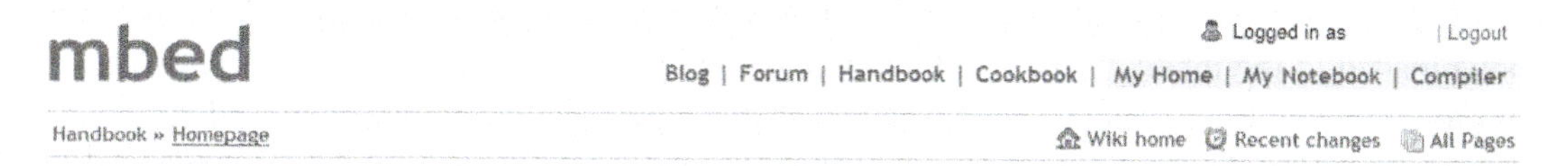

Figure 2.5:
The mbed home page

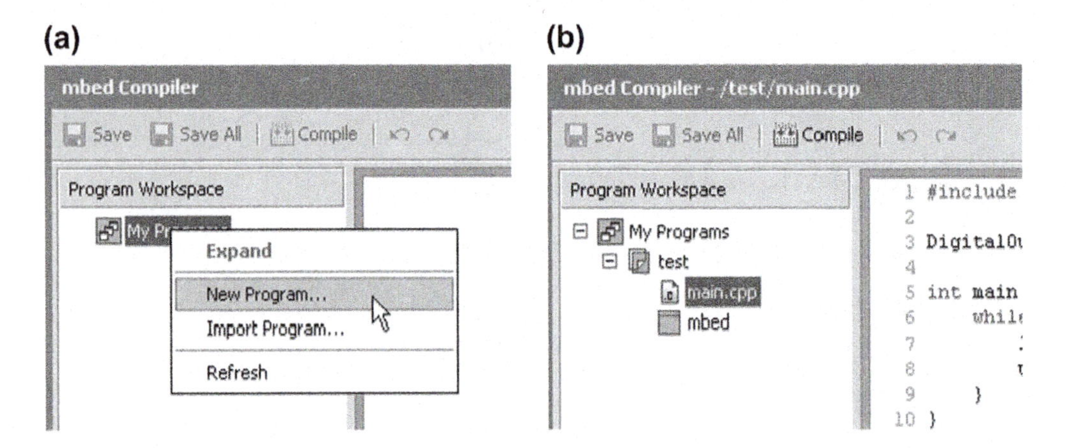

Figure 2.6:
Opening a new program: (a) selecting New Program; (b) opening the main source file

a new program it always contains the same simple code. This is shown here as Program Example 2.1. This program will be examined in the next chapter.

The other item in the program folder is the 'mbed' library. This provides all the functions used to start up and control the mbed, such as the **DigitalOut** interface used in this example.

```
/* Program Example 2.1: Simple LED flashing
*/
#include "mbed.h"
DigitalOut myled(LED1);
int main() {
  while(1) {
    myled = 1;
    wait(0.2);
    myled = 0;
    wait(0.2);
  }
}
```

Program Example 2.1 Simple LED flashing

Step 4. Compiling the Program

To compile the program, click the 'Compile' button in the toolbar. This will compile all the source code files within the program folder to create the binary machine code which will be downloaded to the mbed. Typically, this is the single program you have written, plus the library calls you have almost certainly made. After a successful compile, you will get a 'Success!' message in the compiler output, and a popup will prompt you to download the compiled **.bin** file to the mbed.

Of course, with this given program, it would be most surprising to find it had an error in it; you will not be so lucky with future programs! Try inserting a small error into your source code, for example by removing the semi-colon at the end of a line, and compiling again. Notice how the compiler gives a useful error message at the bottom of the screen. Correct the error, compile again and proceed. The type of error you have just inserted is often called a *syntax error*. This is an error which relates to the rules of writing lines of C code. When a syntax error is found, the compiler is unable to proceed with the compilation, as it perceives that the program has stepped outside the rules of the language, and hence cannot reliably interpret the code that is written.

Step 5. Downloading the Program Binary Code

After a successful compile, the program code, in binary form, can be downloaded to the mbed. Save it to the location of the mbed drive. You should see the Status LED (as seen in Figure 2.1) flash as the program downloads. Once the Status LED has stopped flashing, press the reset button on the mbed to start your program running. You should now see LED1 flashing on and off every 0.2 seconds.

Step 6. Modifying the Program Code

In the main.cpp file, simply change the DigitalOut statement to read:

```
DigitalOut myled(LED4);
```

Now compile and download your code to the mbed. You should now see that LED4 flashes instead of LED1. You can also change the pause between flashes by modifying the values bracketed in the **wait( )** command.

2.3 The Development Environment

As Section 1.3 suggests, there are many different approaches to development in embedded systems. With the mbed there is no software to install, and no extra development hardware needed for program download. All software tools are placed online, so that you can compile and download wherever you have access to the Internet. Notably, there is a C++ compiler and an extensive set of software libraries, used to drive the peripherals. Thus, there is no need to write code to configure peripherals, which in some systems can be very time consuming.

2.3.1 The mbed Compiler and API

The mbed development environment uses the ARM RVDS (RealView Development Suite) compiler, currently Version 4.1. All features of this compiler relevant to the mbed are available through the mbed portal.

One thing that makes the mbed special is that it comes with an application programming interface (API). In brief, this is the set of programming building blocks, appearing as C++ utilities, which allow programs to be devised quickly and reliably. Therefore, we will be writing code in C or C++, but drawing on the features of the API. You will meet most of the features of this as you work through the book. Note that you can see all components of the API by opening the mbed handbook, linked from the mbed home page.

2.3.2 Using C/C++

As just mentioned, the mbed development environment uses a C++ compiler. That means that all files will carry the **.cpp** (C plus plus) extension. C, however, is a subset of C++, and is simpler to learn and apply. This is because it does not use the more advanced 'object-oriented' aspects of C++. In general, C code will compile on a C++ compiler, but not the other way round.

C is usually the language of choice for any embedded program of low or medium complexity, so will suit us well in this book. For simplicity, therefore, we aim to use only C in the programs we develop. It should be recognized, however, that the mbed API is written in C++ and uses the features of that language to the full. We will aim to outline any essential features when we come to them.

This book does not assume that you have any knowledge of C or C++, although you have an advantage if you do. We aim to introduce all new features of C as they come up, flagging this by using the symbol alongside. If you see that symbol and you are a C expert, then it means you can probably skim through that section. If you are not an expert, you will need to read the section with care and refer to Appendix B, which summarizes all the C features used. Even if you are a C expert, you may not have used it in an embedded context. As you work through the book you will see a number of tricks and techniques that are used to optimize the language for this particular environment.

Chapter Review

- The mbed is a compact, microcontroller-based hardware platform, accompanied by a program development environment.
- Communication to the mbed from a host computer is by USB cable; power can also be supplied through this link.
- The mbed has 40 pins, which can be used to connect to an external circuit. On the board it has four user-programmable LEDs, so very simple programs can be run with no external connection to its pins.
- The mbed uses the LPC1768 microcontroller, which contains an ARM Cortex-M3 core. Most mbed connections link directly to the microcontroller pins, and many of the mbed characteristics derive directly from the microcontroller.

- The mbed development environment is hosted on the web. Program development is undertaken while online, and programs are stored on the mbed server.

Quiz

1. What do ADC, DAC and SRAM stand for?
2. What do UART, CAN, I^2C and SPI stand for, and what do these mbed features have in common?
3. How many digital inputs are available on the mbed?
4. Which mbed pins can be used for analog input and output?
5. How many microcontrollers are on the mbed PCB and what specifically are they?
6. What is unique about the mbed compiler software?
7. An mbed is part of a circuit which is to be powered from a 9 V battery. After programming the mbed is disconnected from the USB. One part of the circuit external to the mbed needs to be supplied from 9 V, and another part from 3.3 V. No other battery or power supply is to be used. Draw a diagram which shows how these power connections should be made.
8. An mbed is connected to a system, and needs to connect with three analog inputs, one SPI connection, one analog output and two PWM outputs. Draw a sketch showing how these connections can be made, and indicate mbed pin number.
9. A friend enters the code shown below into the mbed compiler, but when compiling a number of errors are flagged. Find and correct the faults.

```
#include "mbed"
Digital Out myled(LED1);
int man() {
    white(1) {
        myled = 1;
        wait(0.2)
        myled = 0;
        watt(0.2);
    }
```

10. By not connecting all the LPC1768 microcontroller pins to the mbed external pins, a number of microcontroller peripherals are 'lost' for use. Identify which ones these are, for ADC, UART, CAN, I^2C, SPI and DAC.

References

2.1. The mbed home site. http://mbed.org/

2.2. MBED Circuit Diagrams. 26/08/2010. http://mbed.org/media/uploads/chris/mbed-005.1.pdf

2.3. NXP B.V. LPC1768/66/65/64 32-bit ARM Cortex-M3 microcontroller. Objective data sheet. Rev. 6.0. August 2010. http://www.nxp.com/

2.4. NXP B.V. LPC17xx User Manual. Rev. 02. August 2010. http://www.nxp.com/

CHAPTER 3

Digital Input and Output

Chapter Outline

3.1 Starting to Program

This chapter will consider that most basic of microcontroller activity, the input and output of digital signals. Moving beyond this, it will show how we can make simple decisions within a program based on the value of a digital input. Figure 1.1 also predicted the importance of time in the embedded environment; no surprise therefore that at this early stage we meet timing activity as well.

Fast and Effective Embedded Systems Design. DOI: 10.1016/B978-0-08-097768-3.00003-9

You may also be embarking on C programming for the first time. This chapter takes you through quite a few of the key concepts. If you are new to C, we suggest you now read through Sections B1–B5 inclusive of Appendix B.

One thing that is not expected of you in reading this book is a deep knowledge of digital electronics. Some understanding of electronic theory will, however, be useful. If you need support in this area, you might like to have a book like Reference 3.1 available, to access when needed. There are many good websites in this area as well.

3.1.1 Thinking about the First Program

We now take a look at our first program, introduced as Program Example 2.1. This is shown again below for convenience, this time with comments inserted. Compare this with the original appearance of the program, shown in Chapter 2. Remember to check across to Appendix B when you need further C background.

Comments are text messages that you write to yourself within the program; they have no impact on the actual working of the program. It is good practice to introduce comments widely; they help your own thought process as you write, remind you what the program was meant to do as you look back over it, and help others to read and understand the program. This last point is essential for any sort of team-working, or if you are handing in work to be marked!

There are two ways of inserting comments into a program, and both are used in this example. One is to place the comment between the markers /* and */. This is useful for a block of text information running over several lines. Alternatively, when two forward-slash symbols (//) are used, the compiler ignores any text that follows on that line only; this can then be used for comment.

The opening comment in the program, lasting three lines, gives a brief summary of what the program does. We adopt this practice in all subsequent programs in the book. Notice also that we have introduced some blank lines, to make the program a little more readable. We then put a number of in-line comments; these indicate what individual lines of code do.

```
/*Program Example 2.1: A program which flashes mbed LED1 on and off.
Demonstrating use of digital output and wait functions. Taken from the mbed site.
*/

#include "mbed.h"  //include the mbed header file as part of this program

// program variable myled is created, and linked with mbed LED1
DigitalOut myled(LED1);

int main() {         //the      function starts here
  while (1) {        //a continuous loop is created
     myled = 1;      //switch the led on, by setting the output to logic 1
     wait(0.2);      //wait 0.2 seconds
```

```
        myled = 0;        //switch the led off
        wait(0.2);        //wait 0.2 seconds
    }                     //end of while loop
}                         //end of main function
```

Program Example 2.1 Repeated (and commented) for convenience

Let's now identify all the C programming features of Program Example 2.1. The action of *any* C or C++ program is contained within its **main( )** function, so that is always a good place to start looking. By the way, writing **main( )** with those brackets after it reminds us that it is the name of a C function; all function names are written in this way in the book. The *function definition* – what goes on inside the function – is contained within the opening curly bracket or brace, appearing immediately after **main( )**, and goes on until the closing brace. There are further pairs of braces inside these outer ones. Using pairs of braces like this is one of the ways that C groups blocks of code together and creates program structure.

Many programs in embedded systems are made up of an endless loop, i.e. a program which just goes on and on repeating itself indefinitely. Here is our first example. We create the loop by using the **while** keyword; this controls the code within the pair of braces that follow. Section B.7 (in Appendix B) tells us that normally **while** is used to set up a loop, which repeats if a certain condition is satisfied. However, if we write **while (1)**, then we "trick" the **while** mechanism to repeat indefinitely.

The real action of the program is contained in the four lines within the **while** loop. These are made up of two calls to the library function **wait( )**, and two statements, in which the value of **myled** is changed. The **wait( )** function is from the mbed library; further options are shown in Table 3.1. The 0.2 parameter is in seconds, and defines the delay length caused by this function – another value can be chosen. In the two statements containing **myled** we make use of a C operator for the first time. Here, we use the *assign* operator, which applies the conventional equals sign. Note carefully that in C this operator means that a variable is set to the value indicated. Thus,

```
myled = 1;
```

means that the variable **myled** is set to the value 1, whatever its previous value was. Conventional "equals" is represented by a double equals sign, i.e. == . It is used later in this chapter. There are many C operators, and it is worth noticing and learning each one as you meet it for the first time. They are summarized in Section B.5.

Table 3.1: mbed library wait functions

C/C++ function	Action
`wait`	Waits for the number of seconds specified
`wait_ms`	Waits for the number of milliseconds specified
`wait_us`	Waits for the number of microseconds specified

The program starts by including the all-important mbed header file, which is the connecting pathway to all mbed library features. This means that the header file is literally inserted into the program, before compiling starts in earnest. The **#include** compiler directive is used for this (Section B.2.3). The program then applies the mbed application programming interface (API) utility **DigitalOut** to define a digital output, and gives it the name **myled**. Once declared, **myled** can be used as a variable within the program. The name LED1 is reserved by the API for the output associated with that light-emitting diode (LED) on the mbed board, as labeled in Figure 2.1b.

Notice that we indent the code by two spaces within **main( )**, and then two more within the **while** code block. This has no effect on the action of the program, but does help to make it more readable and hence cut down on programming errors; it is a practice we apply for programs in this book. Check Section B.11 for other code layout practices that we adopt. Companies working with C often apply 'house styles', to ensure that programmers write code in a readable and consistent fashion.

Exercise 3.1

Get familiar with Program Example 2.1, and the general process of compiling, by trying the following variations. Observe the effects.

1. Add the comments, or others of your own, given in the example. See that the program compiles and runs without change.
2. Vary the 0.2 parameter in the **wait( )** functions.
3. Replace the **wait( )** functions with **wait_ms( )** functions, with accompanying parameter set to give the same wait time. Also vary this time.
4. Use LED2, 3 or 4 instead of 1.
5. Make a more complex light pattern using two or more LEDs, and more **wait( )** function calls.

3.1.2 Understanding the mbed API

The mbed API has a pattern that we will see repeated many times, so it is important to get used to it. The library is made up of a set of utilities, which are all itemized in the mbed website Handbook. The first that we look at is **DigitalOut,** which we have already used. The API summary for this is given in Table 3.2.

In a pattern that will become familiar, the **DigitalOut** API component creates a C++ *class*, called **DigitalOut.** This appears at the head of the table. The class then has a set of member functions, which are listed below. The first of these is a C++ *constructor*, which must have the same name as the class itself. This can be used to create C++ objects. By

Table 3.2: The mbed digital output API summary (from www.mbed.org)

Function	Usage
`DigitalOut`	Create a DigitalOut connected to the specified pin
`write`	Set the output, specified as 0 or 1 (int)
`read`	Return the output setting, represented as 0 or 1 (int)
`operator=`	A shorthand for write
`operator int()`	A shorthand for read

using the **DigitalOut** constructor, we can create C++ objects. In our first example we create the object **myled**. We can then write to it and read from it, using the functions **write( )** and **read( )**. These are member functions of the class, so their format would be **myled.write( )**, and so on. Having created the **myled** object we can, however, invoke the API operator shorthand, mentioned in Table 3.2, which applies the assign operator = . Hence, when we write

```
myled = 1;
```

the variable value is then not only changed (normal C usage), but also written to the digital output. This replaces **myled.write(1);**. We will find similar shorthands offered for all peripheral API groups in the mbed API.

3.1.3 Exploring the while *Loop*

Program Example 3.1 is the first "original" program in this book, although it builds directly on the previous example. Create a new program in the mbed compiler and copy this example into it.

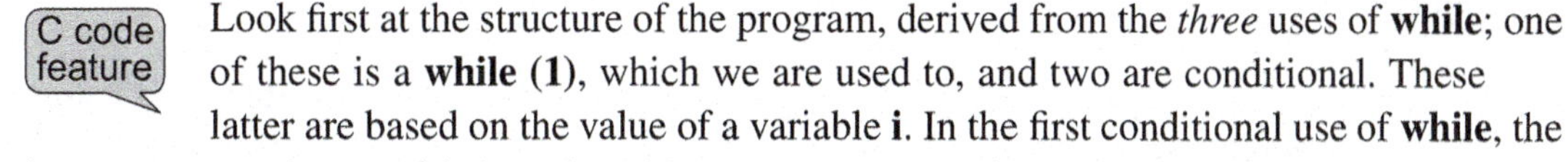

Look first at the structure of the program, derived from the *three* uses of **while**; one of these is a **while (1)**, which we are used to, and two are conditional. These latter are based on the value of a variable **i**. In the first conditional use of **while**, the loop repeats as long as **i** is less than 10. You can see that **i** is incremented within the loop, to bring it up this value. In the second conditional loop, the value is decremented, and the loop repeats as long as **i** is greater than zero.

A very important new C feature seen in this program is the way that the variable **i** is *declared* at the beginning of **main( )**. It is essential in C to declare the data type of any variable before it is used. The possible data types are given in Section B.3. In this example, **i** is declared as a character, effectively an 8-bit number. In the same line, its value is initialized to zero. There are also four new operators introduced: +, −, < and >. The use of these is the same as in conventional algebra, so should be immediately familiar.

```
/*Program Example 3.1: Demonstrates use of while loops. No external connection required
*/
#include "mbed.h"
DigitalOut myled(LED1);
DigitalOut yourled(LED4);

int main() {
  char i=0;          //declare variable i, and set to 0
  while(1){          //start endless loop
  while(i<10) {      //start first conditional while loop
    myled = 1;
    wait(0.2);
    myled = 0;
    wait(0.2);
    i = i+1;         //increment i
  }                  //end of first conditional while loop
  while(i>0)  {      //start second conditional loop
    yourled = 1;
    wait(0.2);
    yourled = 0;
    wait(0.2);
    i = i-1;
  }
    }                //end infinite loop block
}                    //end of main
```

Program Example 3.1 Using *while*

Having compiled Program Example 3.1, download it to the mbed, and run it. You should see LED1 and LED4 flashing 10 times in turn. Make sure you understand why this is so.

Exercise 3.2

1. A slightly more elegant way of incrementing or decrementing a variable is by using the increment and decrement operators seen in Section B.5. Try replacing

   i= i+1; and i = i – 1;

   with i++ ; and i-- ; respectively,

   and run the program again.
2. Change the program so that the LEDs flash only five times each.
3. Try replacing the **myled = 1;** statement with **myled.write(1);**

■

3.2 Voltages as Logic Values

Computers deal with binary numbers made up of lots of binary digits, or bits, and each one of these bits takes a logic value of 0 or 1. Now that we are starting to use the mbed in earnest, it is worth thinking about how those logic values are actually represented inside the mbed's electronic circuit, and at its connection pins.

In any digital circuit, logic values are represented as electrical voltages. Here now is the *big* benefit of digital electronics: we do not need a precise voltage to represent a logical value. Instead, we accept a *range* of voltages as representing a logic value. This means that a voltage can pick up some noise or distortion and still be interpreted as the right logic value. The microcontroller we are using in the mbed, the LPC1768, is powered from 3.3 V. We can find out which range of voltages it accepts as Logic 0, and which as Logic 1, by looking at its technical data. This is found in Reference 2.4 (Chapter 2), with important points summarized in Appendix C. (There will be moments when it will be useful to look at this appendix, but do not rush to it now.) This data shows that, for most digital inputs, the LPC1678 interprets *any* input voltage below 1.0 V (specified as 0.3 × 3.3 V) as Logic 0, and *any* input voltage above 2.3 V (specified as 0.7 × 3.3 V) as Logic 1. This idea is represented in the diagram of Figure 3.1.

If we want to input a signal to the mbed, hoping that it is interpreted correctly as a logic value, then it will need to satisfy the requirements of Figure 3.1. If we are outputting a signal from the mbed, then we can expect it to comply with the same figure. The mbed will normally output Logic 0 as 0 V and Logic 1 as 3.3 V, as long as no electrical current is flowing. If current is flowing, for example into an LED, then we can expect some change in output voltage. We will return to this point. The neat thing is, when we output a logic value, we are also getting a predictable voltage to make use of. We can use this to light LEDs (light emitting diodes), switch on motors, or many other things.

For many applications in this book we do not worry much about these voltages, as the circuits we build meet the requirements of Figure 3.1. There are situations, however, where it is necessary to take some care over logic voltage values, and these are mentioned as they come up.

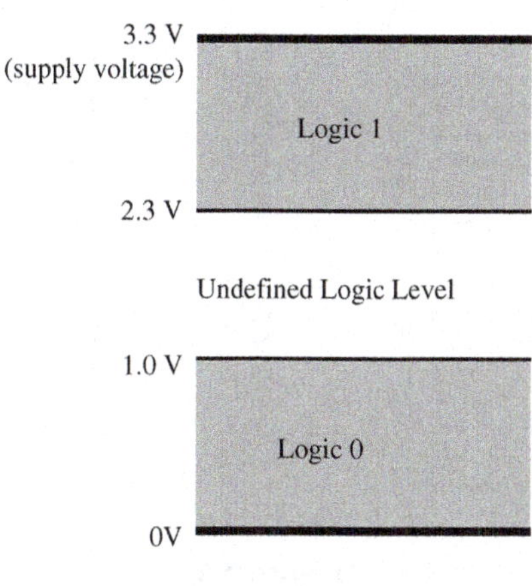

Figure 3.1:
Input logic levels for the mbed

3.3 Digital Output on the mbed

In our first two programs we have just switched the diagnostic LEDs which are fitted on the mbed board. The mbed, however, has 26 digital input/output (I/O) pins (pins 5–30), which can be configured as either digital inputs or outputs. These are shown in Figure 3.2. A comparison of this figure with the master mbed pinout diagram, i.e. Figure 2.1c, shows that these pins are multi-function. While we can use any of them for simple digital I/O, they all have secondary functions, connecting with the microcontroller peripherals. It is up to the programmer to specify, within the program, how they are configured.

3.3.1 Using LEDs

For the first time we are connecting an external device to the mbed. While you may not be an electronics specialist it is important – whenever you connect anything to the mbed – that you have some understanding of what you are doing. The LED is a semiconductor diode, and behaves electrically as one. It will conduct current in one direction, sometimes called the 'forward' direction, but not the other. What makes it so useful is that when it is connected so that it conducts, it emits photons from its semiconductor junction. The LED has the voltage/current characteristic shown in Figure 3.3a. A small forward voltage will cause very little current to flow. As the voltage increases there comes a point where the current suddenly starts flowing rather rapidly. For most LEDs this voltage is in the range shown, typically around 1.8 V.

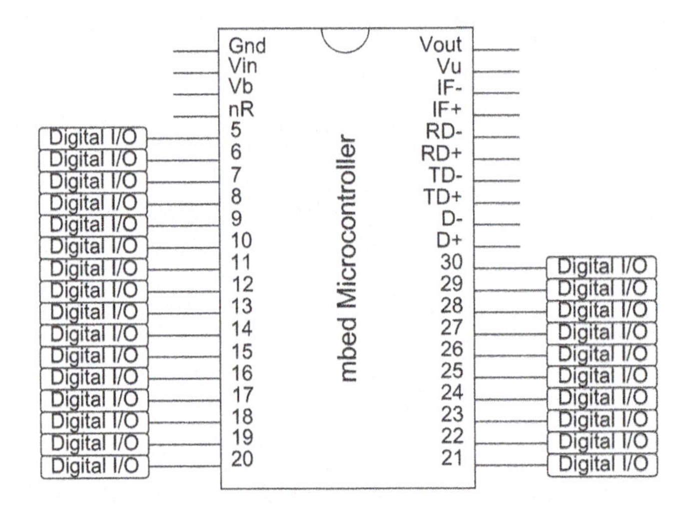

Figure 3.2:
The mbed digital I/O

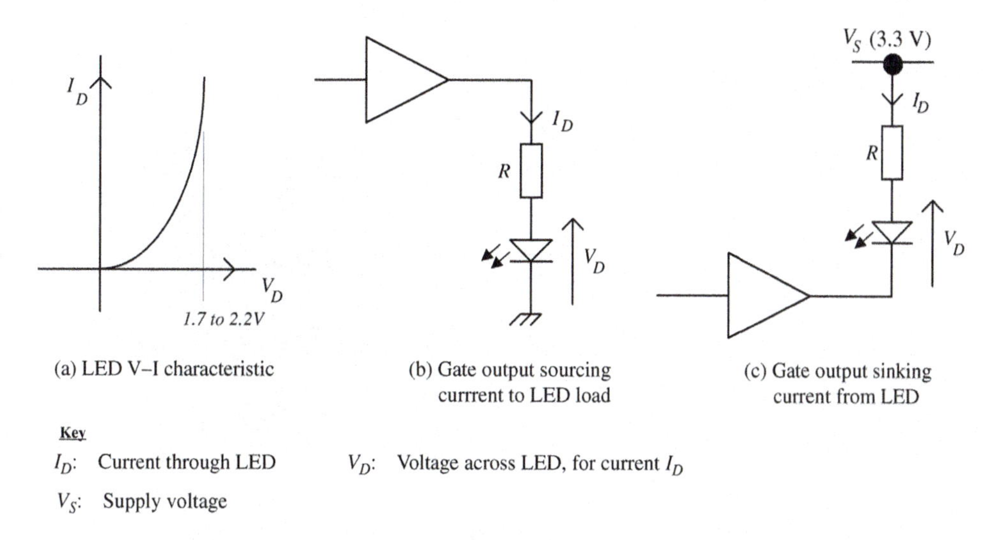

Figure 3.3:
Driving LEDs from logic gates

Figures 3.3b and c show circuits used to make direct connections of LEDs to the output of logic gates, for example an mbed pin configured as an output. The gate is here shown as a logic buffer (the triangular symbol). If the connection of Figure 3.3b is made, the LED lights when the gate is at logic high. Current then flows out of the gate into the LED. Alternatively, the circuit of Figure 3.3c can be made; the LED now lights when the logic gate is at logic zero, with current flowing *into* the gate. Usually a current-limiting resistor needs to be connected in series with the LED, to control how much current flows. The exception to this is if the combination of output voltage and gate internal resistance is such that the current is limited to a suitable value anyway. Some LEDs, such as the ones recommended for the next program, have the series resistance built into them. They therefore do not need any external resistor connected.

3.3.2 Using mbed External Pins

The digital I/O pins are named and configured to output using **DigitalOut**, just as we did in the earlier programs, by defining them at the start of the program code, for example:

```
DigitalOut myname(p5);
```

The **DigitalOut** object created in this way can be used to set the state of the output pin, and also to read back its current state.

We will be applying the circuit of Figure 3.4a. Appendix D gives example part numbers for all parts used; equivalent devices can of course be implemented. Within this circuit we are applying the connection of Figure 3.3b. The particular LED recommended in Appendix D, however, has an internal series resistor, of value around 240 Ω, so an external one is not required.

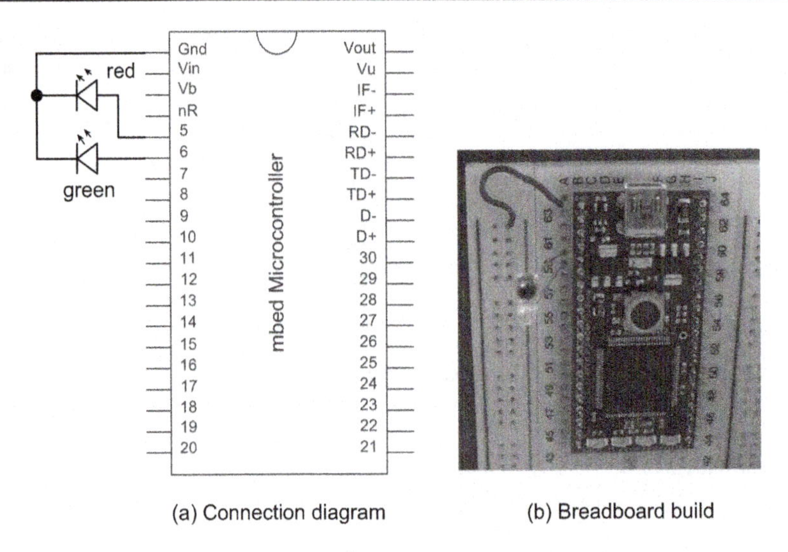

(a) Connection diagram

(b) Breadboard build

Figure 3.4:
mbed set-up for simple LED flashing

Place the mbed in a breadboard, as seen in Figure 3.4b, and connect the circuit shown. The mbed has a common ground on pin 1. Connect this across to one of the outer rows of connections of the breadboard, as seen in Figure 3.4b. You should adopt this as a habit for all similar circuits that you build. Remember to attach the anode of the LEDs (the side with the longer leg) to the mbed pins. The negative side (cathode) should be connected to ground. For this and many circuits we will take power from the universal serial bus (USB).

Create a new program in the mbed compiler, and copy across Program Example 3.2.

```
/*Program Example 3.2: Flashes red and green LEDs in simple time-based pattern
*/
#include "mbed.h"
DigitalOut redled(p5);    //define and name a digital output on pin 5
DigitalOut greenled(p6);  //define and name a digital output on pin 6

int main() {
  while(1) {
    redled = 1;
    greenled = 0;
    wait(0.2);
    redled = 0;
    greenled = 1;
    wait(0.2);
  }
}
```

Program Example 3.2 Flashing external LEDs

Compile, download and run the code on the mbed. The code extends ideas already applied in Program Examples 2.1 and 3.1, so it should be easy to understand that the green and red LEDs are programmed to flash alternately. You should see this happen when the code runs.

Exercise 3.3

Using any digital output pin, write a program which outputs a square wave, by switching the output repeatedly between Logic 1 and 0. Use **wait( )** functions to give a frequency of 100 Hz (i.e. a period of 10 ms). View the output on an oscilloscope. Measure the voltage values for Logic 0 and 1. How do they relate to Figure 3.1? Does the square wave frequency agree with your programmed value?

3.4 Using Digital Inputs

3.4.1 Connecting Switches to a Digital System

Ordinary electromechanical switches can be used to create logic levels, which will satisfy the logic level requirements seen in Figure 3.1. Three commonly used ways are shown in Figure 3.5. The simplest, Figure 3.5a, uses an SPDT (single-pole, double-throw) switch. This is what we will use in the next example. A resistor in series with the logic input is sometimes included in this circuit, as a precautionary measure. However, an SPST (single-pole, single-throw) switch can be lower cost and smaller, and so is very widely used. They are connected with a pull-up or pull-down resistor to the supply voltage, as shown in Figure 3.5b or c. When the switch is open, the logic level is defined by the connection through the resistor. When it is closed, the switch asserts the other logic state. A wasted electrical current then flows through the resistor. This is kept small by having a high-value resistor. On the mbed, as with many microcontrollers, pull-up and pull-down resistors are available within the microcontroller, saving the need to make the external connection.

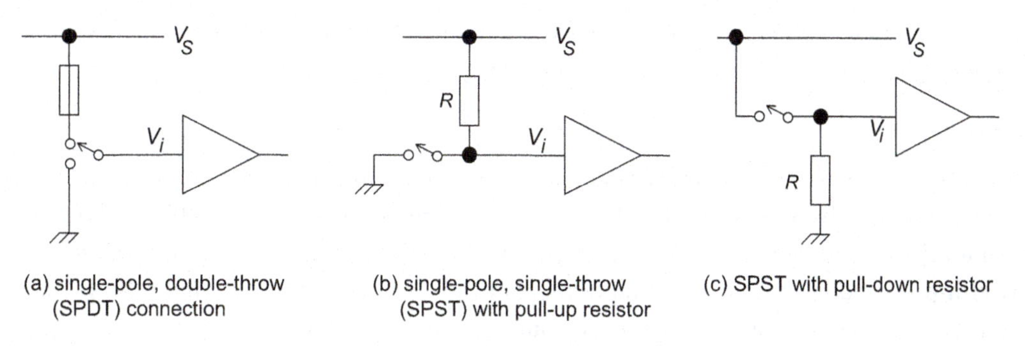

Figure 3.5:
Connecting switches to logic inputs

Table 3.3: The mbed digital input API summary (from www.mbed.org)

Function	Usage
`DigitalIn`	Create a DigitalIn connected to the specified pin
`read`	Read the input, represented as 0 or 1 (int)
`mode`	Set the input pin mode
`operator int()`	An operator shorthand for read()

3.4.2 The DigitalIn API

The mbed API has the digital input functions listed in Table 3.3, with format identical to that of **DigitalOut**, which we have already seen. This section of the API creates a class called **DigitalIn**, with the member functions shown. The **DigitalIn** constructor can be used to create digital inputs, and the **read( )** function used to read the logical value of the input. In practice, the shorthand offered allows us to read input values through use of the digital object name. This will be seen in the program example that follows. As with digital outputs, the same 26 pins (pins 5–30) can be configured as digital inputs. Input voltages will be interpreted according to Figure 3.1. Note that use of **DigitalIn** enables by default the internal pull-down resistor, i.e. the input circuit is configured as Figure 3.5c. This can be disabled, or an internal pull-up enabled, using the **mode( )** function. See the mbed Handbook for details on this.

3.4.3 Using if *to Respond to a Switch Input*

We will now connect to our circuit a digital input, a switch, and use it to control LED states. In so doing, we make a significant step forward. For the first time we will have a program that makes a decision based on an external variable, the switch position. This is the essence of many embedded systems. Program Example 3.3 is applied to achieve this; the digital input is connected to pin 7, and created with the **DigitalIn** constructor.

To make the decision within the program, we use the statement

```
if(switchinput==1).
```

This line sees use of the C equal operator, ==, for the first time. Read Section B.6.1 to review use of the **if** and **else** keywords. The line causes the line or block of code which follows it to execute, if the specified condition is met. In this case, the condition is that the variable **switchinput** is equal to 1. If the condition is not satisfied, then the code that follows **else** is executed. Looking now at the program we can see that if the switch gives a value of Logic 1, the green LED is switched off and the red LED is programmed to flash. If the switch input is 0, the **else** code block is invoked and the roles of the LEDs are reversed. Again, the **while (1)** statement is used to create an overall infinite loop, so the LEDs flash continuously.

```
/*Program Example 3.3: Flashes one of two LEDs, depending on the state of a 2-way switch
*/
#include "mbed.h"
DigitalOut redled(p5);
DigitalOut greenled(p6);
DigitalIn switchinput(p7);
int main() {
 while(1) {
    if (switchinput==1) {   //test value of switchinput
    //execute following block if switchinput is 1
        greenled = 0;       //green led is off
        redled = 1;         // flash red led
        wait(0.2);
        redled = 0;
        wait(0.2);
  }                         //end of if
  else {                    //here if switchinput is 0
   redled = 0;              //red led is off
   greenled = 1;            // flash green led
   wait(0.2);
   greenled = 0;
   wait(0.2);
  }                         //end of else
 }                          //end of while(1)
}                           //end of main
```

Program Example 3.3 Using *if* and *else* to respond to external switch

Adjust the circuit of Figure 3.4 to that of Figure 3.6a, by adding an SPDT switch as shown. The input it connects to is configured in the program as a digital input. The photograph in Figure 3.6b shows the addition of the switch. In this case, wires have been soldered to the switch and connected to the breadboard. It is also possible to find switches that plug in directly. Create a new program and copy across Program Example 3.3. Compile, download and run the code on the mbed.

Exercise 3.4

Write a program that creates a square wave output, as in Exercise 3.3. Include now two possible frequencies, 100 Hz and 200 Hz, depending on an input switch position. Use the same switch connection you have just used. Observe the output on an oscilloscope.

Exercise 3.5

The circuit of Figure 3.6 uses an SPDT switch connected as shown in Figure 3.5a. However, the mbed Handbook tells us that the **DigitalIn** utility actually configures an on-chip pull-down resistor. Reconfigure the circuit using Figure 3.5c, using either the same toggle switch or an SPST push button. Test that the program runs correctly.

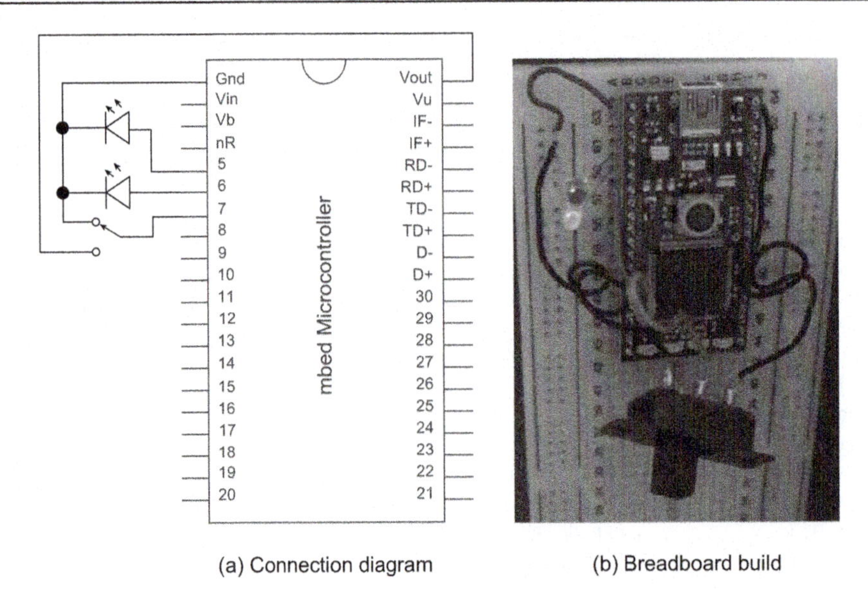

(a) Connection diagram (b) Breadboard build

Figure 3.6:
Controlling the LED with a switch

3.5 Interfacing Simple Opto Devices

Now that we have the ability to input and output single bits of data, a wide range of possibilities opens up. Many simple sensors can interface directly with digital inputs. Others have their own designed-in interfaces, which produce a digital output. This section looks at some simple and traditional sensors and displays that can be interfaced directly to the mbed. In later chapters this is taken further, connecting to some very new and hi-tech devices.

3.5.1 Opto Reflective and Transmissive Sensors

Opto-sensors, such as seen in Figure 3.7a and b, are simple examples of sensors with 'almost' digital outputs. When a light falls on the base of an opto-transistor, it conducts; when there is no light it does not. In the reflective sensor (Figure 3.7a), an infrared LED is mounted in the same package as an opto-transistor. When a reflective surface is placed in front, the light bounces back, and the transistor can conduct. In the transmissive sensor (Figure 3.7b), the LED is mounted opposite the transistor. With nothing in the way, light from the LED falls directly on the transistor, and so it can conduct. When something comes in the way the light is blocked, and the transistor stops conducting. This sensor is also sometimes called a slotted opto-sensor or photo-interrupter. Each sensor can be used to detect certain types of object.

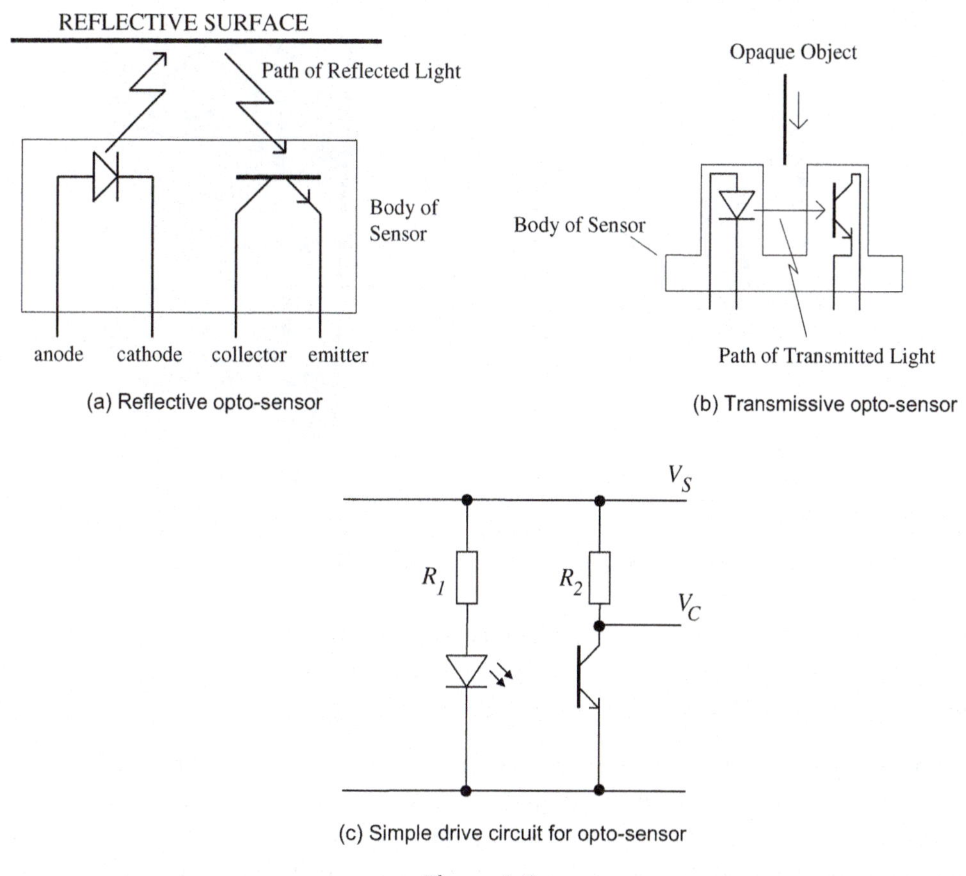

(a) Reflective opto-sensor

(b) Transmissive opto-sensor

(c) Simple drive circuit for opto-sensor

Figure 3.7:
Simple opto-sensors

Either of the sensors shown can be connected in the circuit of Figure 3.7c. Here, R_1 is calculated to control the current flowing in the LED, taking suitable values from the sensor datasheet. The resistor R_2 is chosen to allow a suitable output voltage swing, invoking the voltage thresholds indicated in Figure 3.1. When light falls on the transistor base, current can flow through the transistor. The value of R_2 is chosen so that with that current flowing, the transistor collector voltage V_C falls almost to 0 V. If no current flows, then V_C rises to V_S. In general, the sensor is made more sensitive by either decreasing R_1 or increasing R_2.

3.5.2 Connecting an Opto-Sensor to the mbed

Figure 3.8 shows how a transmissive opto-sensor can be connected to an mbed. The device used is a KTIR0621DS, made by Kingbright; similar devices can be used. The particular sensor used has pins that can plug directly into the breadboard. Take care

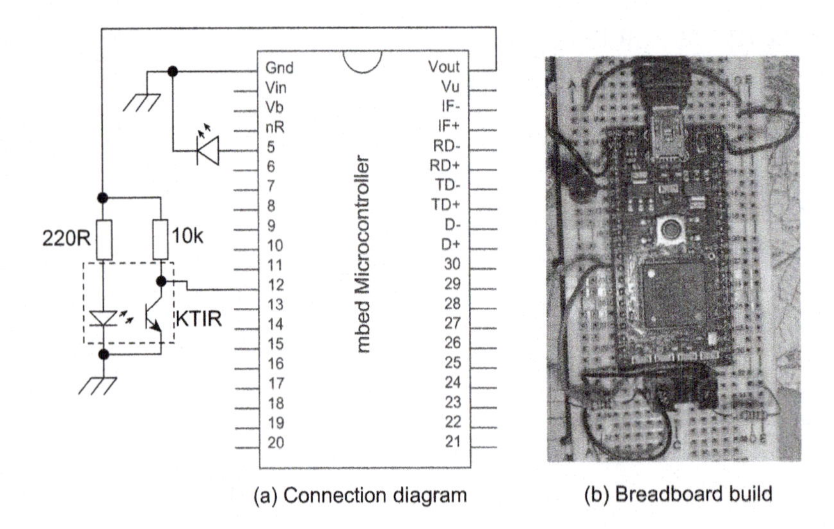

(a) Connection diagram (b) Breadboard build

Figure 3.8:
mbed connected to a transmissive opto-sensor

when connecting these four pins. Connections are indicated on its housing; alternatively, you can look these up in the data that Kingbright provides. This can be obtained from the Kingbright website, Reference 3.2, or from the supplier you use to buy the sensor.

Program Example 3.4 controls this circuit. The output of the sensor is connected to pin 12, which is configured in the program as a digital input. When there is no object in the sensor, then light falls on the photo-transistor. It conducts, and the sensor output is at Logic 0. When the beam is interrupted, then the output is at Logic 1. The program therefore switches the LED on when the beam is interrupted, i.e. an object has been sensed. To make the selection, we use the **if** and **else** keywords, as in the previous example. Now, however, there is just one line of code to be executed for either state. It is not necessary to use braces to contain these single lines.

```
/*Program Example 3.4: Simple program to test KTIR slotted optosensor. Switches an LED
according to state of sensor
*/

#include "mbed.h"
DigitalOut redled(p5);
DigitalIn opto_switch(p12);

int main() {
  while(1) {
     if (opto_switch==1)           //input = 1 if beam interrupted
        redled = 1;                //switch led on if beam interrupted
     else
        redled = 0;           //led off if no interruption
  }                                //end of while
}
```

Program Example 3.4 Applying the photo-interrupter

3.5.3 Seven-Segment Displays

We have now used single LEDs in several programs. In addition, LEDs are often packaged together, to form patterns, digits or other types of display. A number of standard configurations are very widely used; these include bar graph, seven-segment display, dot matrix and 'star-burst'.

The seven-segment display is a particularly versatile configuration. An example single digit, made by Kingbright (Reference 3.2), is shown in Figure 3.9. By lighting different combinations of the seven segments, all numerical digits can be displayed, as well as a surprising number of alphabetic characters. A decimal point is usually included, as shown. This means that there are eight LEDs in the display, needing 16 connections. To simplify matters, either all LED anodes are connected together, or all LED cathodes. This is seen in Figure 3.9b; the two possible connection patterns are called *common cathode* or *common anode*. Now instead of 16 connections being needed, there are only nine, one for each LED and one for the common connection. The actual pin connections in the example shown lie in two rows, at the top and bottom of the digit. There are 10 pins in all, with the common anode or cathode taking two pins.

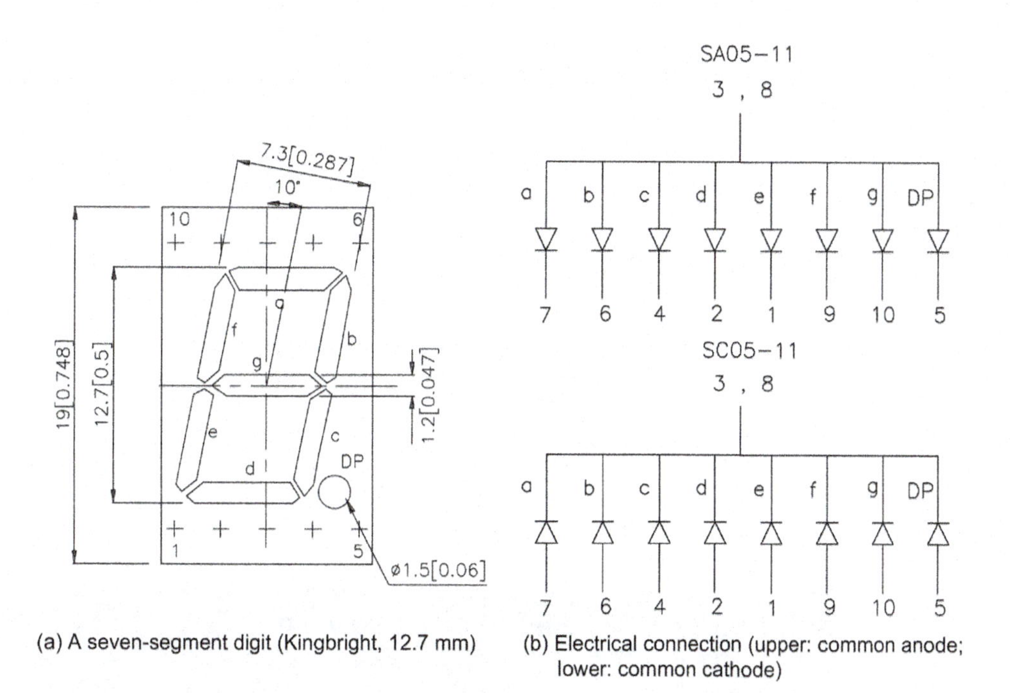

(a) A seven-segment digit (Kingbright, 12.7 mm)

(b) Electrical connection (upper: common anode; lower: common cathode)

Figure 3.9:
The seven-segment display. *(Image reproduced with permission of Kingbright Elec. Co. Ltd.)*

A small seven-segment display as seen in Figure 3.9 can be driven directly from a microcontroller. In the case of common cathode, the cathode is connected to ground, and each segment is connected to a port pin. If the segments are connected in this sequence to form a byte,

(MSB) DP g f e d c b a (LSB)

then the values shown in Table 3.4 apply. For example, if 0 is to be displayed, then all outer segments, i.e. abcdef, must be lit, with the corresponding bits from the microcontroller set to 1. If 1 is to be displayed, then only segments b and c need to be lit. Note that larger displays have several LEDs connected in series, for each segment. In this case, a higher voltage is needed to drive each series combination and, depending on the supply voltage of the microcontroller, it may not be possible to drive the display directly.

3.5.4 Connecting a Seven-segment Display to the mbed

As we know, the mbed runs from a 3.3 V supply. The datasheet of our display (accessed from Reference 3.2), shows that each LED requires around 1.8 V across it in order to light. This is within the mbed capability. If there were two LEDs in series in each segment however, the mbed would barely be able to switch them into conduction. But can we just connect the mbed output directly to the segment, or do we need a current-limiting resistor, as seen in Figure 3.3b? Looking at the LPC1768 data summarized in Appendix C, we see that the output voltage of a port pin drops around 0.4 V, when 4 mA is flowing. This implies an output resistance of 100 Ω. Let's not get into the electronics of all of this; suffice it to say, this value is approximate, and only applies in this region of operation. Applying Ohm's law, the current flow in an LED connected directly to a port pin of this type is given by

$$I_D \cong (3.3 - 1.8)/100 = 15 \text{ mA} \tag{3.1}$$

Table 3.4: Example seven-segment display control values

Display value	0	1	2	3	4	5	6	7	8	9
Segment drive (B)										
(MSB)	0011	0000	0101	0100	0110	0110	0111	0000	0111	0110
(LSB)	1111	0110	1011	1111	0110	1101	1101	0111	1111	1111
B (hex)	0x3F	0x06	0x5B	0x4F	0x66	0x6D	0x7D	0x07	0x7F	0x6F
Actual display										

This current, 15 mA, will light the segments very brightly, but is acceptable. It can be reduced, for example for power-conscious applications, by inserting a resistor in series with each segment.

Connect a seven-segment display to an mbed, using the circuit of Figure 3.10. In this simple application the common cathode is connected direct to ground, and each segment is connected to one mbed output.

The circuit can be driven by Program Example 3.5. This first of all applies the **BusOut** mbed API class. **BusOut** allows you to group a set of digital outputs into one bus, so that you can write a digital word direct to it. The equivalent input is the **Busin** object, though that is not used here. Applying **BusOut** simply requires you to specify a name, in this case **display**, and then list in brackets the pins which will be members of that bus.

C code feature

Within this program we see for the first time a **for** loop. This is an alternative to **while**, for creating a conditional loop. Review its format in Section B.7.2. In this example, the variable **i** is initially set to 0, and on each iteration of the loop it is incremented by 1. The new value of **i** is applied within the loop. When **i** reaches 4, the

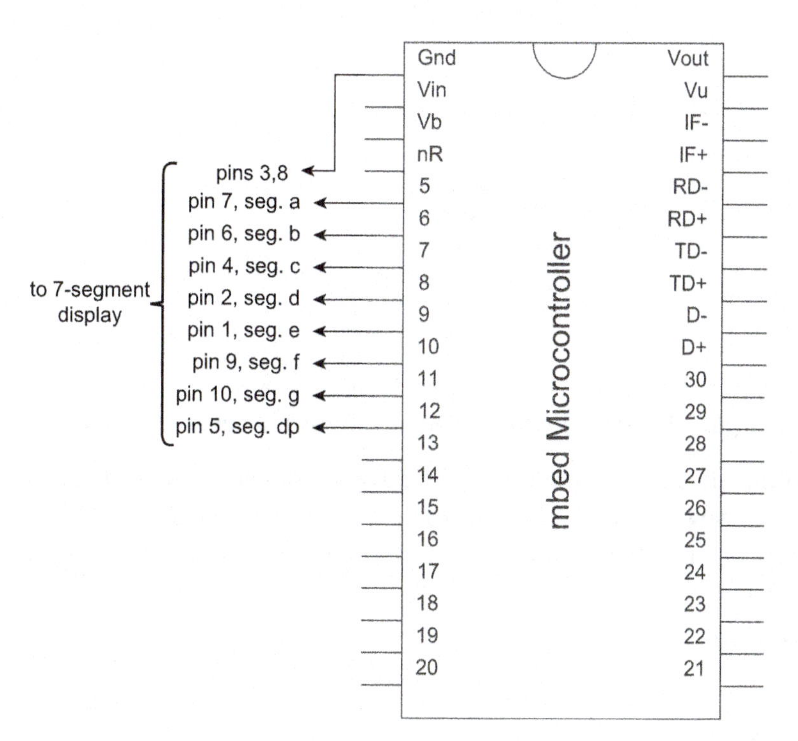

Figure 3.10:
The mbed connected to a common cathode seven-segment display

loop terminates. However, as the **for** loop is the only code within the endless **while** loop, it simply starts again.

The program goes on to apply the **switch**, **case** and **break** keywords. Used together, these words provide a mechanism to allow one item to be chosen from a list, as described in Section B.6.2. In this case, as the variable **i** is incremented, its value is used to select the word that must be sent to the display, in order for the required digit to illuminate. This method of choosing one value from a list is one way of achieving a *look-up table*, which is an important programming technique.

There are now quite a few nested blocks of code in this example. The **switch** block lies within the **for** block which lies inside the **while** block, which finally lies within the **main** block. The closing brace for each is commented in the program listing. When writing more complex C programs it becomes very important to ensure that each block is ended with a closing brace, at the right place.

```
/*Program Example 3.5: Simple demonstration of 7-segment display. Display digits 0, 1,
2, 3 in turn.
*/

#include "mbed.h"
BusOut display(p5,p6,p7,p8,p9,p10,p11,p12);  // segments a,b,c,d,e,f,g,dp

int main() {
  while(1) {
    for(int i=0; i<4; i++) {
      switch (i){
          case 0: display = 0x3F; break;     //display 0
          case 1: display = 0x06; break;     //display 1
          case 2: display = 0x5B; break;
          case 3: display = 0x4F; break;
      }                                        //end of switch
    wait(0.2);
    }                                        //end of for
  }                                          //end of while
}                                            //end of main
```

Program Example 3.5 Using *for* to sequence values to a seven-segment display

Compile, download and run the program. The display should appear similar to Figure 3.11. Notice carefully, however, by looking at where pin 8 is connected, that this is a common anode connection. It does not exactly replicate Figure 3.10. Pin 3 has been left unconnected.

Exercise 3.6

Write a program which flashes the letters H E L P in turn on a seven-segment display, using the circuit of Figure 3.10.

Figure 3.11:
mbed connected to a common anode seven-segment display

3.6 Switching Larger DC Loads

3.6.1 Applying Transistor Switching

The mbed can drive simple DC loads directly with its digital I/O pins, as we saw with the LEDs. The mbed summary data (repeated in Appendix C) tells us that a port pin can source up to around 40 mA. However, this is a short circuit current, so we are unlikely to be able to benefit from it with an actual electrical load connected to the port.

If it is necessary to drive a load – say a motor – which needs more current than an mbed port pin can supply, or which needs to run from a higher voltage, then an interface circuit will be needed. Three possibilities, which allow DC loads to be switched, are shown in Figure 3.12. Each has an input labeled V_L, which is the logic voltage supplied from the port pin. The first two circuits show how a resistive load, such as a motor or heater, can be switched, using a bipolar transistor and a metal oxide semiconductor field effect transistor (MOSFET – a mouthful, but an important device in today's electronics), respectively. In the case of the bipolar transistor, the simple formulae shown can be applied to calculate R_B, starting with knowledge of the required load current and the value of the current gain (β) of the transistor. In the case of the MOSFET, there is a threshold gate-to-source voltage, above which the transistor switches on. This is a particularly useful configuration, as the MOSFET gate can be readily driven by a microcontroller port bit output.

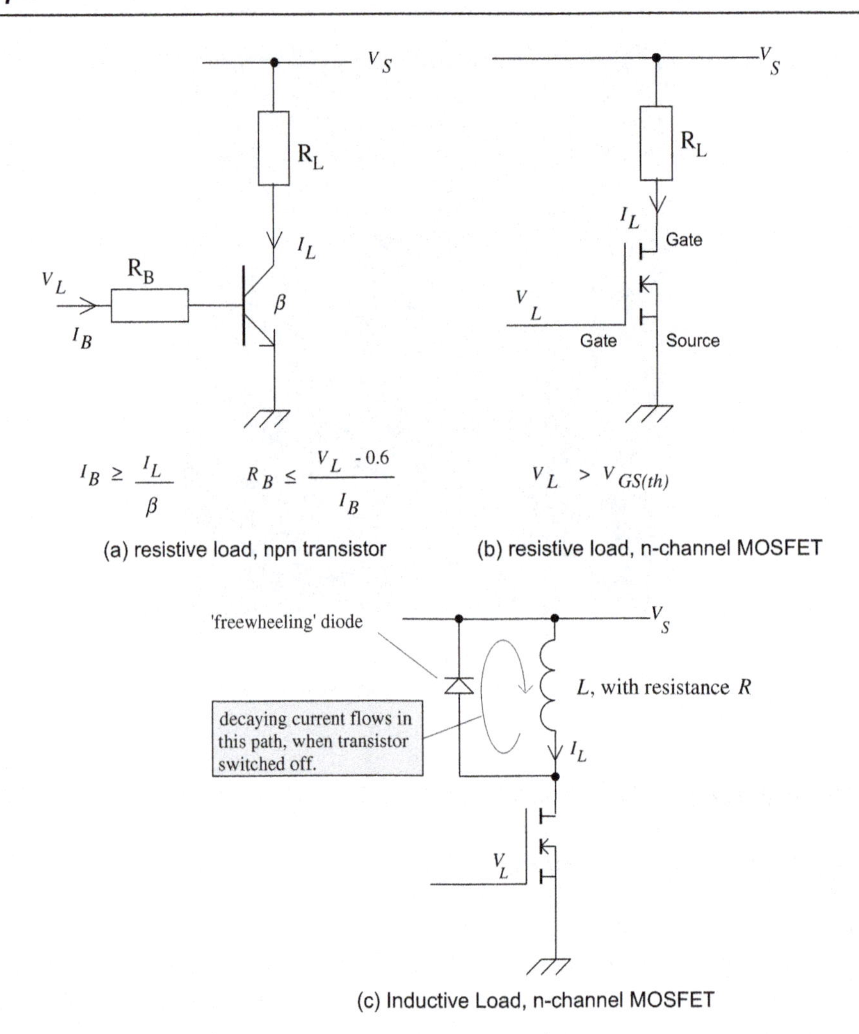

Figure 3.12:
Transistor switching of DC loads

Figure 3.12c shows an inductive load, such as a solenoid or DC motor, being switched. An important addition here is the *freewheeling diode.* This is needed because any inductance with current flowing in it stores energy in the magnetic field that surrounds it. When that current is interrupted, in this case by the transistor being switched off, the energy has to be returned to the circuit. This happens through the diode, which allows decaying current to continue to circulate. If the diode is not included then a high-voltage transient occurs, which can/will destroy the field effect transistor.

3.6.2 Switching a Motor with the mbed

A good switching transistor for small DC loads is the ZVN4206A, whose main characteristics are listed in Table 3.5. An important value is the maximum V_{GS} threshold

Table 3.5: Characteristics of the ZVN4206A n-channel MOSFET

Characteristic	ZVN4206A
Maximum drain-source voltage V_{DS}	60 V
Maximum gate-source threshold $V_{GS(th)}$	3 V
Maximum drain-source resistance when 'On'. $R_{DS(on)}$	1.5 Ω
Maximum continuous drain current I_D	600 mA
Maximum power dissipation	0.7 W
Input capacitance	100 pF

value, shown as 3 V. This means that the MOSFET will respond, just, to the 3.3 V Logic 1 output level of the mbed.

Exercise 3.7

Connect the circuit of Figure 3.13, using a small DC motor. The 6 V for the motor can be supplied from an external battery pack or bench power supply. Its exact voltage is non-critical, and depends on the motor you are using. Write a program so that the motor switches on and off continuously, say 1 s on, 1 s off. Increase the frequency of the switching until you can no longer detect that it is switching on and off. How does the motor speed compare to when the motor was just left switched on?

3.6.3 Switching Multiple Seven-Segment Displays

We saw earlier in the chapter how a single seven-segment display could be connected to the mbed. Each display required eight connections, and if we wanted many displays, then we

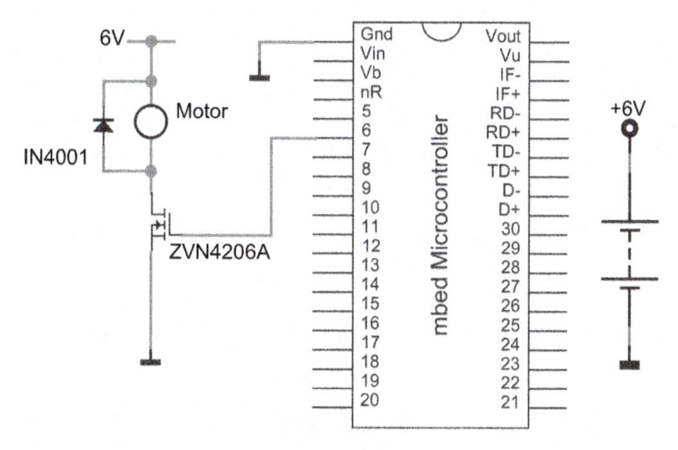

Figure 3.13:
Switching a DC motor

would quickly run out of I/O pins. There is a very useful technique to get around this problem, shown in Figure 3.14. Each segment type on each display is wired together, as shown, and connected back to a microcontroller pin configured as digital output. The common cathode of each digit is then connected to its own drive MOSFET. The timing diagram shown then applies. The segment drives are configured for Digit 1, and that digit's drive transistor is activated, illuminating the digit. A moment later the segment drives are configured for Digit 2, and that digit's drive transistor is activated. This continues endlessly with each digit in turn. If it is done rapidly enough, then the human eye perceives all digits as being continuously illuminated; a useful rate is for each digit to be illuminated in turn for around 5 ms.

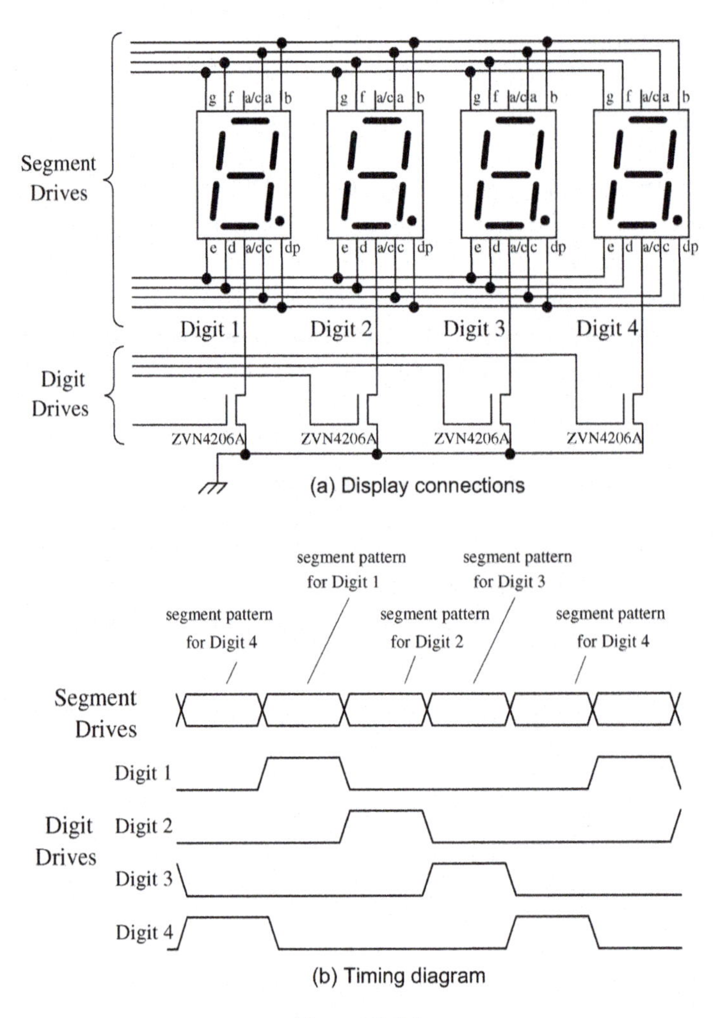

Figure 3.14:
Multiplexing seven-segment displays

3.7 Mini-Project: Letter Counter

Use the slotted opto-sensor, push-button switch and one seven-segment LED display to create a simple letter counter. Increment the number on the display by one every time a letter passes the sensor. Clear the display when the push button is pressed. Use an LED to create an extra 'half' digit, so you can count from 0 to 19.

Chapter Review

- Logic signals, expressed mathematically as 0 or 1, are represented in digital electronic circuits as voltages. One range of voltages represents 0, another represents 1.
- The mbed has 26 digital I/O pins, which can be configured as either input or output.
- LEDs can be driven directly from the mbed digital outputs. They are a useful means of displaying a logic value, and of contributing to a simple human interface.
- Electromechanical switches can be connected to provide logic values to digital inputs.
- A range of simple opto-sensors have almost digital outputs, and can with care be connected directly to mbed pins.
- Where the mbed pin cannot provide enough power to drive an electrical load directly, interface circuits must be used. For simple on/off switching a transistor is often all that is needed.

Quiz

1. Complete Table 3.6, converting between the different number types. The first row of numbers is an example.
2. Is it possible to display unambiguously all of the capitalized alphabet characters A, B, C, D, E and F on the seven-segment display shown in Figure 3.9? For those that can usefully be displayed, determine the segment drive values. Give your answers in both binary and hexadecimal formats.
3. A loop in an mbed program is untidily coded as follows:

```
    while (1)
{redled = 0;
    wait_ms(12);
greenled = 1;
    wait(0.002);
    greenled = 0;
    wait_us(24000);
}
```

   What is the total period of the loop, expressed in seconds, milliseconds and microseconds?

Table 3.6:

Binary	Hexadecimal	Decimal
0101 1110	5E	94
1101		
	77	
		129
	6F2	
1101 1100 1001		
		4096

4. The circuit of Figure 3.5b is used eight times over to connect eight switches to 8 mbed digital inputs. The pull-up resistors have a value of 10 kΩ, and are connected to the mbed supply of 3.3 V. What current is consumed due to this circuit configuration when all switches are closed simultaneously? If this current drain must be limited to 0.5 mA, to what value must the pull-up resistors be increased? Reflect on the possible impact of pull-up resistors in low-power circuits.
5. What is the current taken by the display connected in Figure 3.10, when the digit 3 is showing?
6. If in Figure 3.10 a segment current of approximately 4 mA was required, what value of resistor would need to be introduced in series with each segment?
7. A student builds an mbed-based system. To one port he connects the circuit of Figure 3.15a, using LEDs of the type used in Figure 3.4, but is then disappointed that the

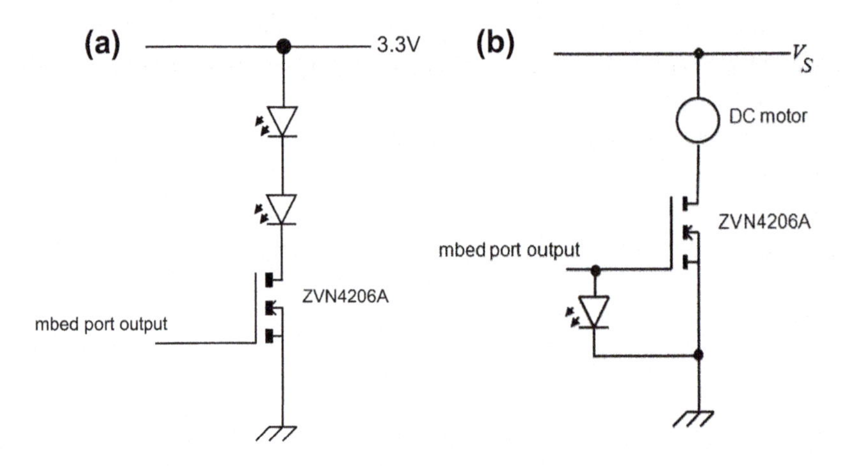

Figure 3.15:
Trial switching circuits

LEDs do not appear to light when expected. Explain why this is so, and suggest a way of changing the circuit to give the required behavior.

8. Another student wants to control a DC motor from the mbed, and therefore builds the circuit of Figure 3.15b, where V_S is a DC supply of appropriate value. As a further indication that the motor is running, she connects a standard LED, as seen in Figure 3.3, directly to the port bit. She then complains of unreliable circuit behavior. Explain any necessary changes that should be made to the circuit.
9. Look at the mbed circuit diagram, Reference 2.2, and find the four onboard LEDs. Estimate the current each one takes when 'on', assuming a forward voltage of 1.8 V.

References

3.1. Floyd, T. (2008). Digital Fundamentals. 10th edition. Pearson Education.
3.2. The Kingbright home site. http://www.kingbright.com/

CHAPTER 4

Analog Output

Chapter Outline

4.1 Introducing Data Conversion

Microcontrollers are digital devices, but they spend most of their time dealing with a world which is analog, as illustrated in Figure 1.1. To make sense of incoming analog signals, for example from a microphone or temperature sensor, they must be able to convert them into digital form. After processing the data, they may then need to convert digital data back to analog form, for example to drive a loudspeaker or direct current (DC) motor. We give these processes the global heading 'data conversion'. Techniques applied in data conversion form a huge and fascinating branch of electronics. They are outside the subject matter of this book; however, if you wish to learn more about them, consider any major electronics text. To get deep into the topic, read Reference 4.1.

While conversion in both directions between digital and analog is necessary, it is conversion from analog to digital form that is the more challenging task; therefore, in this chapter the

Fast and Effective Embedded Systems Design. DOI: 10.1016/B978-0-08-097768-3.00004-0

easier option is considered: conversion from digital to analog. (Analog-to-digital conversion is covered in Chapter 5.)

4.1.1 The Digital-to-Analog Converter

A digital-to-analog converter (DAC) is a circuit that converts a binary input number into an analog output. The actual circuitry inside a DAC is complex, and need not concern us. We can, however, represent the DAC as a block diagram, as in Figure 4.1. This has a digital input, represented by D, and an analog output, represented by V_o. The 'yardstick' by which the DAC calculates its output voltage is a *voltage reference*, a precise, stable and known voltage.

Most DACs have a simple relationship between their digital input and analog output, with many (including the one inside the LPC1768) applying Equation 4.1:

$$V_o = \frac{D}{2^n} V_r \tag{4.1}$$

Here, V_r is the value of the voltage reference, D is the value of the binary input word, n is the number of bits in that word, and V_o is the output voltage. Figure 4.2 shows this equation represented graphically. For each input digital value, there is a corresponding analog output. It is as if we are creating a voltage staircase with the digital inputs. The number of possible output values is given by 2^n, and the step size by $V_r/2^n$; this is called the *resolution.* The maximum possible output value occurs when $D = (2^n - 1)$, so the value of V_r as an output is never quite reached. The *range* of the DAC is the difference between its maximum and minimum output values. For example, a 6-bit DAC will have 64 possible output values; if it has a 3.2 V reference, it will have a resolution (step size) of 50 mV.

Figure 2.3 shows that the LPC1768 has a 10-bit DAC; there will therefore be 2^{10} steps in its output characteristic, i.e. 1024. Reference 2.4 further tells us that it normally uses its own power supply voltage, i.e. 3.3 V, as voltage reference. The step size, or resolution, will therefore be 3.3/1024, i.e. 3.22 mV.

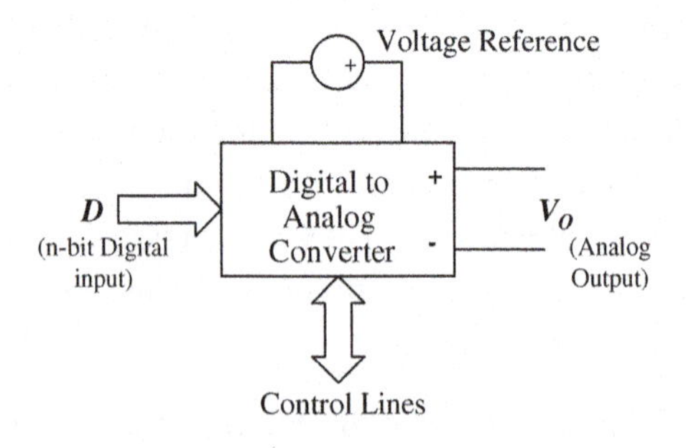

Figure 4.1:
The digital-to-analog converter

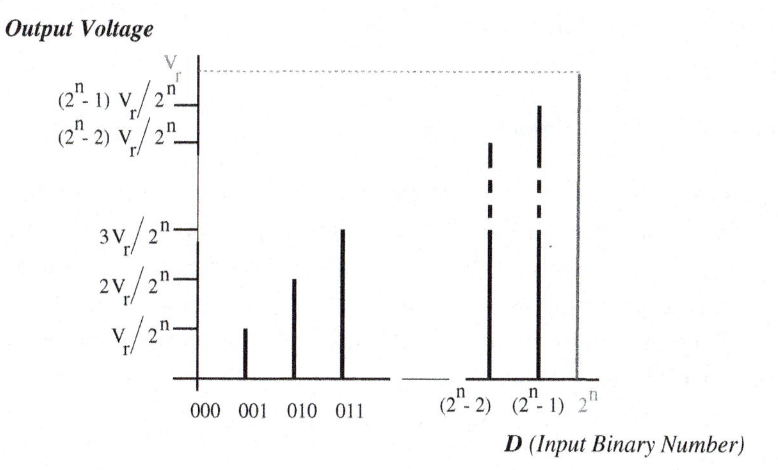

Figure 4.2:
The DAC input/output characteristic

4.2 Analog Outputs on the mbed

The mbed pin connection diagram, Figure 2.1c, has already shown us that the mbed is rich in analog input/output capabilities. Pins 15–20 can be analog input, while pin 18 is the only analog output.

The application programming interface (API) summary for analog output is shown in Table 4.1. It follows a pattern similar to the digital input and output utilities we have already seen. Here, with **AnalogOut**, we can initialize and name an output; using **write( )** or **write_u16( )**, we can set the output voltage either with a floating point number or with a hexadecimal number. Finally, we can simply use the = sign as a shorthand for write, which is what we will mostly be doing.

C code feature

Notice in Table 4.1 the way that data types *float*, and *unsigned short* are invoked. In C/C++ all data elements have to be declared before use; the same is true for the return type of a function and the parameters it takes. A review of the concept of *floating point* number representation can be found in Appendix A. Table B.4 in Appendix B summarizes the different data types that are available. The DAC itself requires an unsigned binary number as input, so the **write_u16( )** function represents the more direct approach of writing to it.

Table 4.1: API summary for mbed analog output

Function	Usage
`AnalogOut`	Create an AnalogOut object connected to the specified pin
`write`	Set the output voltage, specified as a percentage (float)
`write_u16`	Set the output voltage, represented as an unsigned short in the range [0x0, 0xFFFF]
`read`	Return the current output voltage setting, measured as a percentage (float)
`operator=`	An operator shorthand for write()

We try now a few simple programs which apply the mbed DAC, creating first fixed voltages and then waveforms.

4.2.1 Creating Constant Output Voltages

Create a new program using Program Example 4.1. In this we create an analog output labeled **Aout**, by using the **AnalogOut** utility. It is then possible to set the analog output simply by setting **Aout** to any permissible value; we do this three times in the program. By default, **Aout** takes a floating point number between 0.0 and 1.0 and outputs this to pin 18. The actual output voltage on pin 18 is between 0 V and 3.3 V, so the floating point number that you output is scaled to this.

```
/*Program Example 4.1: Three values of DAC are output in turn on Pin 18.
Read the output on a DVM.
*/
#include "mbed.h"
AnalogOut Aout(p18); //create an analog output on pin 18
int main() {
  while(1) {
  Aout=0.25;            // 0.25*3.3V = 0.825V
  wait(2);
  Aout=0.5;             // 0.5*3.3V = 1.65V
  wait(2);
  Aout=0.75;            // 0.75*3.3V = 2.475V
  wait(2);
  }
}
```

Program Example 4.1 Trial DAC output

Compile the program in the usual way, and let it run. Connect a digital voltmeter (DVM) between pins 1 and 18 of the mbed. You should see the three output voltages named in the comments of Program Example 4.1 being output in turn.

Exercise 4.1

Adjust Program Example 4.1 so that the **write_u16( )** function is used to set the analog output, giving the same output voltages.

4.2.2 Sawtooth Waveforms

We will now make a sawtooth wave and view it on an oscilloscope. Create a new program and enter the code of Program Example 4.2.

As in many cases previously, the program is made up of an endless **while(1)** loop. Within this there is a **for** loop. In this example, the variable **i** is initially set to 0, and on each iteration of the loop it is incremented by 0.1. The new value of **i** is applied within the loop. When **i** reaches 1, the loop terminates. As the **for** loop is the only code within the endless **while** loop, it repeats continuously.

```
/*Program Example 4.2: Sawtooth waveform on DAC output. View on oscilloscope
*/
#include "mbed.h"
AnalogOut Aout(p18);
float i;
int main() {
  while (1){
    for (i=0;i<1;i=i+0.1){    // i is incremented in steps of 0.1
      Aout=i;
      wait(0.001);             // wait 1 millisecond
    }
  }
}
```

Program Example 4.2 Sawtooth waveform

Connect an oscilloscope probe to pin 18 of the mbed, with its earth connection to pin 1. Check that you get a sawtooth waveform similar to that shown in Figure 4.3. Ensure that the duration

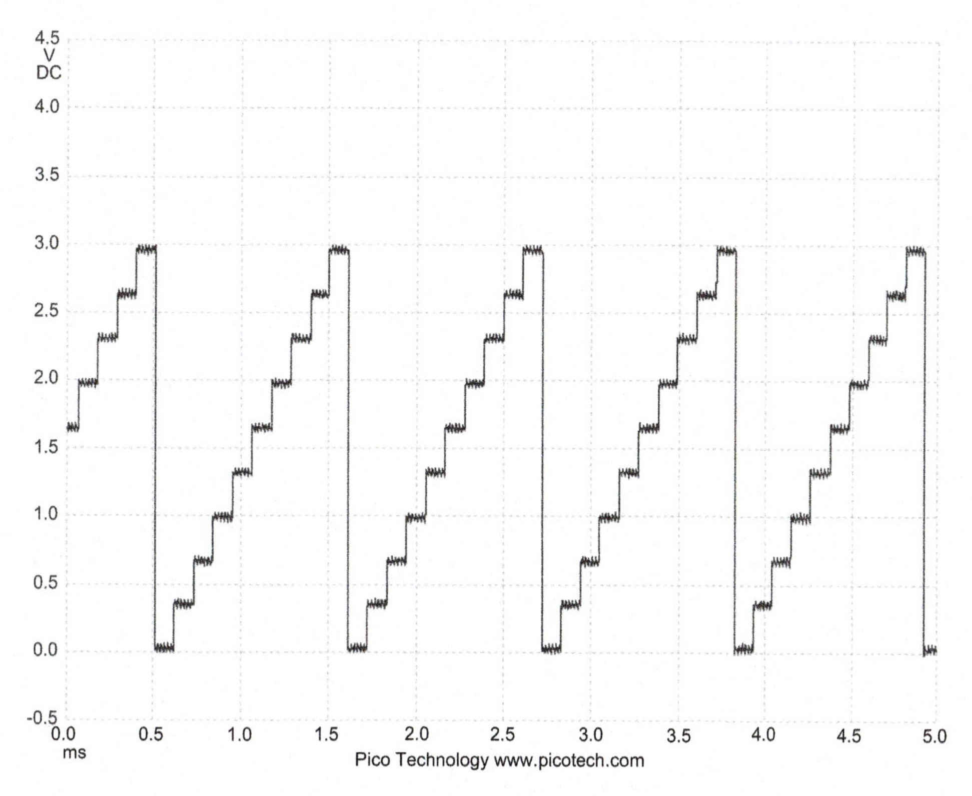

Figure 4.3:
A stepped sawtooth waveform

of each step is the 1 ms you define in the program, and try varying this. The waveform should start from 0 V and not go higher than 3.3 V. Explain the maximum value you see.

If you do not have an oscilloscope you can set the wait parameter to be much longer (say 100 ms) and use the DVM; you should then see the voltage step up from 0 V to 3.3 V and then reset back to 0 V again.

Exercise 4.2

Improve the resolution of your sawtooth by having more but smaller increments, i.e. reduce the value by which **i** increments. The result should be as seen in Figure 4.4.

Exercise 4.3

Create a new project and devise a program which outputs a triangular waveform (i.e. one that counts down as well as up). The oscilloscope output should look like Figure 4.5.

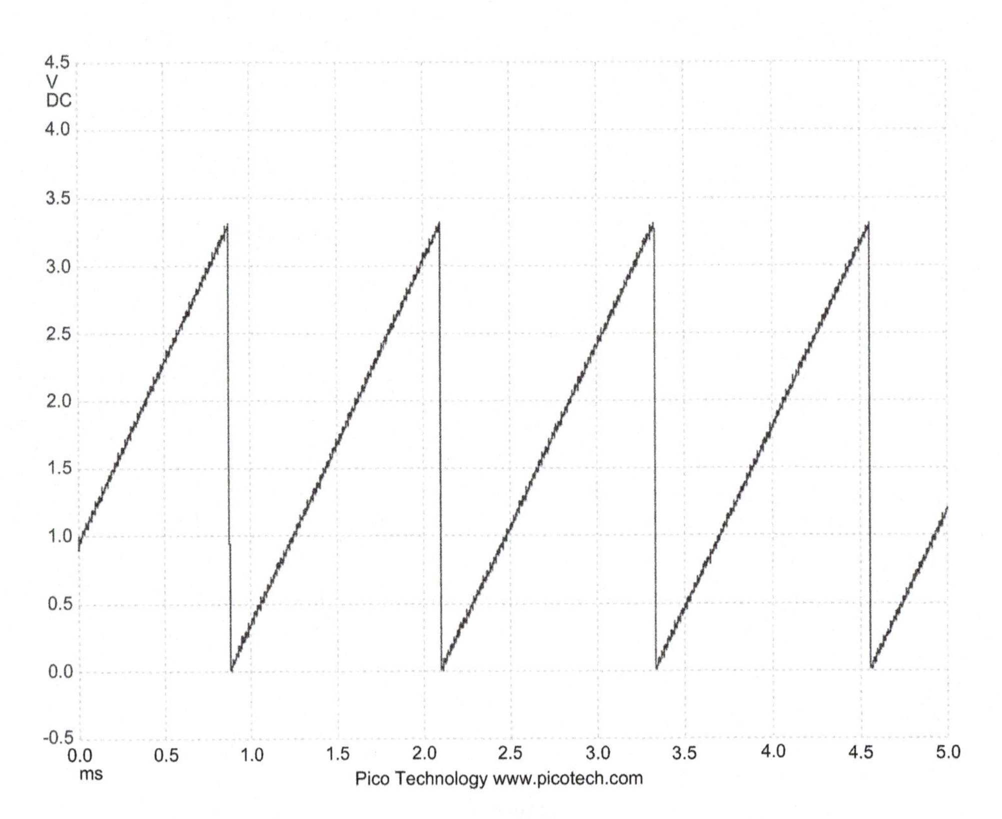

Figure 4.4:
A smooth sawtooth waveform

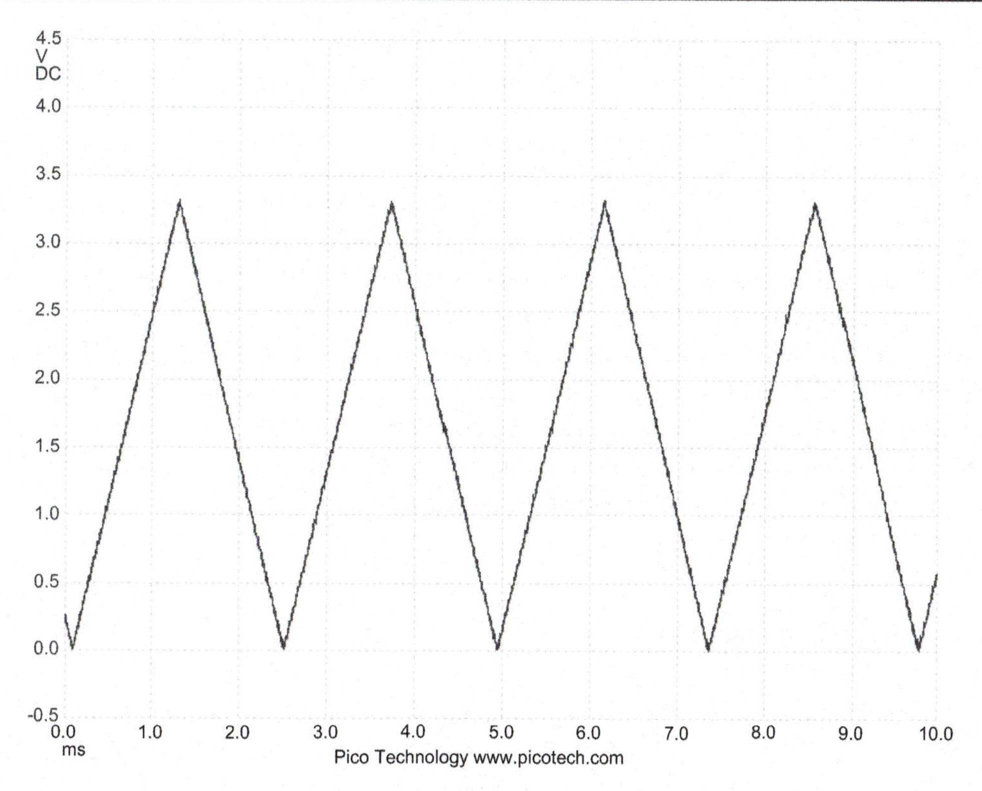

Figure 4.5:
A triangular waveform

4.2.3 Testing the DAC Resolution

Now let us return to the question of the DAC resolution, which we touched on in Section 4.1.1. Try changing the **for** loop in your version of Program Example 4.2 to this:

```
for (i=0;i<1;i=i+0.0001){
  Aout=i;
  wait(1);
  led1=!led1;
}
```

We have also included a little LED indication here, so add this line before **main( )** to set it up:

```
DigitalOut led1(LED1);
```

C code feature

This adjusted program will produce an extremely slow sawtooth waveform, which will take 10 000 steps to reach the maximum value, each one taking a second (hence the period of the waveform is 10 000 seconds, or two and three quarter hours!). Our purpose is not, however, to view the waveform, but to explore carefully the DAC characteristic of Figure 4.2. Notice the use of the NOT operator, in the form of an exclamation mark (!), for the first time. This causes logical inversion, so a Logic 0 is replaced by 1, and vice versa.

Switch your DVM to its finest voltage range, so that you have millivolt resolution on the scale; the 200 mV scale is useful here. Connect the DVM between pins 1 and 18, and run the program. The LED changes state every time a new value is output to the DAC. However, you will notice an interesting thing. Your DVM reading does not change with every LED change. Instead it changes state in distinct steps, with each change being around 3 mV. You will notice also that it takes around five LED 'blinks', i.e. 10 updates of the DAC value, before each DAC change. All this was anticipated in Section 4.1.1, where a step size of 3.22 mV was predicted; each step size is equal to the DAC resolution. The float value is rounded to the nearest digital input to the DAC, and it takes around 10 increments of the float value for the DAC digital input to be incremented by one.

4.2.4 Generating a Sine Wave

C code feature

It is an easy step from here to generate a sine wave. We will apply the **sin( )** function, which is part of the C standard library (see Section B.9.2). Take a look at Program Example 4.3. To produce one cycle of the sine wave, we want to take sine values of a number which increases from 0 to 2π radians. In this program we use a **for** loop to increment variable **i** from 0 to 2 in small steps, and then multiply that number by π when the sine value is calculated. There are other ways of getting this result. The final challenge is that the DAC cannot output negative values. Therefore, we add a fixed value (an *offset*) of 0.5 to the number being sent to the DAC; this ensures that all output values lie within its available range. Notice that we are using the multiply operator, *, for the first time.

```
/*Program Example 4.3: Sine wave on DAC output. View on oscilloscope
*/
#include "mbed.h"
AnalogOut Aout(p18);
float i;
int main() {
   while(1) {
      for (i=0;i<2;i=i+0.05) {
         Aout=0.5+0.5*sin(i*3.14159); // Compute the sine value, + half the range
         wait(.001);                  // Controls the sine wave period
      }
   }
}
```

Program Example 4.3 Generating a sinusoidal waveform

Exercise 4.4

Observe on the oscilloscope the sine wave that Program Example 4.3 produces. Estimate its frequency from information in the program, and then measure it. Do the two values agree? Try varying the frequency by varying the wait parameter. What is the maximum frequency you can achieve with this program?

4.3 Another Form of Analog Output: Pulse Width Modulation

The DAC is a fine circuit and it has many uses. Yet it adds complexity to a microcontroller, and sometimes moves us to the analog domain before we are ready to go. Pulse width modulation (PWM) is an alternative; indeed, the mbed has six PWM outputs but only one analog output. PWM represents a neat and remarkably simple way of getting a rectangular digital waveform to control an analog variable, usually voltage or current. PWM control is used in a variety of applications, ranging from telecommunications to robotic control.
A PWM signal is shown in Figure 4.6. The period is normally kept constant, and the pulse width, or 'on' time, is varied, hence the name. The *duty cycle* is the proportion of time that the pulse is 'on' or 'high', and is expressed as a percentage, i.e.:

$$\text{duty cycle} = \frac{\text{pulse on time}}{\text{pulse period}} * 100\% \qquad 4.2$$

A 100% duty cycle therefore means 'continuously on' and a 0% duty cycle means 'continuously off'. PWM streams are easily generated by digital counters and comparators, which can readily be designed into a microcontroller. They can also be produced simply by program loops and a standard digital output, with no dedicated hardware at all. This is seen later in the chapter.

Whatever duty cycle a PWM stream has, there is an average value, as indicated by the dotted line in Figure 4.6. If the on time is small, the average value is low; if the on time is large, the average value is high. By controlling the duty cycle, therefore, this average value is controlled. When using PWM, it is this average that we are usually interested in. It can be extracted from the PWM

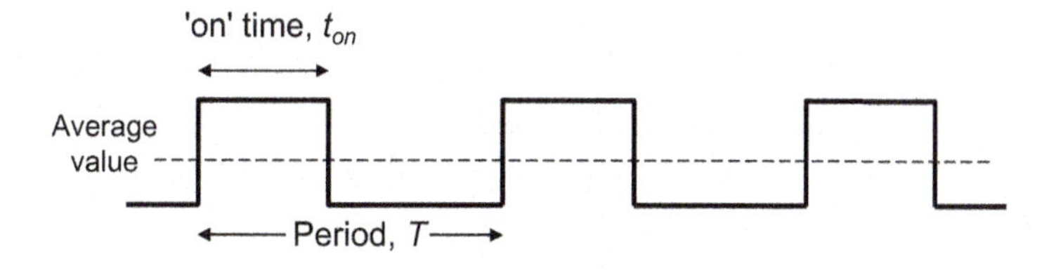

Figure 4.6:
A PWM waveform

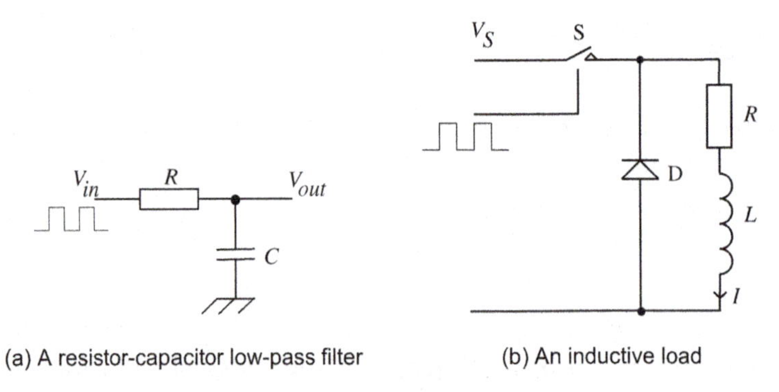

Figure 4.7:
Simple averaging circuits

stream in a number of ways. Electrically, we can use a low-pass filter, e.g. the resistor-capacitor combination of Figure 4.7a. In this case, and as long as PWM frequency and values of R and C are appropriately chosen, V_{out} becomes an analog output, with a bit of ripple, and the combination of PWM and filter acts just like a DAC. Alternatively, if we switch the current flowing in an inductive load, as seen in Figure 4.7b, then the inductance has an averaging effect on the current flowing through it. This is very important, as the windings of any motor are inductive, so this technique can be used for motor control. The switch in Figure 4.7b is controlled by the PWM stream, and can be a transistor. We have incidentally to introduce the freewheeling diode, just as we did in Figure 3.12c, to provide a current path as the switch opens.

In practice, this electrical filtering is not always required. Many physical systems have internal inertias which, in reality, act like low-pass filters. We can, for example, dim a conventional filament light bulb with PWM. In this case, varying the pulse width varies the average temperature of the bulb filament, and the dimming effect is achieved.

As an example, the control of a DC motor is a very common task in robotics; the speed of a DC motor is proportional to the applied DC voltage. We could use a conventional DAC output, drive it through an expensive and bulky power amplifier, and use the amplifier output to drive the motor. Alternatively, a PWM signal can be used to drive a power transistor directly, which replaces the switch in Figure 4.7b; the motor is the inductor/resistor combination in the same circuit. This technique is taken much farther in the field of power electronics, with the PWM concept being taken far beyond these simple but useful applications.

4.4 Pulse Width Modulation on the mbed

4.4.1 Using the mbed PWM Sources

As mentioned, it is easy to generate PWM pulse streams using simple digital building blocks. We do not explore how that is done in this book, but you can find the information

Table 4.2: API summary for PWM output

Function	Usage
`PwmOut`	Create a PwmOut object connected to the specified pin
`write`	Set the output duty cycle, specified as a normalized float (0.0–1.0)
`read`	Return the current output duty-cycle setting, measured as a normalized float (0.0–1.0)
`period`	Set the PWM period, specified in seconds (float), keeping the duty cycle the same
`period_ms`	Set the PWM period, specified in milliseconds (int), keeping the duty cycle the same
`period_us`	Set the PWM period, specified in microseconds (int), keeping the duty cycle the same
`pulsewidth`	Set the PWM pulse width, specified in seconds (float), keeping the period the same
`pulsewidth_ms`	Set the PWM pulse width, specified in milliseconds (int), keeping the period the same
`pulsewidth_us`	Set the PWM pulse width, specified microseconds (int), keeping the period the same
`operator =`	An operator shorthand for write()

elsewhere if you wish, for example in Reference 1.1. As for the mbed, Figure 2.1c shows that there are six PWM outputs available, from pin 21 to 26 inclusive. Note that on the LPC1768 microcontroller, and hence the mbed, the PWM sources all share the same period/ frequency; if the period is changed for one, then it is changed for all.

As with all peripherals, the mbed PWM ports are supported by library utilities and functions, as shown in Table 4.2. This is rather more complex than the API tables we have seen so far, so we can expect a little more complexity in its use. Notably, instead of having just one variable to control (e.g. a digital or an analog output), we now have two, period and pulse width, or duty cycle derived from this combination. Similar to previous examples, a PWM output can be established, named and allocated to a pin using **PwmOut**. Subsequently, it can be varied, by setting its period, duty cycle or pulse width. As shorthand, the **write( )** function can simply be replaced by =.

4.4.2 Some Trial PWM Outputs

As a first program using an mbed PWM source, let's create a signal that can be seen on an oscilloscope. Make a new project, and enter the code of Program Example 4.4. This will generate a 100 Hz pulse with 50% duty cycle, i.e. a perfect square wave.

```
/*Sets PWM source to fixed frequency and duty cycle. Observe output on oscilloscope.
*/
#include "mbed.h"
PwmOut PWM1(p21);        //create a PWM output called PWM1 on pin 21
int main() {
   PWM1.period(0.010);   // set PWM period to 10 ms
   PWM1=0.5;             // set duty cycle to 50%
}
```

Program Example 4.4 Trial PWM output

In this program example we first set the PWM period. There is no shorthand for this, so we have to use the full **PWM1.period(0.010);** statement. We then define the duty cycle as a decimal number, between the values of 0 and 1. We could also set the duty cycle as a pulse time with the following:

```
PWM1.pulsewidth_ms(5);          // set PWM pulsewidth to 5 ms
```

When you run the program you should be able to see the square wave on the oscilloscope, and verify the output frequency.

Exercise 4.5

1. Change the duty cycle of Program Example 4.4 to some different values, say 0.2 (20%) and 0.8 (80%), and check the correct display is seen on the oscilloscope, for example as shown in Figure 4.8.
2. Change the program to give the same output waveforms, but using **period_ms( )** and **pulsewidth_ms( )**.

4.4.3 Speed Control of a Small Motor

We will now apply the mbed PWM source to control motor speed. Use the simple motor circuit we used in the previous chapter (Figure 3.13), and create a project using Program Example 4.5. This ramps the PWM up from a duty cycle of 0% to 100%, using programming features that we know.

```
/*Program Example 4.5: PWM control to DC motor is repeatedly ramped
*/
#include "mbed.h"
PwmOut PWM1(p21);
float i;
int main() {
  PWM1.period(0.010);          //set PWM period to 10 ms
  while(1) {
    for (i=0;i<1;i=i+0.01) {
      PWM1=i;                  // update PWM duty cycle
      wait(0.2);
    }
  }
}
```

Program Example 4.5 Controlling motor speed with mbed PWM source

Compile, download and run the program. See how your motor performs and observe the waveform on the oscilloscope. You are seeing one of the classic applications of PWM, controlling motor speed through a simple digital pulse stream.

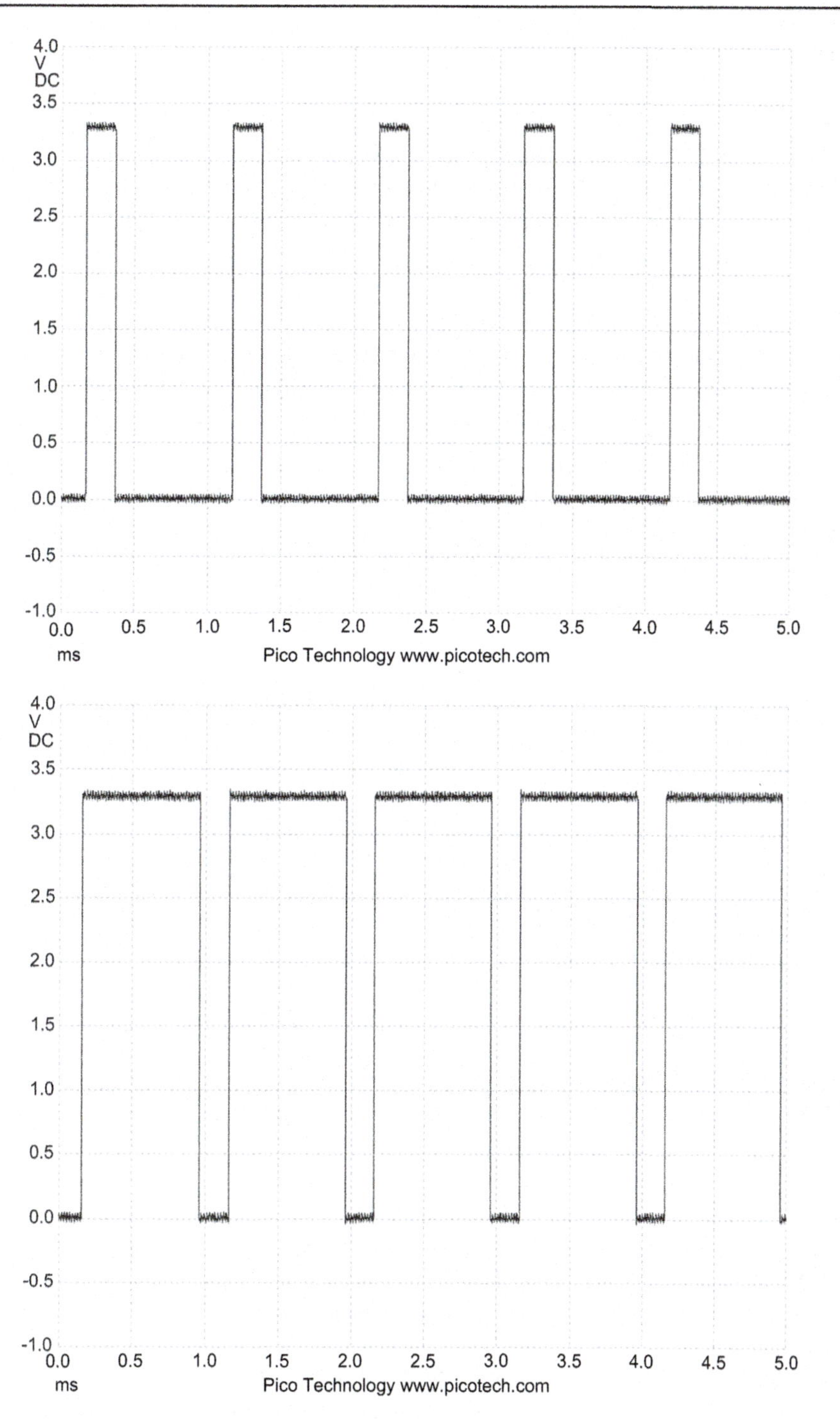

Figure 4.8:
PWM observed from the mbed output

4.4.4 Generating PWM in Software

Although we have just used PWM sources that are available on the mbed, it is useful to realize that these are not essential for creating PWM; we can actually do it just with a digital output and some timing. In Exercise 3.8 you were asked to write a program that switched a small DC motor on and off continuously, 1 second on, 1 second off. If you speed this switching up, to say 1 ms on, 1 ms off, you will immediately have a PWM source.

Create a new project with Program Example 4.6 as source code. Notice carefully what the program does. There are two **for** loops in the main **while** loop. The motor is initially switched off for 5 s. The first **for** loop then switches the motor on for 400 μs and off for 600 us; it does this 5000 times. This results in a PWM signal of period 1 ms and duty cycle 40%. The second **for** loop switches the motor on for 800 μs and off for 200 us; the period is still 1 ms but the duty cycle is 80%. The motor is then switched full on for 5 s. This sequence continues indefinitely.

```
/*Program Example 4.6: Software generated PWM. 2 PWM values generated in turn, with full
on and off included for comparison.
*/
#include "mbed.h"
DigitalOut motor(p6);
int i;
int main() {
 while(1) {
      motor = 0;                    //motor switched off for 5 secs
      wait (5);
      for (i=0;i<5000;i=i+1) {      //5000 PWM cycles, low duty cycle
        motor = 1;
        wait_us(400);               //output high for 400us
        motor = 0;
        wait_us(600);               //output low for 600us
      }
      for (i=0;i<5000;i=i+1) {      //5000 PWM cycles, high duty cycle
        motor = 1;
        wait_us(800);               //output high for 800us
        motor = 0;
        wait_us(200);               //output low for 200us
      }
      motor = 1;                    //motor switched fully on for 5 secs
      wait (5);
 }
}
```

Program Example 4.6 Generating PWM in software

Compile and download this program, and apply the circuit of Figure 3.12. You should find that the motor runs with the speed profile indicated. PWM period and duty cycle can be readily verified on the oscilloscope. You may wonder whether it is necessary to have dedicated PWM ports, when it seems quite easy to generate PWM with software. Remember, however, that in this example the central processing unit (CPU) becomes totally committed to this task and can do nothing else. With the hardware ports, we can set the PWM running and the CPU can then get on with some completely different activity.

Exercise 4.6

Vary the motor speeds in Program Example 4.6 by changing the duty cycle of the PWM, initially keeping the frequency constant. Depending on the motor you use, you will probably find that for small values of duty cycle the motor will not run at all, owing to its own friction. This is particularly true of geared motors. Observe the PWM output on the oscilloscope, and confirm that on and off times are as indicated in the program. Try also at much higher and lower frequencies.

4.4.5 Servo Control

A servo is a small rotary position control device, often used in radio-controlled cars and aircraft to control the angular position of variables such as steering, elevators and rudders. Servos are now popular in certain robotic applications. The servo shaft can be positioned at specific angular positions by sending the servo a PWM signal. As long as the modulated signal exists on the input line, the servo will maintain the angular position of the shaft. As the modulated signal changes, the angular position of the shaft changes. This is illustrated in Figure 4.9. Many servos use a PWM signal with a 20 ms period, as is shown here. In this example, the pulse width is modulated from 1.25 ms to 1.75 ms to give the full 180 degree range of the servo.

Connect a servo to the mbed as indicated in Figure 4.10. The servo requires a higher current than the universal serial bus (USB) standard can provide, so it is essential that you power

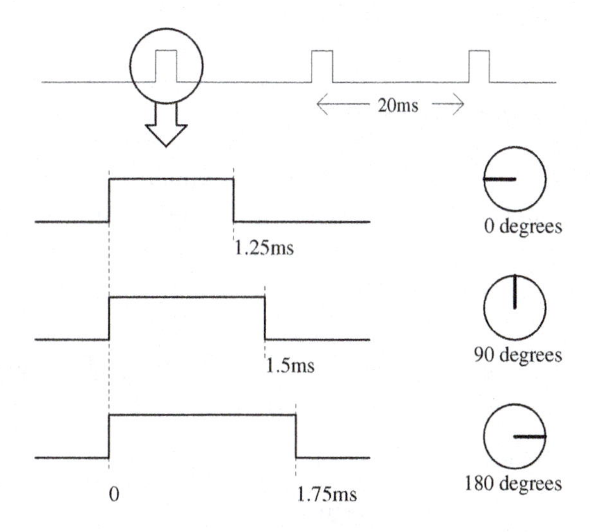

Figure 4.9:
Servo characteristics

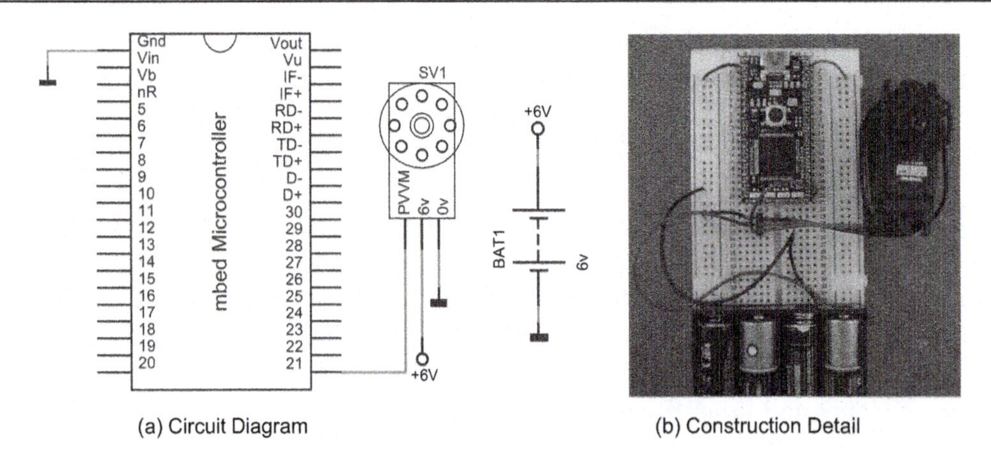

(a) Circuit Diagram (b) Construction Detail

Figure 4.10:
Using PWM to drive a servo

it using an external supply. A 4 × AA (6 V) battery pack meets the supply requirement of the servo. Note that the mbed itself can still be supplied from the USB. Alternatively, because the pack voltage lies within the permissible input voltage range for pin 2 mentioned in Figure 2.1, the mbed could be supplied from the battery pack, through this pin. The mbed then regulates the incoming 6 V to the 3.3 V that it requires.

Exercise 4.7

Create a new project and write a program that sets a PWM output on pin 21. Set the PWM period to 20 ms. Try a number of different duty periods, taking values from Figure 4.9, and observe the servo's position. Then write a program that continually moves the servo shaft from one limit to the other.

4.4.6 Outputting to a Piezo Transducer

We can use the PWM source simply as a variable frequency signal generator. In this example we use it to sound a piezo transducer and play the start of an old London folk song called 'Oranges and Lemons'. This song imagines that the bells of each church in London call a particular message. If you can read music you may recognize the tune in

Figure 4.11:
The 'Oranges and Lemons' tune

Table 4.3: Frequencies of notes used in tune

Word/syllable	Musical note	Frequency (Hz)	Beats
Oran-	E	659	1
ges	C#	554	1
and	E	659	1
le-	C#	554	1
mons,	A	440	1
say	B	494	½
the	C#	554	½
bells	D	587	1
of	B	494	1
St	E	659	1
Clem-	C#	554	1
ent's	A	440	2

Figure 4.11; if you cannot, don't worry! You just need to know that any note that is a minim ('half note' in the USA) lasts twice as long as a crotchet ('quarter note' in the USA), which in turn lasts twice as long as a quaver ('eighth note' in the USA). Put another way, a crotchet lasts one beat, a minim two, and a quaver a half. The pattern for the music is as shown in Table 4.3. Note that here we are simply using the PWM as a variable frequency signal source, and not actually modulating the pulse width as a proportion of frequency at all.

C code feature

Create a new program and enter Program Example 4.7. This introduces an important new C feature, the *array*. If you are unfamiliar with this, then read the review in Section B.8.1. The program uses two arrays, one defined for frequency data, the other for beat length. There are 12 values in each, for each of the 12 notes in the tune. The program is structured round a **for** loop, with variable **i** as counter. As **i** increments, each array element is selected in turn. Notice that **i** is set just to reach the value 11; this is because the value 0 addresses the first element in each array, and the value 11 hence addresses the twelfth. From the frequency array the PWM period is calculated and set, always with a 50% duty ratio. The beat array determines how long each note is held, using the **wait** function.

```
/*Program Example 4.7: Plays the tune "Oranges and Lemons" on a piezo buzzer, using PWM
*/
#include "mbed.h"
PwmOut buzzer(p21);
```

```
                                                  //frequency array
float frequency[]={659,554,659,554,440,494,554,587,494,659,554,440};
float beat[]={1,1,1,1,1,0.5,0.5,1,1,1,1,2};      //beat array
int main() {
  while (1) {
    for (int i=0;i<=11;i++) {
      buzzer.period(1/(2*frequency[i]));          // set PWM period
    buzzer=0.5;                    // set duty cycle
    wait(0.4*beat[i]);             // hold for beat period
    }
  }
}
```

Program Example 4.7 'Oranges and Lemons' program

Compile the program and download. Connect the piezo transducer between pins 1 and 21, and the sequence should play on reset. The transducer seems very quiet when held in air. You can increase the volume significantly by fixing or holding it to a flat surface such as a table top.

Exercise 4.8

Try the following:

1. Make the 'Oranges and Lemons' sequence play an octave higher by doubling the frequency of each note.
2. Change the tempo by modifying the multiplier in the wait command.
3. (For the more musically inclined.) Change the tune, so that the mbed plays the first line of 'Twinkle Twinkle Little Star'. This uses the same notes, except it also needs F#, of frequency 699 Hz. The tune starts on A. Because there are repeated notes, consider putting a small pause between each note.

Chapter Review

- A digital-to-analog converter (DAC) converts an input binary number to an output analog voltage, which is proportional to that input number.
- DACs are widely used to create continuously varying voltages, for example to generate analog waveforms.
- The mbed has a single DAC, and an associated set of library functions.
- Pulse width modulation (PWM) provides a way of controlling certain analog quantities, by varying the pulse width of a fixed frequency rectangular waveform.
- PWM is widely used for controlling flow of electrical power, for example LED brightness or motor control.
- The mbed has six possible PWM outputs. They can all be individually controlled, but must all share the same frequency.

Quiz

1. A 7-bit DAC obeys Equation 4.1, and has a voltage reference range of 2.56 V.
   (a) What is its resolution?
   (b) What is its output if the input is 100 0101?
   (c) What is its output if the input is 0x2A?
   (d) What is its digital input in decimal and binary if its output reads 0.48 V?
2. What is the mbed's DAC resolution and what is the smallest analog voltage step increase or decrease which can be output from the mbed?
3. What is the output of the LPC1768 DAC, if its input digital word is:
   (a) 00 0000 1000
   (b) 0x80
   (c) 10 1000 1000 ?
4. What output voltages will be read on a DVM while this program loop runs on the mbed?

```
while(1){
  for (i=0;i<1;i=i+0.2){
     Aout=i;
     wait(0.1);
  }
}
```

5. The waveform in Question 5 gives a crude sawtooth. What is its period?
6. What are the advantages of using pulse width modulation (PWM) for control of analog actuators?
7. A PWM data stream has a frequency of 4 kHz and a duty cycle of 25%. What is its pulse width?
8. A PWM data stream has period of 20 ms and an on time of 1 ms. What is its duty cycle?
9. The PWM on an mbed is set up with these statements. What is the on time of the waveform?

```
PWM1.period(0.004);    // set PWM period
PWM1=0.75;             // set duty cycle
```

Reference

4.1. Kestner, W., Ed. (2005). The Data Conversion Handbook. Analog Devices Inc. Newnes.

CHAPTER 5

Analog Input

Chapter Outline

5.1 Analog-to-Digital Conversion

The world around the embedded system is a largely analog one, and sensors – of temperature, sound, acceleration, and so on – mostly have analog outputs. Yet it is essential for the microcontroller to have these signals available in digital form. This is where the analog-to-digital converter (ADC) comes in. Analog signals can be repeatedly converted into digital representations, with a resolution and at a rate determined by the ADC. Following this analog-to-digital conversion, the microcontroller can be used to process or analyze this information, based on the value(s) of the analog input.

Fast and Effective Embedded Systems Design. DOI: 10.1016/B978-0-08-097768-3.00005-2

5.1.1 The Analog-to-Digital Converter

An ADC is an electronic circuit whose digital output is proportional to its analog input. Effectively, it 'measures' the input voltage and gives a binary output number proportional to its size. The list of possible analog input signals is endless, including such diverse sources as audio and video, medical or climatic variables, and a host of industrially generated signals. Of these, some, such as temperature, have a very low frequency content. Others, video for example, are very high frequency. Owing to this huge range of applications, it is not surprising to discover that many types of ADC have been developed, with characteristics optimized for these differing applications.

The ADC almost always operates within a larger environment, often called a data acquisition system. Some features of a general purpose data acquisition system are shown in Figure 5.1. To the right of the diagram is the ADC itself. This has an analog input and a digital output. It is under computer control; the computer can start a conversion. The conversion takes finite time, maybe some microseconds, so the ADC needs to signal when it has finished. The output data can then be read. The ADC works with a voltage reference, which may be thought of as a ruler or tape measure. In one way or another the ADC compares the input voltage with the voltage reference, and comes up with the output number depending on this comparison. As with so

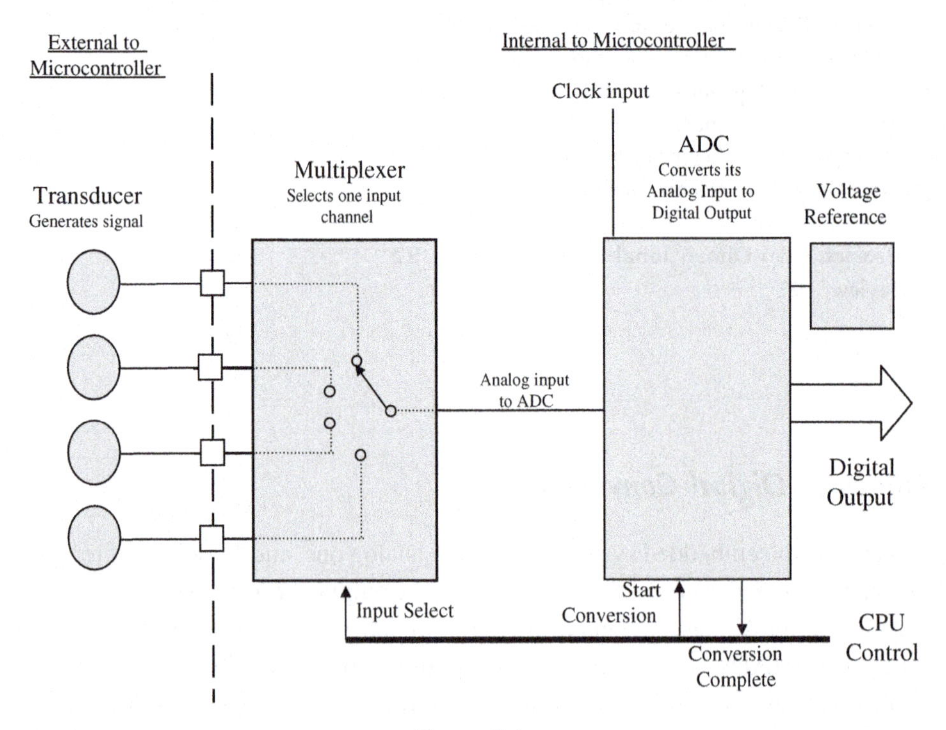

Figure 5.1:
An example data acquisition system

many digital or digital/analog subsystems, there is also a clock input, a continuously running square wave, which sequences the internal operation of the ADC. The frequency set here determines how rapidly the ADC operates.

Once we start working with an ADC, we usually find that we want to work with more than one signal. We could just use more ADCs, but this is costly and takes up semiconductor space. Instead, the usual practice is to put an analog multiplexer in front of the ADC. This acts as a selector switch. The user can then select any one of several inputs to the ADC. If this is done quickly enough, it is as if all inputs are being converted at the same time. Many microcontrollers, including the NXP1768, include an ADC and multiplexer on chip. The inputs to the multiplexer are connected to microcontroller pins and multiple inputs can be used. This is as shown in Figure 5.1 for a 4-bit multiplexer. We return to some of the detail of Figure 5.1 in Chapter 14.

5.1.2 Range, Resolution and Quantization

Many ADCs obey Equation 5.1, where V_i is the input voltage, V_r the reference voltage, n the number of bits in the converter output, and D the digital output value. The output binary number D is an integer, and for an n-bit number can take any value from 0 to $(2^n - 1)$. The internal ADC process effectively rounds or truncates the calculation in Equation 5.1 to produce an integer output. Clearly, the ADC cannot just convert any input voltage, but has maximum and minimum permissible input values. The difference between this maximum and minimum is called the *range.* Often the minimum value is 0 V, so the range is then just the maximum possible input value. Analog inputs that exceed the maximum or minimum permissible input values are likely to be digitized as the maximum and minimum values respectively, i.e. a limiting (or 'clipping') action takes place. The input range of the ADC is directly linked to the value of the voltage reference; in many ADC circuits the range is actually equal to the reference voltage.

$$D = \frac{V_i}{V_r} \times 2^n \quad 5.1$$

Equation 5.1 is represented in graphical form in Figure 5.2, for a 3-bit converter. If the input voltage is gradually increased from 0 V and the converter is running continuously, then the ADC's output is initially 000. If the input slowly increases, there comes a point when the output will take the value 001. As the input increases further, the output changes to 010, and so on. At some point it reaches 111, i.e. 7 in decimal, or $(2^3 - 1)$. This is the maximum possible output value. The input may be increased further, but it cannot force any increase in output value.

As Figure 5.2 demonstrates, by converting an analog signal to digital, we run the risk of approximating it. This is because any one digital output value has to represent a small range of analog input voltages, i.e. the width of any of the steps on the 'staircase' of Figure 5.2. For

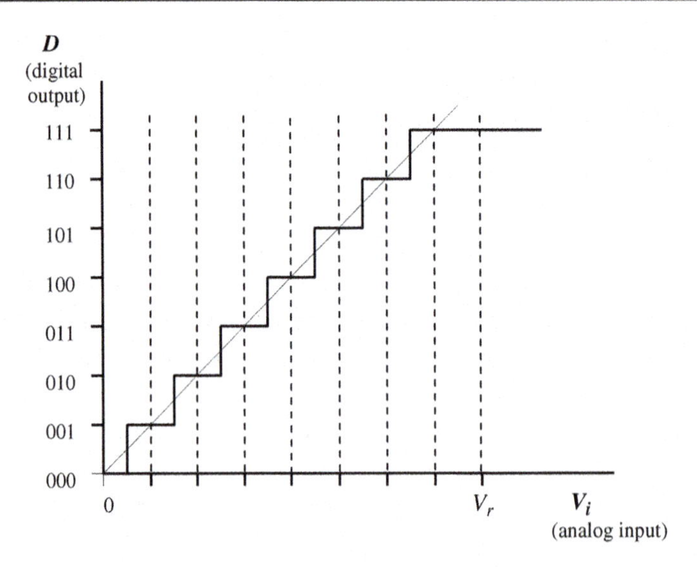

Figure 5.2:
A 3-bit ADC characteristic

example, the digital output 001 in the figure must represent *any* analog voltage along the step that it represents. If the output value of 001 is precisely correct for the input voltage at the middle of the step, then the greatest error occurs at either end of the step. This is called *quantization error.* Following this line of reasoning, the greatest quantization error is one half of the step width, or half of one least significant bit (LSB) equivalent of the voltage scale. Clearly, the more steps there are representing the range, the narrower they will be, and hence the quantization error is reduced. More steps are obtained by increasing the number of bits in the ADC process. This inevitably increases the complexity and cost of the ADC, and also often the time it takes to complete a conversion.

As an example, if we want to convert an analog signal that has a range 0–3.3 V to an 8-bit digital signal, then there are 256 (i.e. 2^8) distinct output values. Each step has a width of $3.3/256 = 12.89$ mV, and the worst case quantization error is 6.45 mV. As far as the mbed is concerned, Figure 2.3 shows that the LPC1768 ADC is 12 bit. This leads to a step width of $3.3/2^{12}$, or 0.8 mV; the worst case quantization error is therefore 0.4 mV.

For many applications an 8, 10 or 12-bit ADC allows sufficient resolution; it all depends on the required accuracy. In certain audio applications listening tests have demonstrated that 16-bit resolution is adequate; improved quality can, however, be noticed using 24-bit conversion.

All of this assumes that all other aspects of the ADC are perfect, which they are not. The reference voltage can be inaccurate or can drift with temperature, and the staircase pattern can have non-linearities. Furthermore, different methods of analog-to-digital conversion each bring their unique inaccuracies to the equation. Reference 5.1 discusses in detail a number of analog-to-digital conversion designs, including, 'successive approximation', 'flash', 'dual slope' and 'delta-sigma' methods, which all have their own advantages and inaccuracies. Discussing the specific design of the ADC system, however, is not necessary for us to understand the concepts of data conversion and implementing the mbed's ADC.

5.1.3 Sampling Frequency

When converting an analog signal to digital, a 'sample' is taken repeatedly and quantized to the accuracy defined by the resolution of the ADC. The more samples taken, the more accurate the digital data will be. Sampling is generally done at a fixed frequency, called the sampling frequency. This is illustrated in Figure 5.3.

The sampling frequency depends on the maximum frequency of the signal being digitized. If the sampling frequency is too low then rapid changes in the analog signal may not be represented in the resulting digital data. The Nyquist sampling criterion states that the sampling frequency must be at least double that of the highest signal frequency. For example, the human auditory system is known to extend up to approximately 20 kHz, so standard audio CDs are sampled and played back at 44.1 kHz in order to adhere to the Nyquist sampling criterion. If the sampling criterion is not satisfied, then a phenomenon called *aliasing* occurs – a new lower frequency is generated. This is illustrated in Figure 5.4 and demonstrated later in the chapter. Aliasing is very damaging to a signal and must always be avoided. A common

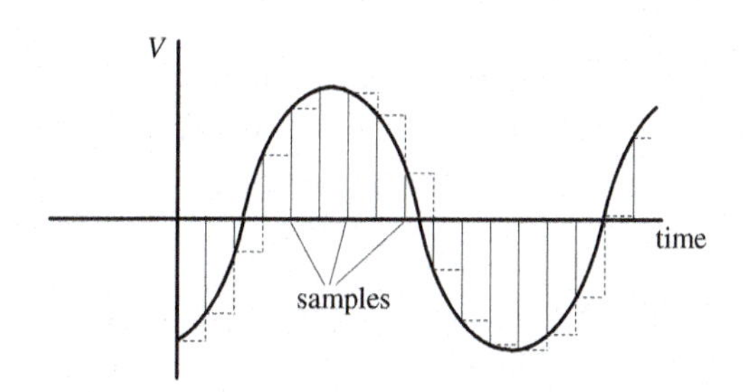

Figure 5.3:
Digitizing a sine wave

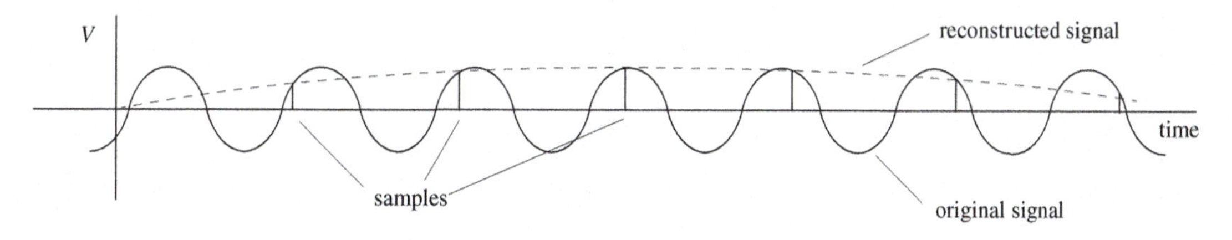

Figure 5.4:
The effect of aliasing

approach is to use an *anti-aliasing filter*, which limits all signal components to those that satisfy the sampling criterion.

5.1.4 Analog Input with the mbed

Figure 2.1 shows us that the mbed has up to six analog inputs, on pins 15–20. The application programming interface (API) summary, following a pattern that is quite familiar, is shown in Table 5.1. It is useful to note that the ADC output is available either in unsigned binary form (as it would be at the ADC output) or as a floating point number.

5.2 Combining Analog Input and Output

The ADC is an input device, which transfers data *into* the microcontroller. If we use it on its own, we will have no idea what values it has created. We now therefore go on to do two things to make that data visible. We will first use the ADC output values to immediately control an output variable, for example the digital-to-analog converter (DAC) or pulse width modulation (PWM). We will later transfer its output values to the PC screen, and explore some measurement applications.

5.2.1 Controlling LED Brightness by Variable Voltage

Let us start with a simple program which reads the analog input, and uses it to control the brightness of a light-emitting diode (LED) by varying the voltage drive to the LED. Here, we will use a potentiometer to generate the analog input voltage, and will then pass the value read straight to the analog output.

Connect up the circuit of Figure 5.5a. This uses pin 20 as the analog input, connecting the potentiometer between 0 and 3.3 V. The LED is connected to pin 18, the analog output. Start a new program and copy into it the very simple code of Program Example 5.1. This just sets up the analog input and output, and then continuously transfers the input to the output.

```
/*Program Example 5.1: Uses analog input to control LED brightness, through DAC output
*/
#include "mbed.h"
AnalogOut Aout(p18);  //defines analog output on Pin 18
AnalogIn Ain(p20)     //defines analog input on Pin 20
int main() {
  while(1) {
    Aout=Ain;         //transfer analog in value to analog out, both are type float
  }
}
```

Program Example 5.1 Controlling LED brightness by variable voltage

Table 5.1: API summary for analog input

Function	Usage
`AnalogIn`	Create an AnalogIn object, connected to the specified pin
`read`	Read the input voltage, represented as a float in the range (0.0–1.0)
`read_u16`	Read the input voltage, represented as an unsigned short in the range (0x0–0xFFFF)

Compile the program and download to the mbed. With the program running, the potentiometer should control the brightness of the LED. You will probably find, however, that there is a range of the potentiometer rotation where the LED is off. Remember that the LED will be following the curve of Figure 3.3a, and there will be very little illumination when the drive voltage is low.

Exercise 5.1

Measure the DAC output voltage at pin 18, as you adjust the potentiometer. You will find that when this is above around 1.8 V, the LED will be lit, with varying levels of brightness. When it is below 1.8 V, the LED no longer conducts, and there is no illumination.

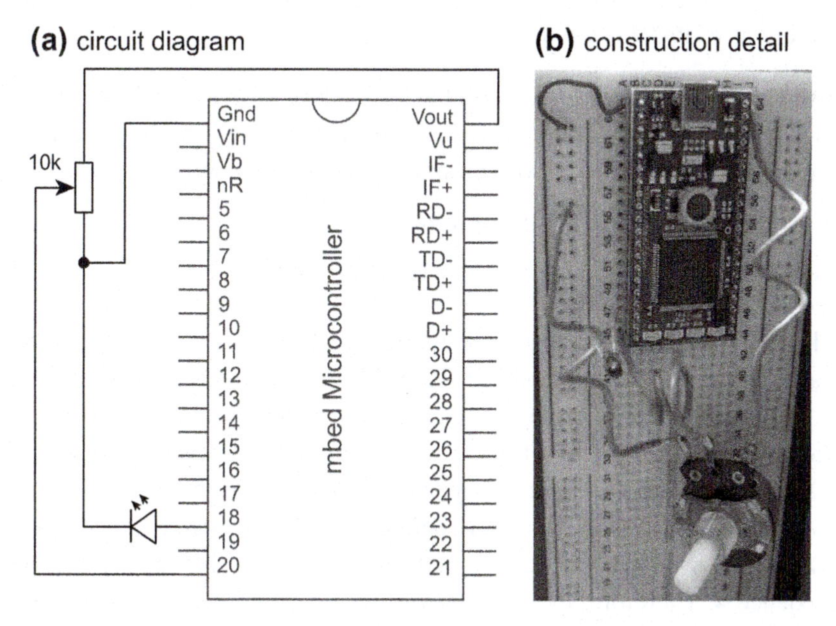

Figure 5.5:
A potentiometer controlling LED brightness

5.2.2 Controlling LED Brightness by PWM

We can use the potentiometer input to alter the PWM duty cycle, using the same approach as we did for the analog output. The circuit of Figure 5.5 is used again, except that the LED should now be connected to the PWM output on pin 21. Create a new program, and enter the code of Program Example 5.2. Here, we see the analog input value being transferred to the PWM duty cycle.

```
/*Program Example 5.2: Uses analog input to control PWM duty cycle, fixed period
*/
#include "mbed.h"
PwmOut PWM1(p21);
AnalogIn Ain(p20);          //defines analog input on Pin 20

int main() {
  while(1){
    PWM1.period(0.010); // set PWM period to 10 ms
    PWM1=Ain;           //Analog in value becomes PWM duty, both are type float
    wait(0.1);
  }
}
```

Program Example 5.2 Controlling PWM pulse width with a potentiometer

The LED brightness should again be controlled by the potentiometer. While the outcome is very similar to the previous program, the means of doing it is quite different. In practice, we would normally not wish to commit a whole DAC to controlling the brightness of an LED; we would be more ready to commit a (simpler and more readily available) PWM source.

5.2.3 Controlling PWM Frequency

Instead of using the potentiometer to control the PWM duty cycle, it can be used to control the PWM frequency. Leave the LED connected to Pin 21. Create a new program and enter the code of Program Example 5.3. No extra components need be connected to the mbed.

Notice that the PWM period is calculated in the line:

```
PWM1.period(Ain/10+0.001);     // set PWM period
```

It is first worth noting that a calculation is placed where one might have expected a simple parameter to be placed. This is not a problem for C. The program will first evaluate the expression inside the brackets, and then call the **period( )** function. This calculation invokes for the first time the divide operator, /. It also raises the thorny little question, which all children face when they learn arithmetic, of the order in which operators should be evaluated. In C this is very clearly defined, and can be seen by checking Table B.5 in Appendix B. This shows a precedence for each operator, with / having precedence 3, and + having precedence 4. Therefore, the division will be evaluated before the addition, and there will be no uncertainty in evaluating the expression. The values used mean that the minimum period, when **Ain** is zero, is 0.001 s (i.e. 1000 Hz). The maximum is when **Ain** is 1, leading to a period of 0.101 s, i.e. around 10 Hz.

```
/*Program Example 5.3: Uses analog input to control PWM period
*/
#include "mbed.h"
PwmOut PWM1(p21);
AnalogIn Ain(p20);

int main() {
  while(1){
    PWM1.period(Ain/10+0.001);    // set PWM period
    PWM1=0.5;                     // set duty cycle
    wait(0.5);
  }
}
```

Program Example 5.3 Controlling PWM frequency with a potentiometer

Observe the waveform on an oscilloscope, setting the time base initially to 5 ms/div. You should be able to see the frequency change as the potentiometer is adjusted.

Exercise 5.2

1. Adjust the values in the PWM period calculation to give different ranges of frequency output.
2. At what frequency does the LED appear not to flash, but seems to be continuously on? Measure this as carefully as you can, and see if the perceived frequency varies between different people. This is an important question, as knowing its value allows us to know at what frequency we can 'trick' the eye into thinking that a flashing image is continuous, for example in conventional raster scan television screens, oscilloscope traces and multiplexed LED displays.

The 0.5 s delay in Program Example 5.3 is added to the loop so that each new PWM period is allowed to be implemented before the next update. If it was omitted then the PWM would potentially be updated repeatedly within each cycle, which would lead to a large amount of jitter in the output (try removing the delay to see the effect). Notice that there can be discontinuities in the PWM output as the frequency values are updated. With care (and especially if you are using a storage oscilloscope) you can see this as the PWM is updated every 0.5 s. This is one reason why it is good to fix the frequency for a PWM signal.

Exercise 5.3

Connect the servo to the mbed as indicated in Figure 4.10a, with the potentiometer connected as in Figure 5.5a. Write a program which allows the potentiometer to control servo position. Scale values so that the full range of potentiometer adjustment leads to the full range of servo position changes.

5.3 Processing Data from Analog Inputs

5.3.1 Displaying Values on the Computer Screen

We turn now to the second way of making use of the ADC output, promised at the beginning of Section 5.2. It is possible to read analog input data through the ADC, and then print the value to the PC screen. This is a very important step forward, as it gives the possibility of displaying any data we are working with on the computer screen. To do this, we need to configure both mbed and host computer to be ready to send and receive data, and we need the host computer to be able to display that data. For the computer a terminal emulator is required. The mbed site recommends use of Tera Term for this purpose; Appendix E describes how to set this up. The mbed can be made to appear to the computer as a serial port, communicating through the universal serial bus (USB) connection. It links up with the USB through one of its own asynchronous serial ports. This is set up simply by adding this program line:

```
Serial pc(USBTX, USBRX);
```

Writing to the computer screen is then achieved with the **pc.printf( )** function. Explanations for this are given in Section 7.9.3.

Start a new mbed project and enter the code of Program Example 5.4. Leave the potentiometer connected as in Figure 5.5a. Note that the program uses the **printf( )** function for the first time, along with some of its far-from-friendly format specifiers. Check Section B.9 in Appendix B for some background on this.

```
/*Program Example 5.4: Reads input voltage through the ADC, and transfers to PC terminal
*/
#include "mbed.h"
Serial pc(USBTX, USBRX);                    //enable serial port which links to USB
AnalogIn Ain(p20);
float ADCdata;
int main() {
  pc.printf("ADC Data Values...\n\r");  //send an opening text message
  while(1){
    ADCdata=Ain;
    wait(0.5);
    pc.printf("%1.3f \n\r",ADCdata);    //send the data to the terminal
  }
}
```

Program Example 5.4 Logging data to the PC

You should now be able to compile and run your code to give an output on Tera Term. If you have problems, check from Appendix E or the mbed site that you have set up Tera Term correctly.

5.3.2 Scaling ADC Outputs to Recognized Units

The data displayed through Program Example 5.4 is just a set of numbers proportional to the voltage input. These values can readily be scaled to give a voltage reading, by multiplying by 3.3. Substitute the code lines below into the **while** loop of Program Example 5.4 to do just this, and to place a unit after the voltage value.

```
ADCdata=Ain*3.3;
wait(0.5);
pc.printf("%1.3f",ADCdata);
pc.printf("V\n\r");
```

Run the adjusted program; its output should appear similar to Figure 5.6. View the measured voltage on the PC screen, and read the actual input voltage on a digital voltmeter. How well do they compare?

5.3.3 Applying Averaging to Reduce Noise

If you leave Program Example 5.4 running, with a fixed input and values displayed on Tera Term, you may be surprised to see that the measured value is not always the same, but varies around some average value. You may already have noticed that the PWM value in Section 5.2.2 or 5.2.3 also appeared to vary, even when the potentiometer was not being moved. Several effects may be at play here, but almost certainly you are seeing the effect of some interference, and all the problems it can bring. If you look with the oscilloscope at the ADC input (i.e. the 'wiper' of the potentiometer) you are likely to see some high-frequency noise superimposed on this; exactly how much will depend on what equipment is running nearby, how long your interconnecting wires are, and a number of other things.

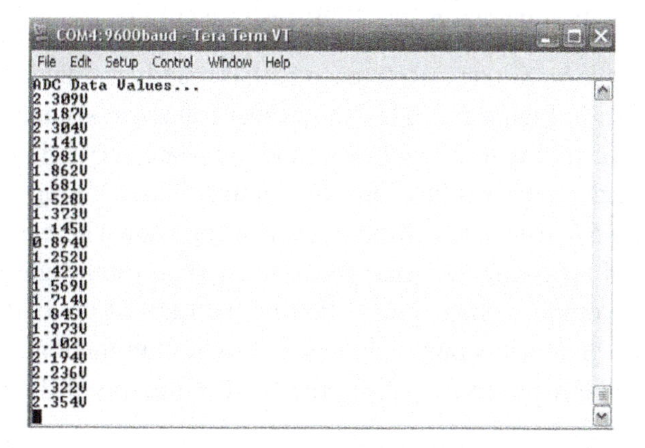

Figure 5.6:
Logged data on Tera Term

A very simple first step to improve this situation is to average the incoming signal. This should help to find the underlying average value and remove the high-frequency noise element. Try inserting the **for** loop shown below, replacing the `ADCdata=Ain;` line in Program Example 5.4. You will see that this code fragment sums 10 ADC values and takes their average. Try running the revised program and see whether a more stable output results. Note that while this sort of approach gives some benefit, the actual measurement now takes 10 times as long. This is a very simple example of digital signal processing.

```
for (int i=0;i<=9;i++) {
  ADCdata=ADCdata+Ain*3.3;     //sum 10 samples
}
ADCdata=ADCdata/10;            //divide by 10
```

5.4 Some Simple Analog Sensors

Now that we are equipped with analog input, it is appropriate to explore some simple analog sensors. These tend to be the simpler and more traditional ones, which have an analog output voltage that can be connected to the mbed ADC input. Later in the book we will come across sensors that can communicate with the mbed via a digital interface.

5.4.1 The Light-Dependent Resistor

The light-dependent resistor (LDR) is made from a piece of exposed semiconductor material. When light falls on it, its energy flips some electrons out of the crystalline structure; the brighter the light, the more electrons are released. These electrons are then available to conduct electricity, with the result that the resistance of the material falls. If the light is removed the electrons pop back into their place and the resistance goes up again. The overall effect is that as illumination increases, the LDR resistance falls.

The NORP12 LDR, made by Silonex (Reference 5.2), is readily available and inexpensive. Its summary data is shown in Figure 5.7. This shows that it has a resistance when completely dark of at least 1.0 MΩ, falling to a few hundred ohms when very brightly illuminated. An easy way to connect such a sensor is in a simple potential divider, giving a voltage output; this is also shown in Figure 5.7. The value of the series resistor, shown here as 10 kΩ, is chosen to give an output value of approximately mid-range for normal room light levels. It can be adjusted to modify the output voltage range. Putting the LDR at the bottom of the potential divider, as shown here, gives a low output voltage in bright illumination and a high one in low illumination. This can be reversed by putting the LDR at the top of the divider.

The LDR is a simple, effective and low-cost light sensor. Its output is not, however, linear, and each device tends to give slightly different output from another. Hence, it is not usually used for precision measurements.

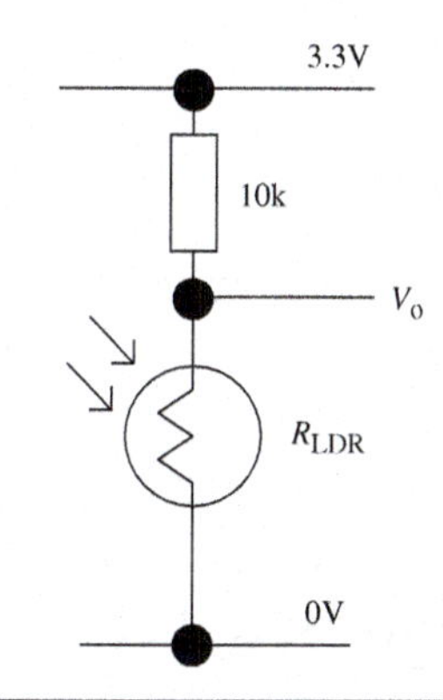

Illumination (lux)	R_{LDR} (Ω)	V_o
Dark	≥ 1.0 M	≥ 3.27 V
10	9k	1.56 V
1,000	400	0.13 V

Figure 5.7:
The NORPS12 LDR connected in a potential divider, with indicative output values

Exercise 5.4

Using the circuit of Figure 5.7, connect a NORP12 LDR to the mbed. Write a program to display light readings on the Tera Term screen. You will not be able to scale these into any useful unit. Try reversing resistor and LDR, and note the effect.

5.4.2 Integrated Circuit Temperature Sensor

Semiconductor action is highly dependent on temperature, so it is not surprising that semiconductor temperature sensors are made. A very useful form of sensor is one which is contained in an integrated circuit, such as the LM35 (Figure 5.8). This allows the output to be scaled to an immediately useful value. This device has an output of 10 mV/°C, with operating temperature from −55°C to 150°C. It is thus immediately useful for a range of temperature-sensing applications. The simplest connection for the LM35, which can be used with the mbed, is shown in Figure 5.8. A range of more advanced connections, for example to obtain an output for temperatures below 0°C, is shown in the datasheet (Reference 5.3).

Exercise 5.5

Design, build and program a simple embedded system using an LM35 sensor, which displays temperature on the computer screen and flashes an LED whenever the

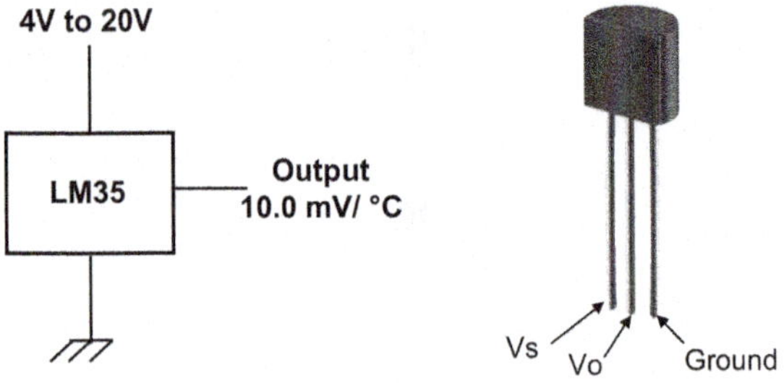

Figure 5.8:
The LM35 integrated circuit temperature sensor

temperature exceeds 30°C. The V_S pin of the sensor can be connected to pin 39 of the mbed. When connecting the sensor, it can be plugged directly into a suitable location in the mbed breadboard. Each terminal can, however, also be soldered to a wire, so that remote sensing can be undertaken. If these wires are insulated appropriately (e.g. with silicone rubber at the sensor end), then the sensor can be used to measure liquid temperatures. How well does this arrangement exploit the input range of the mbed ADC?

■

5.5 Exploring Data Conversion Timing

Nyquist's sampling theorem suggests that a slow ADC will only be able to convert low-frequency signals. When designing a carefully specified system it is therefore very important to know how long each data conversion takes. For this reason, we will make a measurement of mbed ADC and DAC conversion times, and then put Nyquist to the test.

5.5.1 Estimating Conversion Time and Applying Nyquist

Program Example 5.5 provides a very simple mechanism for measuring conversion times and then viewing Nyquist's sampling theorem in action. It adapts Program Example 5.1, but pulses a digital output between each stage. Enter this as a new program, compile and run.

```
/*Program Example 5.5: Inputs signal through ADC, and outputs to DAC. View DAC output on
oscilloscope. To demonstrate Nyquist, connect variable frequency signal generator to
ADC input. Allows measurement of conversion times, and explores Nyquist limit.
*/
#include "mbed.h"
AnalogOut Aout(p18);      //defines analog output on Pin 18
AnalogIn Ain(p20);        //defines analog input on Pin 20
DigitalOut test(p5);
float ADCdata;
```

```
int main() {
  while(1) {
     ADCdata=Ain;     //starts A-D conversion, and assigns analog value to ADCdata
     test=1;           //switch test output, as time marker
     test=0;
     Aout=ADCdata;    // transfers stored value to DAC, and forces a D-A conversion
     test=1;          //a double pulse, to mark the end of conversion
     test=0;
     test=1;
     test=0;
     //wait(0.001);   //optional wait state, to explore different cycle times
  }
}
```

Program Example 5.5 Estimating data conversion times

This program allows us to make a number of measurements which are of very great importance, and which require careful use of the oscilloscope. The measurements are presented as Exercises 5.6 and 5.7.

Exercise 5.6

Running Program Example 5.5, observe carefully the waveform displayed by the 'test' output (i.e. pin 5) on an oscilloscope – a digital storage oscilloscope will give the best results. This may require some patience, as they are very narrow pulses. You can widen the pulses if necessary by inserting a wait while the output is high. Note that for this test you do not need anything connected to the ADC input.

You will be able to detect the single pulse at the end of the analog-to-digital conversion, and the double one at the end of the loop. Measure the time duration of the analog-to-digital conversion, and the digital-to-analog conversion. What comment can you make on these? Note that the conversion times you measure are not the actual conversion times of the ADC and DAC; they include all associated programming overheads. Keep a note of these values as we aim to account for them later, in Exercises 5.7 and 14.8.

■

Exercise 5.7

Armed with the knowledge of the conversion times, connect a signal generator as an input to the ADC. Set the signal amplitude so that it is just under 3.3 V peak to peak, and apply a DC offset so that the voltage value never goes below 0 V. This facility is available on most signal generators. Insert the **wait(0.001);** line at the end of the loop ('commented out' in Program Example 5.5). This will give a sampling frequency of a little below 1 kHz. Nyquist's sampling theorem predicts that the maximum signal frequency that can be digitized will be 500 Hz, for this sampling frequency. Let's test it.

Start with an input signal of around 200 Hz. Observe input signal and DAC output on the two beams of the oscilloscope. You should see the input signal and a reconstructed version of it, something like Figure 5.3, with a new conversion approximately every millisecond. Now gradually increase the signal frequency towards 500 Hz. As you approach Nyquist's limit the output becomes a square wave. When input frequency equals sampling frequency, a straight line on the oscilloscope should occur, although in practice it may be difficult to find this condition exactly. As the input frequency increases further, an *alias* signal (as illustrated in Figure 5.4) appears at the output.

Decrease the duration of the wait state, and predict and observe the new Nyquist frequency. Finally, remove the wait state altogether. The data conversion should now be taking place at the fastest possible rate, with conversion time corresponding to your earlier measurement. The Nyquist limit that you now find is the limit for this particular hardware/software configuration.

■

5.6 Mini-Project: Two-Dimensional Light Tracking

Light-tracking devices are very important for the capture of solar energy. Often they operate in three dimensions, and tilt a solar panel so that it is facing the sun as accurately as possible. To start rather more simply, create a two-dimensional light tracker by fitting two LDRs, angled away from each other by around 90°, to a servo. Connect the LDRs using the circuit of Figure 5.7 to two ADC inputs. Write a program that reads the light value sensed by the two LDRs and rotates the servo so that each is receiving equal light. The servo can only rotate 180°. This is not, however, unreasonable, as a sun-tracking system will be located to track the sun from sunrise to sunset, i.e. not more than 180°. Can you think of a way of meeting this need using only one ADC input?

Chapter Review

- An ADC is available in the mbed; it can be used to digitize analog input signals.
- It is important to understand ADC characteristics, in terms of input range, resolution and conversion time.
- Nyquist's sampling theorem must be understood, and applied with care when sampling AC signals. The sampling frequency must be at least twice that of the highest frequency component in the sampled analog signal.
- Aliasing occurs when the Nyquist criterion is not met; this can introduce false frequencies to the data. Aliasing can be avoided by introducing an anti-aliasing filter to the analog signal before it is sampled.
- Data gathered by the ADC can be further processed, and displayed or stored.

- There are many sensors available which have an analog output; in many cases this output can be directly connected to the mbed ADC input.

Quiz

1. Give three types of analog signal which might be sampled through an ADC.
2. An ideal 8-bit ADC has an input range of 5.12 V. What is its resolution, and what is its greatest quantization error?
3. Give an example of how a single ADC can be used to sample four different analog signals.
4. An ideal 10-bit ADC has a reference voltage of 2.048 V and behaves according to Equation 5.1. For a particular input its output reads 10 1110 0001. What is the input voltage?
5. What will be the result if an mbed is required to sample an analog input value of 4.2 V?
6. An ultrasound signal of 40 kHz is to be digitized. Recommend the minimum sampling frequency.
7. The conversion time of an ADC is found to be 7.5 μs. The ADC is set to convert repeatedly, with no other programming requirements. What is the maximum frequency signal it can digitize?
8. The ADC in Question 7 is now used with a multiplexer, so that four inputs are repeatedly digitized in turn. A further time of 2500 ns per sample is required to save the data and switch the input. What is the maximum frequency signal that can now be digitized?
9. An LM35 temperature sensor is connected to an mbed ADC input, and senses a temperature of 30°C. What is the binary output of the ADC?
10. What will be the value of integer *x* for 1.5 V and 2.5 V signals sampled on the mbed using the following program code?

```
#include "mbed.h"
AnalogIn Ain(p20);
int main(){
  int x=Ain.read_u16();
}
```

References

5.1. Horowitz, P. and Hill, W. (1989). The Art of Electronics. 2nd edition. Cambridge University Press.
5.2. The Silonex home site. http://www.silonex.com/
5.3. LM35 Precision Centigrade Temperature Sensors. November 2000. National Semiconductor Corporation. http://www.national.com/

CHAPTER 6

Further Programming Techniques

Chapter Outline

6.1 The Benefits of Considered Program Design and Structure

There are numerous challenges when tackling an embedded system design project. It is usually wise first to consider the software design structure, particularly with large and multi-functional projects. It is not possible to program all functionality into a single control loop, so the approach for breaking up code into understandable features should be well thought out. In particular, it helps to ensure that the following can be achieved:

- that code is readable, structured and documented
- that code can be tested for performance in a modular form
- that development reuses existing code utilities to keep development time short
- that code design supports multiple engineers working on a single project
- that future upgrades to code can be implemented efficiently.

There are various C/C++ programming techniques that enable these design requirements to be considered, as discussed in this chapter.

Fast and Effective Embedded Systems Design. DOI: 10.1016/B978-0-08-097768-3.00006-4

6.2 Functions

A function is a portion of code within a larger program. The function performs a specific task and is relatively independent of the main code. Functions can be used to manipulate data; this is particularly useful if several similar data manipulations are required in the program. Data values can be input to the function and the function can return the result to the main program. Functions, therefore, are particularly useful for coding mathematical algorithms, look-up tables and data conversions, as well as control features that may operate on a number of different parallel data streams. It is also possible to use functions with no input or output data, simply to reduce code size and to improve readability of code. Figure 6.1 illustrates a function call.

There are several advantages when using functions. First, a function is written once and compiled into one area of memory, irrespective of the number of times that it is called from the main program, so program memory is reduced. Functions also allow clean and manageable code to be designed, allowing software to be well structured and readable at a number of levels of abstraction. The use of functions also enables the practice of modular coding, where teams of software engineers are often required to develop large and advanced applications. Writing code with functions therefore allows one engineer to develop a particular software feature, while another engineer may take responsibility for something else.

Using functions is not always completely beneficial, however. There is a small execution time overhead in storing program position data and jumping and returning from the function, but this should only be an issue for consideration in the most time-critical systems. Furthermore, it is possible to 'nest' functions within functions, which can sometimes make software challenging to follow. A limitation of C functions is that only a single value can be returned from the function, and arrays of data cannot be passed to or from a function (only single-value

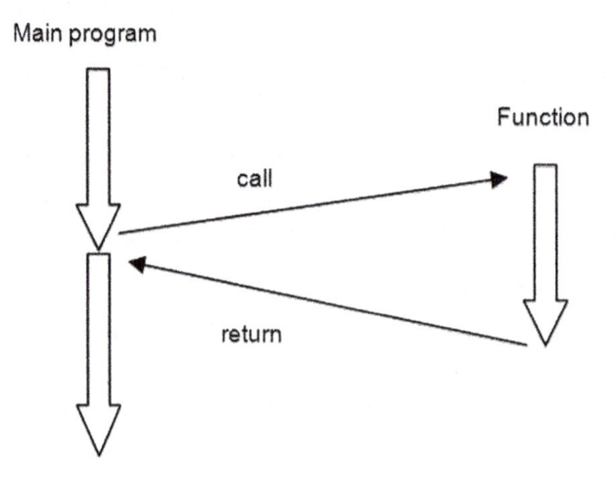

Figure 6.1:
A function call

variables can be used). Working with functions and modular techniques therefore requires a considered software structure to be designed and evaluated before programming is started.

6.3 Program Design

6.3.1 Using Flowcharts to Define Code Structure

It is often useful to use a flowchart to indicate the operation of program flow and the use of functions. Code flow can be designed using a flowchart prior to coding. Figure 6.2 shows some of the flowchart symbols that are used.

For example, take the following software design specification:

> Design a program to increment continuously the output of a seven-segment numerical light-emitting diode (LED) display (as shown in Figure 6.3, and similar to the one used in Section 3.5) through the numbers 0 to 9, then reset back to 0 to continue counting. This includes:
>
> - Use a function to convert a hexadecimal counter byte A to the relevant seven-segment LED output byte B.
> - Output the LED output byte to light the correct segment LEDs.
> - If the count value is greater than 9, then reset to zero.
> - Delay for 500 ms to ensure that the LED output counts up at a rate that is easily visible.

The output of the seven-segment display has been discussed previously in Chapter 3 and in particular in Table 3.4. A feasible software design is shown in Figure 6.4.

Flowcharts allow us to visualize the order of operations of code and to make judgments on which sections of a program may require the most attention or take the most effort to develop.

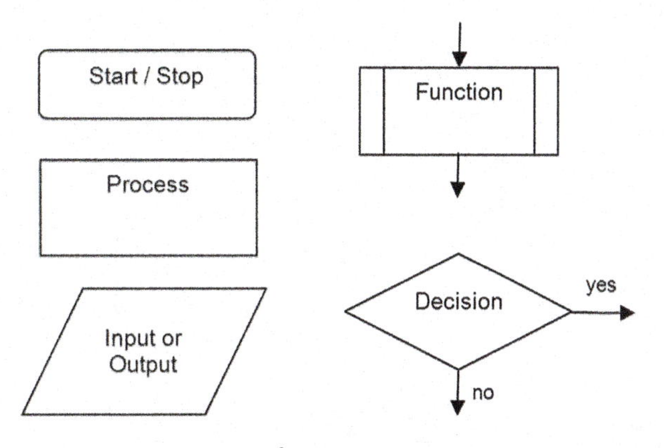

Figure 6.2:
Example flowchart symbols

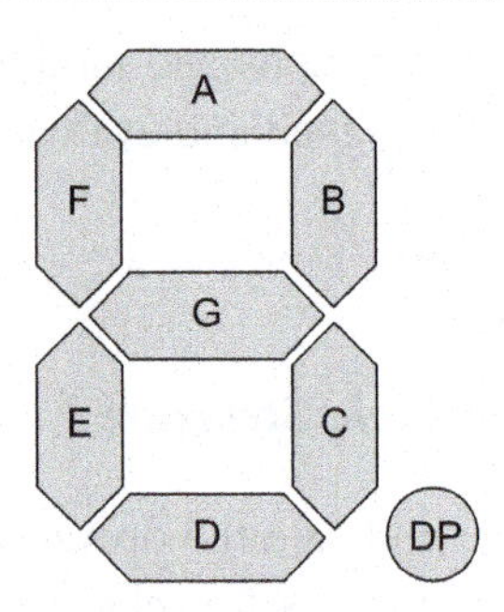

Figure 6.3:
Seven-segment display

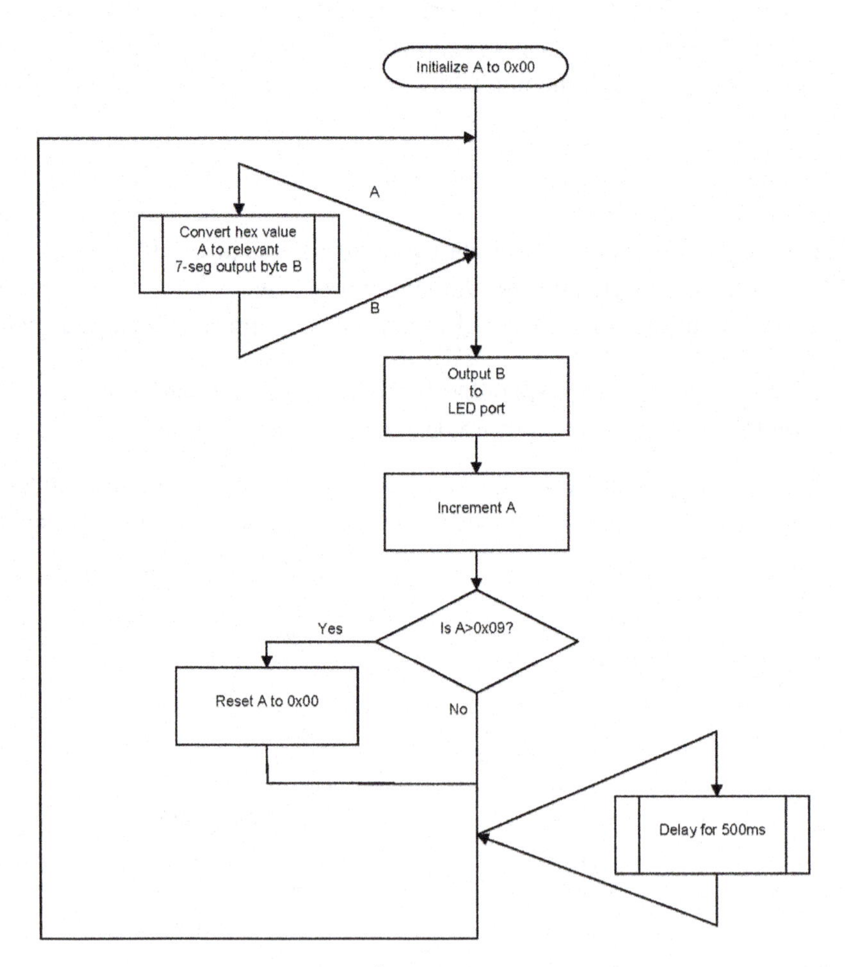

Figure 6.4:
Example flowchart design for a seven-segment display counter

They also help with communicating a potential design with non-engineers, which may hold the key to designing a system that meets a very detailed specification.

6.3.2 Pseudocode

Pseudocode consists of short, English phrases used to explain specific tasks within a program. Ideally, pseudocode should not include keywords in any specific computer language. Pseudocode should be written as a list of consecutive phrases; we can even draw arrows to show looping processes. Indentation can be used to show the logical program flow in pseudocode.

Writing pseudocode saves time later during the coding and testing stage of a program's development and also helps communication among designers, coders and project managers. Some projects may use pseudocode for design, others may use flowcharts, and some a combination of both.

The software design shown by the flowchart in Figure 6.4 could also be described in pseudocode as shown in Figure 6.5.

Note that the functions **SegConvert( )** and **Delay( )** are defined elsewhere, for example in a separate 'utilities' file, authored by a different engineer. Function **SegConvert( )** could implement a simple look-up table or number of if statements that assigns the suitable value to B.

```
Program start
Initialise variable A=0
Initialise variable B
Start infinite loop
    Call function SegConvert withinput A
    SegConvert returns value B
    Output B to LED port
    Increment A
    If A > 9
        A=0
    Call function Delay for 500ms
End infinite loop
```

Figure 6.5:
Example pseudocode for seven-segment display counter

6.4 Working with Functions on the mbed

The implementation and syntax of functions in C/C++ are discussed in detail in Section B.4 of Appendix B with some simple examples. It is important to remember that, as with variables, all functions (except the **main()** function) must be declared at the start of a program. The declaration statements for functions are called *prototypes*. So each function in the code must have an associated prototype for it to compile and run. The actual function code needs defining within a C/C++ file also, in order for it to be called from within the main code. This is done by specifying the function in the same way as the prototype, followed by the actual function code. A number of examples will be presented during this chapter, so in each case make a point of identifying the function prototype and the actual function definition. A further point to note is that if a function is defined in code prior to the **main( )** C function, then its definition serves as its prototype also.

6.4.1 Implementing a Seven-Segment Display Counter

Program Example 6.1 shows a program which implements the designs described by the flowchart in Figure 6.4 and the pseudocode shown in Figure 6.5. It applies some of the techniques first used in Program Example 3.5, but goes beyond these. The main design requirement is that a seven-segment display is used to count continuously from 0 to 9 and loop back to 0. Declarations for the **BusOut** object and the A and B variables, as well as the **SegConvert( )** function prototype, appear early in the program. It can be seen that the **main( )** program function is followed by the **SegConvert( )** function, which is called regularly from within the main code. Notice in the line

```
B=SegConvert(A);          // Call function to return B
```

that B can immediately take on the return value of the **SegConvert( )** function.

Notice in the **SegConvert( )** function the final line immediately below, which applies the **return** keyword:

```
return SegByte;
```

This line causes program execution to return to the point from which the function was called, carrying the value **SegByte** as its return value. It is an important technique to use once you start writing functions that provide return values. Notice that **SegByte** has been declared as part of the function prototype early in the program listing.

```
/* Program Example 6.1: seven-segment display counter                              */
#include "mbed.h"
BusOut Seg1(p5,p6,p7,p8,p9,p10,p11,p12);      // A,B,C,D,E,F,G,DP
char SegConvert(char SegValue);               // function prototype
char A=0;                                     // declare variables A and B
char B;
int main() {                                  // main program
   while (1) {                                // infinite loop
   B=SegConvert(A);                           // Call function to return B
   Seg1=B;                                    // Output B
   A++;                                       // increment A
   if (A>0x09){                               // if A > 9 reset to zero
   A=0;
        }
   wait(0.5);                                 // delay 500 milliseconds
     }
}
char SegConvert(char SegValue) {              // function 'SegConvert'
  char SegByte=0x00;
  switch (SegValue) {                           //DP G F E D C B A
       case 0 : SegByte = 0x3F;break;           // 0 0 1 1 1 1 1 1 binary
       case 1 : SegByte = 0x06;break;           // 0 0 0 0 0 1 1 0 binary
       case 2 : SegByte = 0x5B;break;           // 0 1 0 1 1 0 1 1 binary
       case 3 : SegByte = 0x4F;break;           // 0 1 0 0 1 1 1 1 binary
       case 4 : SegByte = 0x66;break;           // 0 1 1 0 0 1 1 0 binary
       case 5 : SegByte = 0x6D;break;           // 0 1 1 0 1 1 0 1 binary
       case 6 : SegByte = 0x7D;break;           // 0 1 1 1 1 1 0 1 binary
       case 7 : SegByte = 0x07;break;           // 0 0 0 0 0 1 1 1 binary
       case 8 : SegByte = 0x7F;break;           // 0 1 1 1 1 1 1 1 binary
       case 9 : SegByte = 0x6F;break;           // 0 1 1 0 1 1 1 1 binary
     }
  return SegByte;
}
```

Program Example 6.1 Seven-segment display counter

Connect a seven-segment display to the mbed and implement Program 6.1. The wiring diagram for a seven-segment LED display was shown previously in Figure 3.10. Verify that the display output continuously counts from 0 to 9 and then resets back to 0. Ensure that you understand how the program works by cross-referencing with the flowchart and pseudocode designs shown previously.

Exercise 6.1

Change Program Example 6.1 so that the display counts up in hexadecimal from 0 to F. You will need to work out the display patterns for A to F using the seven-segment display. For a few you will need lower case, for others upper case.

6.4.2 Function Reuse

Now that we have a function to convert a decimal value to a seven-segment display byte, we can build projects using multiple seven-segment displays with little extra effort. For example, we can implement a second seven-segment display (see Figure 6.6) by simply defining its mbed **BusOut** declaration and calling the same **SegConvert( )** function as before.

It is possible to implement a counter program that counts from 00 to 99 by simply modifying the main program code to that shown in Program Example 6.2. Note that the **SegConvert( )** function previously defined in Program Example 6.1 is also required to be copied (reused) in this example. Note also that a slightly different programming approach is used; here we use two **for** loops to count each of the tens and units values.

```
/* Program Example 6.2: Display counter for 0-99
*/
#include "mbed.h"
BusOut Seg1(p5,p6,p7,p8,p9,p10,p11,p12);        // A,B,C,D,E,F,G,DP
BusOut Seg2(p13,p14,p15,p16,p17,p18,p19,p20);
char SegConvert(char SegValue);                 // function prototype
int main() {                                    // main program
   while (1) {                                  // infinite loop
   for (char j=0;j<10;j++) {                    // counter loop 1
         Seg2=SegConvert(j);                    // tens column
         for (char i=0;i<10;i++) {              // counter loop 2
             Seg1=SegConvert(i);                // units column
             wait(0.2);
         }
   }
  }
}
 // add SegConvert function here...
```

Program Example 6.2 Two digit seven-segment display counter

Using two seven-segment displays, with pin connections shown in Figure 6.6, implement Program Example 6.2 and verify that the display output counts continuously from 00 to 99 and then resets back to 0. Review the program design and familiarize yourself with the method used to count the tens and units digits each from 0 to 9.

Exercise 6.2

Write and test mbed programs to perform the following action with a dual seven-segment display setup:

1. Count 0 to FF, in hexadecimal format
2. Create a one-minute timer, count 0–59, with a precise increment rate of 1 s. Flash an LED every time the count overflows back to zero (i.e. every minute)

3. Introduce two LEDs to make a simple '1' digit, and count 0 to 199. This is called a two and a half digit display.

In each case set the count rate to a different speed and ensure that the program loops back to 0x00, so that the counter operates continuously.

■

6.4.3 A More Complex Program Using Functions

A more advanced program could read two numerical values from a host terminal application and display these on two seven-segment displays connected to the mbed. The program can therefore display any integer number between 00 and 99, as required by user key presses.

An example program design uses four functions to implement the host terminal output on seven-segment displays. The four functions are as follows:

- **SegInit( )** – to set up and initialize the seven-segment displays
- **HostInit( )** – to set up and initialize the host terminal communication
- **GetKeyInput( )** – to get keyboard data from the terminal application
- **SegConvert( )** – function to convert a decimal integer to a seven-segment display data byte.

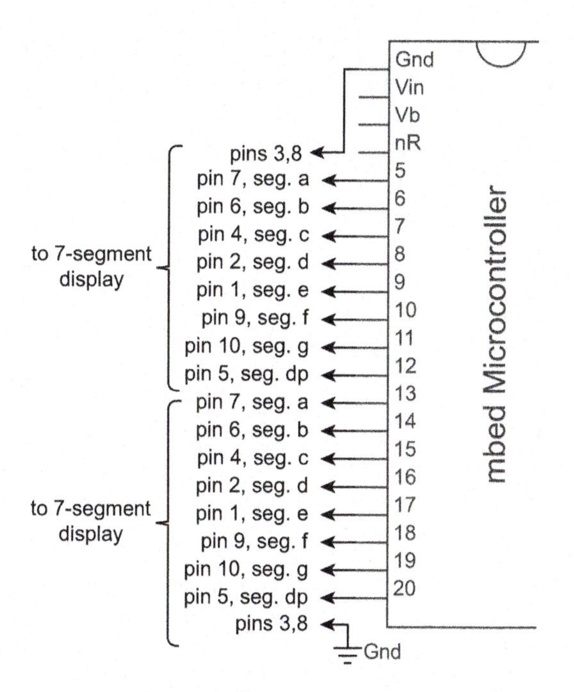

Figure 6.6:
Two seven-segment display control with the mbed

We will use the mbed universal serial bus (USB) interface to communicate with the host PC, as we did in Section 5.3, and two seven-segment displays, as in the previous exercise.

For the first time now we come across a method for communicating keyboard data and display characters, using ASCII codes. The term ASCII refers to the American Standard Code for Information Interchange method for defining alphanumeric characters as 8-bit values. Each alphabet character (lower and upper case), number (0–9) and a selection of punctuation characters are all described by a unique identification byte, i.e. the 'ASCII value'. Therefore, for example, when a key is pressed on a computer keyboard, its ASCII byte is communicated to the PC. The same applies when communicating with displays. The use of ASCII characters will be developed further in Chapter 8.

The ASCII byte for numerical characters has the higher four bits set to value 0x3 and the lower four bits represent the value of the numerical key which is pressed (0x0 to 0x9). Numbers 0–9 are therefore represented in ASCII as 0x30 to 0x39. To convert the ASCII byte returned by the keyboard to a regular decimal digit, the higher four bits need to be removed. We do this by logically ANDing the ASCII code with a bitmask, a number with bits set to 1 where we want to keep a bit in the ASCII, and set to 0 where we want to force the bit to 0. In this case, we apply a bitmask of 0x0F. The logical AND applies the operator '**&**' from Table B.5, and appears in the line:

```
return (c&0x0F);      // apply bit mask to convert to decimal, and return
```

Example functions and program code are shown in Program Example 6.3. Once again, the function **SegConvert( )**, as shown in Program Example 6.1, should be added to compile the program.

```
/* Program Example 6.3: Host keypress to 7-seg display
*/
#include "mbed.h"
Serial pc(USBTX, USBRX);                        // comms to host PC
BusOut Seg1(p5,p6,p7,p8,p9,p10,p11,p12);        // A,B,C,D,E,F,G,DP
BusOut Seg2(p13,p14,p15,p16,p17,p18,p19,p20);   // A,B,C,D,E,F,G,DP

void SegInit(void);                             // function prototype
void HostInit(void);                            // function prototype
char GetKeyInput(void);                         // function prototype
char SegConvert(char SegValue);                 // function prototype
char data1, data2;                              // variable declarations

int main() {                        // main program
   SegInit();                       // call function to initialize the 7-seg displays
   HostInit();                      // call function to initialize the host terminal
   while (1) {                               // infinite loop
   data2 = GetKeyInput();                    // call function to get 1st key press
   Seg2=SegConvert(data2);                   // call function to convert and output
   data1 = GetKeyInput();                    // call function to get 2nd key press
   Seg1=SegConvert(data1);                   // call function to convert and output
```

```
    pc.printf(" ");                              // display spaces between numbers
   }
  }
// functions
void SegInit(void) {
 Seg1=SegConvert(0);               // initialize to zero
 Seg2=SegConvert(0);               // initialize to zero
}

void HostInit(void) {
 pc.printf("\n\rType two digit numbers to be displayed\n\r");
}

char GetKeyInput(void) {
 char c = pc.getc();               // get keyboard data (ascii 0x30-0x39)
 pc.printf("%c",c);                // print ascii value to host PC terminal
 return (c&0x0F);                  // apply bit mask to convert to decimal, and return
}
// copy SegConvert function here too...
```

Program Example 6.3 Two digit seven-segment display based on host key presses

Implement Program Example 6.3 and verify that numerical keyboard presses are displayed on the seven-segment displays. Familiarize yourself with the program design and understand the input and output features of each program function.

6.5 Using Multiple Files in C/C++

Large embedded projects in C/C++ benefit from being split into a number of different files, usually so that a number of engineers can take responsibility for different parts of the code. This approach also improves readability and maintenance. For example, the code for a processor in a vending machine might have one C/C++ file for the control of the actuators delivering the items and a different file for controlling the user input and liquid crystal display. It does not make sense to combine these two code features in the same source file as they each relate to different peripheral hardware. Furthermore, if a new batch of vending machines were to be built with an updated keypad and display, only that piece of the code would need to be modified. All the other source files could be carried over without change.

Modular coding uses header files to join multiple files together. In general, a main C/C++ file (**main.c** or **main.cpp**) is used to contain the high-level code, but all functions and global variables (variables that are available to all functions) are defined in feature-specific C files. It is good practice therefore for each C/C++ feature file to have an associated header file (with a .h extension). Header files typically include declarations only, for example compiler directives, variable declarations and function prototypes.

A number of header files also exist from within C/C++, which can be used for more advanced data manipulation or arithmetic. Many of the built-in C/C++ header files are already linked to through the **mbed.h** header, so we do not need to worry too much about those here. They are discussed in more detail in Section B.9 of Appendix B.

6.5.1 Overview of the C/C++ Program Compilation Process

To understand further the design approach to modular coding, it helps to understand the way programs are preprocessed, compiled and linked to create a binary execution file for the microprocessor. A simplified version of this process is shown in Figure 6.7 and described in detail through Section 6.6.

In summary, first a preprocessor looks at a particular source file and implements any preprocessor directives and associated header files. The compiler then generates an object file for the particular source code. In doing so, the compiler ensures that the source files do not

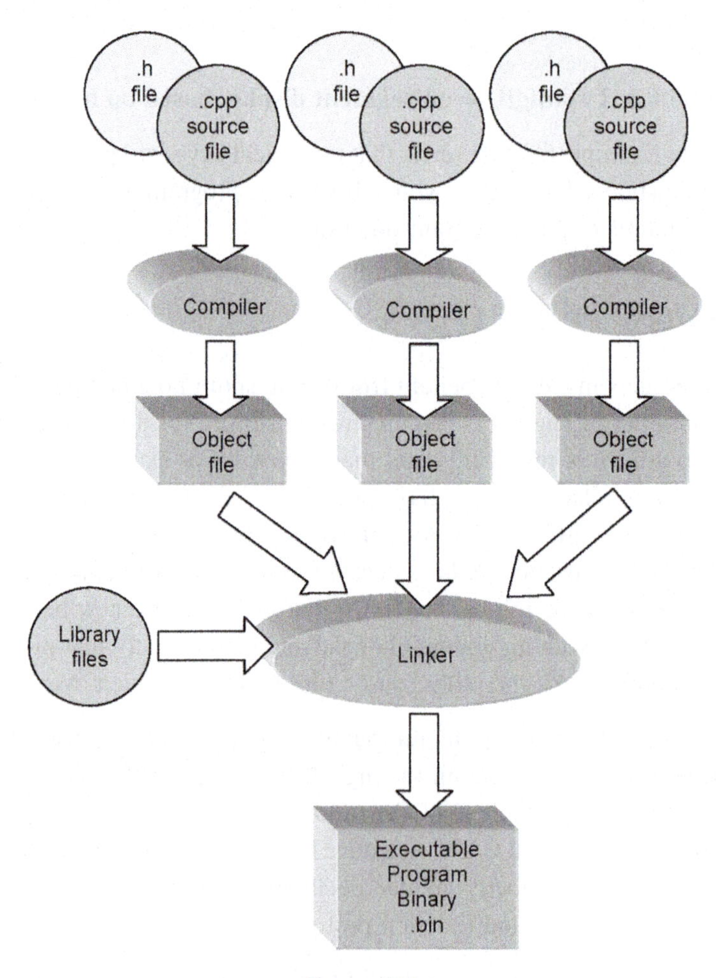

Figure 6.7:
C program compile and link process

contain any syntax errors – note that a program can have no syntax errors, but still be quite useless. The compiler then generates correct object code for each file and ensures that the object and library files are formatted correctly for the linker.

Object files and built-in library files are then linked together to generate an executable binary (.bin) file, which can be downloaded to the microprocessor. The linker also manages the allocation of memory for the microprocessor application and it ensures that all object files are located into memory and linked to each other correctly. In undertaking the task, the linker may uncover programming faults associated with memory allocation and capacity.

6.5.2 The C/C++ Preprocessor and Preprocessor Directives

C code feature

The C/C++ preprocessor modifies code before the program is compiled. Preprocessor directives are denoted with a '#' symbol, as described in Section B.2 of Appendix B. These may also be called 'compiler directives' as the preprocessor is essentially a subprocess of the compiler. The **#include** (usually referred to as 'hash-include') directive is commonly used to tell the preprocessor to include any code or statements contained within an external header file. Indeed, we have seen this ubiquitous **#include** statement in every complete program example so far in this book, as this is used to connect our programs with the core mbed libraries.

Figure 6.8 shows three header files to explain how the **#include** statement works. It is important to note that **#include** essentially just acts as a copy-and-paste feature. If we compile

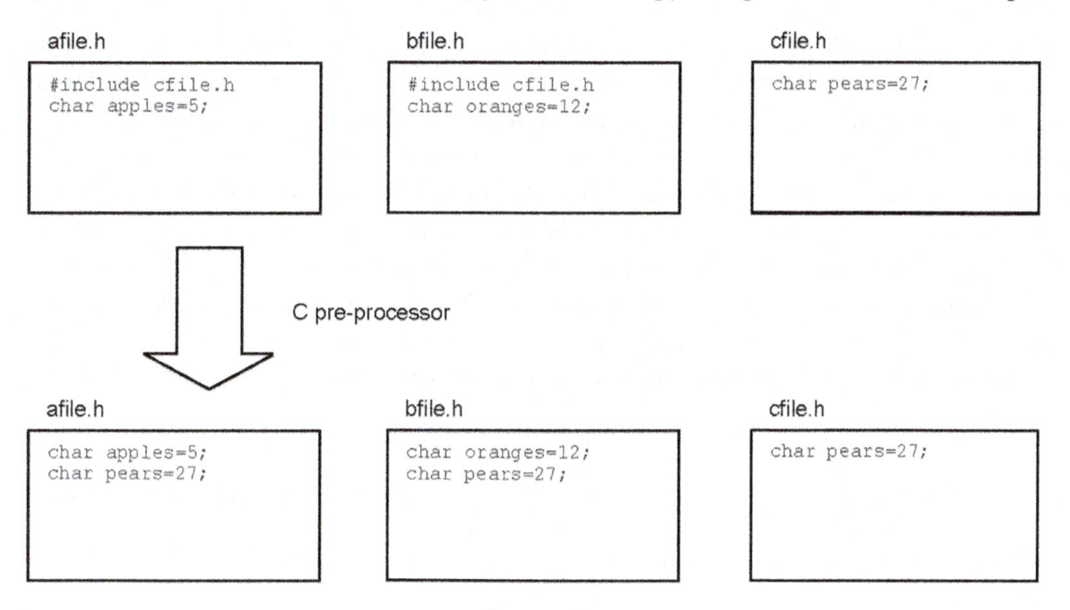

Figure 6.8:
C preprocessor example with multiple declaration error for variable 'pears'

afile.h and **bfile.h** where both files also **#include cfile.h**, we will have two copies of the contents of **cfile.h** (hence variable **pears** will be defined twice). The compiler will therefore highlight an error, as multiple declarations of the same variables or function prototypes are not allowed.

We can also use **#define** statements (usually referred to as 'hash-define'), which allow us to use meaningful names for specific numerical values that never change. Here are some examples:

```
#define  SAMPLEFREQUENCY      44100
#define  PI                   3.141592
#define  MAX_SIZE             255
```

The preprocessor replaces **#define** with the actual value associated with that name, so using **#define** statements does not actually increase the memory load on the processor.

6.5.3 The #ifndef Directive

We have just seen that multiple declarations of a variable or function prototype are not allowed. However, it is important for header files to include all the variables and features that are required for the linker to build the project successfully. It is therefore necessary to ensure that all variables and function prototypes are preprocessed on just one occasion.

When using header files it is therefore good practice to use a conditional statement to define variables and prototypes only if they have not previously been defined. For this the **#ifndef** directive, which means 'if not defined', can be used. In order to use the **#ifndef** directive effectively, a conditional statement based on the existence of a **#define** value is required. If the **#define** value has not previously been defined then that value and all of the header file's variables and prototypes are defined. If the **#define** value has previously been declared then the header file's contents are not implemented by the preprocessor (as they must certainly have already been implemented). This ensures that all header file declarations are added only once to the project. The example code shown in Program Example 6.4 represents a template header file structure using the **#ifndef** condition, avoiding the error highlighted in Figure 6.8.

```
/* Program Example 6.4: Template for .h header file
*/

#ifndef VARIABLE_H                  // if VARIABLE_H has not previously been defined
#define VARIABLE_H                  // define it now

// header declarations here...

#endif                              // end of the if directive
```

Program Example 6.4 Example header file template

6.5.4 Using mbed Objects Globally

All mbed objects must be defined in an 'owner' source file. However, we may wish to use those objects from within other source files in the project, i.e. 'globally'. This can be done by also defining the mbed object in the associated owner's header file. When an mbed object is defined for global use, the **extern** specifier should be used. For example, a bespoke file called **my_functions.cpp** may define and use a **DigitalOut** object called **'RedLed'** as follows:

```
DigitalOut RedLed(p5);
```

If any other source files need to manipulate **RedLed**, the object must also be declared in the **my_functions.h** header file using the **extern** specifier, as follows:

```
extern DigitalOut RedLed;
```

Note that the specific mbed pins do not need to be redefined in the header file, as these will have already been specified in the object declaration in **my_functions.cpp**.

6.6 Modular Program Example

A modular program example can now be built from the non-modular code given in Program Example 6.3. Here, the functional features are separated to different source and header files. Therefore, a keyboard-controlled seven-segment display project with multiple source files is created as follows:

- **main.cpp** – contains the main program function
- **HostIO.cpp** – contains functions and objects for host terminal control
- **SegDisplay.cpp** – contains functions and objects for seven-segment display output.

The following associated header files are also required:

- **HostIO.h**
- **SegDisplay.h**

Note that, by convention, the main.cpp file does not need a header file. The program file structure in the mbed compiler should be similar to that shown in Figure 6.9. Note that modular files can be created by right-clicking on the project name and selecting 'New File'.

The main.cpp file holds the same main function code as before, but with **#include** directives to the new header files. Program Example 6.5 details the source code for **main.cpp**.

```
/* Program Example 6.5: main.cpp file for modular 7-seg keyboard controller
*/
#include "mbed.h"
#include "HostIO.h"
```

```
#include "SegDisplay.h"
char data1, data2;                   // variable declarations
int main() {                         // main program
  SegInit();                         // call init function
  HostInit();                        // call init function
  while (1) {                        // infinite loop
    data2 = GetKeyInput();             // call to get 1st key press
    Seg2 = SegConvert(data2);          // call to convert and output
    data1 = GetKeyInput();             // call to get 2nd key press
    Seg1 = SegConvert(data1);          // call to convert and output
    pc.printf(" ");                    // display spaces on host
  }
}
```

Program Example 6.5 Source code for main.cpp

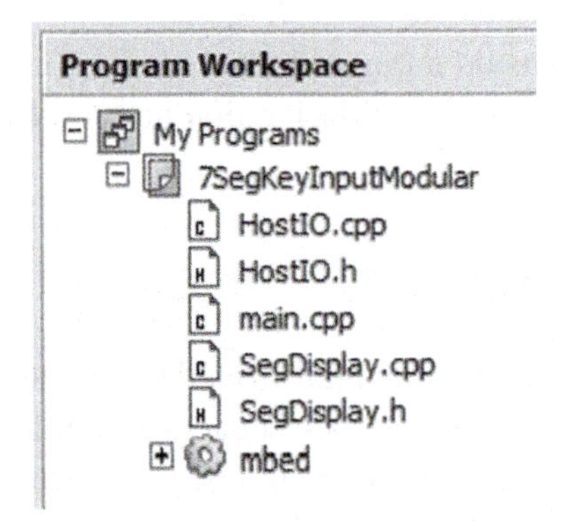

Figure 6.9:
File structure for modular seven-segment display program

The **SegInit( )** and **SegConvert( )** functions are to be 'owned' by the **SegDisplay.cpp** source file, as are the **BusOut** objects named **'Seg1'** and **'Seg2'**. The resulting **SegDisplay.cpp** file is shown in Program Example 6.6.

```
/* Program Example 6.6: SegDisplay.cpp file for modular 7-seg keyboard controller
*/
#include "SegDisplay.h"
BusOut Seg1(p5,p6,p7,p8,p9,p10,p11,p12);      // A,B,C,D,E,F,G,DP
BusOut Seg2(p13,p14,p15,p16,p17,p18,p19,p20); // A,B,C,D,E,F,G,DP

void SegInit(void) {
  Seg1=SegConvert(0);     // initialize to zero
  Seg2=SegConvert(0);     // initialize to zero
}
char SegConvert(char SegValue) {       // function 'SegConvert'
  char SegByte=0x00;
  switch (SegValue) {                  //DP G F E D C B A
    case 0 : SegByte = 0x3F; break;    // 0 0 1 1 1 1 1 1 binary
    case 1 : SegByte = 0x06; break;    // 0 0 0 0 0 1 1 0 binary
```

```
    case 2 : SegByte = 0x5B; break;  // 0 1 0 1 1 0 1 1 binary
    case 3 : SegByte = 0x4F; break;  // 0 1 0 0 1 1 1 1 binary
    case 4 : SegByte = 0x66; break;  // 0 1 1 0 0 1 1 0 binary
    case 5 : SegByte = 0x6D; break;  // 0 1 1 0 1 1 0 1 binary
    case 6 : SegByte = 0x7D; break;  // 0 1 1 1 1 1 0 1 binary
    case 7 : SegByte = 0x07; break;  // 0 0 0 0 0 1 1 1 binary
    case 8 : SegByte = 0x7F; break;  // 0 1 1 1 1 1 1 1 binary
    case 9 : SegByte = 0x6F; break;  // 0 1 1 0 1 1 1 1 binary
  }
  return SegByte;
}
```

Program Example 6.6 Source code for SegDisplay.cpp

Note that **SegDisplay.cpp** file has an **#include** directive to the **SegDisplay.h** header file. This is given in Program Example 6.7.

```
/* Program Example 6.7: SegDisplay.h file for modular 7-seg keyboard controller
*/
#ifndef SEGDISPLAY_H
#define SEGDISPLAY_H

#include "mbed.h"

extern BusOut Seg1;   // allow Seg1 to be manipulated by other files
extern BusOut Seg2;   // allow Seg2 to be manipulated by other files

void SegInit(void);                // function prototype
char SegConvert(char SegValue);    // function prototype

#endif
```

Program Example 6.7 Source code for SegDisplay.h

The **DisplaySet** and **GetKeyInput** functions are to be 'owned' by the **HostIO.cpp** source file, as is the serial USB interface object named '**pc**'. The **HostIO.cpp** file should therefore be as shown in Program Example 6.8.

```
/* Program Example 6.8: HostIO.cpp code for modular 7-seg keyboard controller
*/
#include "HostIO.h"
Serial pc(USBTX, USBRX);    // communication to host PC

void HostInit(void) {
  pc.printf("\n\rType two digit numbers to be \n\r");
}

char GetKeyInput(void) {
  char c = pc.getc();    // get keyboard ascii data
  pc.printf("%c",c);     // print ascii value to host PC terminal
  return (c&0x0F);       // return value as non-ascii
}
```

Program Example 6.8 Source code for HostIO.cpp

```
/* Program Example 6.9: HostIO.h code for modular 7-seg keyboard controller
*/
#ifndef HOSTIO_H

#define HOSTIO_H
#include "mbed.h"

extern Serial pc;         // allow pc to be manipulated by other files
void HostInit(void);      // function prototype
char GetKeyInput(void);   // function prototype

#endif
```

Program Example 6.9 Source code for HostIO.h

The HostIO header file, **HostIO.h**, is shown in Program Example 6.9.

Create the modular seven-segment display project given by Program Examples 6.5–6.9. You will need to create a new project in the mbed compiler and add the required modular files by right clicking on the project and selecting 'New File'. Hence, create a file structure which replicates Figure 6.9.

You should now be able to compile and run your modular program. Use the circuit of Figure 6.6.

Exercise 6.3

Create an advanced modular project which uses a host terminal application and a servo.

The user inputs a value between 1 and 9 from the keyboard which moves the servo to a specified position. An input of 1 moves the servo to 90 degrees left and a user input of 9 moves the servo to 90 degrees right. Numbers between 1 and 9 move the servo to a relative position, for example the value 5 points the servo to the center.

You can reuse the **GetKeyInput( )** function from the previous examples.

You may also need to create a look-up table function to convert the numerical input value to a suitable pulse width modulation (PWM) duty cycle value associated with the desired servo position.

■

This chapter has shown that functions are useful for allowing us to write clean and readable code while allowing manipulation of data. This, in turn, has enabled us to create modular programs which allow large multi-functional projects to be programmed. The program development can also be managed through a team of engineers and with a mechanism that enables reuse of code and a simple approach to updating and upgrading software features.

Chapter Review

- We use functions to allow code to be reusable and easier to read.
- Functions can take input data values and return a single data value as output; however, it is not possible to pass arrays of data to or from a function.
- Flowcharts and pseudocode can be used to assist program design.
- The technique of modular programming involves designing a complete program as a number of source files and associated header files. Source files hold the function definitions, whereas header files hold function and variable declarations.
- The C/C++ compilation process compiles all source and header files and links those together with predefined library files to generate an executable program binary file.
- Preprocessor directives are required to ensure that compilation errors due to multiple variable declarations are avoided.
- Modular programming enables a number of engineers to work on a single project, each taking responsibility for a particular code feature.

Quiz

1. List the advantages of using functions in a C program.
2. What are the limitations associated with using functions in a C program?
3. What is pseudocode and how is it useful at the software design stage?
4. What is a function 'prototype' and where can it be found in a C program?
5. How much data can be input to and output from a function?
6. What is the purpose of the preprocessor in the C program compilation process?
7. At what stage in the program compilation process are predefined library files implemented?
8. When would it be necessary to use the **extern** storage class specifier in an mbed C program?
9. Why is the **#ifndef** preprocessor directive commonly used in modular program header files?
10. Draw a program flowchart which describes a program that continuously reads an analog temperature sensor output once per second and displays the temperature in degrees Celsius on a three-digit seven-segment display.

CHAPTER 7

Starting with Serial Communication

Chapter Outline

7.1 Introducing Synchronous Serial Communication

There is an unending need in computer systems to move data around – lots of it. In Chapters 1 and 2 we came across the idea of data buses, on which data flies backwards and forwards between different parts of a computer. In these buses, data is transferred in *parallel.* There is

Fast and Effective Embedded Systems Design. DOI: 10.1016/B978-0-08-097768-3.00007-6

one wire for each bit of data, and one or two more to provide synchronization and control; data is transferred a whole word at a time. This works well, but it requires a lot of wires and a lot of connections on each device that is being interconnected. It is bad enough for an 8-bit device, but for 16 or 32 bits the situation is far worse. An alternative to parallel communication is *serial.* Here, a single wire is effectively used for data transfer, with bits being sent in turn. A few extra connections are almost inevitably needed, for example for earth return, and synchronization and control.

Once we start applying the serial concept, a number of challenges arise. How does the receiver know when each bit begins and ends, and how does it know when each word begins and ends? There are several ways of responding to these questions. A straightforward approach is to send a clock signal alongside the data, with one clock pulse per data bit. The data is *synchronized* to the clock. This idea, called synchronous serial communication, is represented in Figure 7.1. When no data is being sent, there is no movement on the clock line. Every time the clock pulses, however, one bit is output by the transmitter and should be read in by the receiver. In general, the receiver synchronizes its reading of the data with one edge of the clock. In this example it is the rising edge, highlighted by a dotted line.

A simple serial data link is shown in Figure 7.2. Each device that connects to the data link is sometimes called a *node.* In this figure, Node 1 is designated *Master*; it controls what is going on, as it controls the clock. The *Slave* is similar to the master, but receives the clock signal from the master.

An essential feature of the serial link shown, and indeed of most serial links, is a *shift register.* This is made up of a string of digital flip-flops, connected so that the output of one is connected to the input of the next. Each flip-flop holds one bit of information. Every time the shift register is pulsed by the clock signal, each flip-flop passes its bit on to its neighbor on one side and receives a new bit from its other neighbor. The one at the input end clocks in data received from the outside world, and the one of the output end outputs its bit. Therefore, as the clock pulses the shift register can be both feeding in external data and outputting data. The data held by all the flip-flops in the shift register can, moreover, be read all at the same time, as a parallel word, or a new value can be loaded in. In summary, the shift register is an incredibly useful subsystem: it can convert serial data to parallel data, and vice versa, and it can act as a serial transmitter and/or a serial receiver.

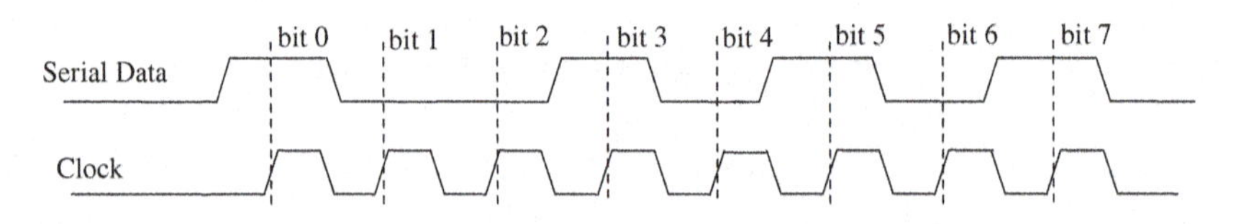

Figure 7.1:
Synchronous serial data

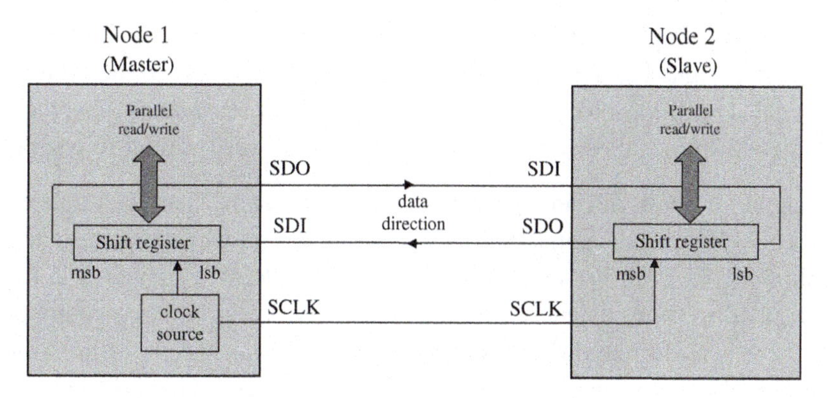

Figure 7.2:
A simple serial link. SDI: serial data in; SDO: serial data out; SCLK: serial clock

As an example, suppose both shift registers in Figure 7.2 are 8-bit, i.e. each has eight flip-flops. Each register is loaded with a new word, and the master then generates eight clock pulses. For each clock cycle, one new bit of data appears at the output end of each shift register, indicated by the SDO (serial data output) label. Each SDO output is, however, connected to the input (SDI: serial data input) of the other register. Therefore as each bit is clocked out of one register, it is clocked in to the other. After eight clock cycles, the word that was in the master shift register is now in the slave, and the word that was in the slave shift register is now held in the master.

The circuitry for the master is usually placed within a microcontroller. The slave might be another microcontroller or some other peripheral device. In general, the hardware circuitry that allows serial data to be sent and received, and interfaces between the microcontroller central processing unit (CPU) and the outside world, is called a serial port.

7.2 Serial Peripheral Interface

To ensure that serial data links are reliable, and can be applied by different devices in different places, several standards or *protocols* have been defined. One that is very widely used these days is the universal serial bus (USB). A protocol defines details of timing and signals, and may go on to define other things, such as the type of connector used. Serial peripheral interface (SPI) is a simple protocol that has had a large influence in the embedded world.

7.2.1 Introducing SPI

In the early days of microcontrollers, both National Semiconductors and Motorola started introducing simple serial communication, based on Figure 7.2. Each formulated a set of rules governing how their microcontrollers worked, and allowed others to develop devices that could interface correctly. These became *de facto* standards, in other words they were never initially formally designed as standards, but were adopted by others to the point where they acted as a formal standard. Motorola called its standard *Serial Peripheral Interface (SPI)* and National Semiconductors called theirs *Microwire*. They are very similar to each other.

It was not long before both SPI and Microwire were adopted by manufacturers of other integrated circuits (ICs), who wanted their devices to be able to work with the new generation of microcontrollers. The SPI has become one of the most durable standards in the world of electronics, applied to short-distance communications, typically within a single piece of equipment. There is no formal document defining SPI, but datasheets for the Motorola 68HC11 (now an old microcontroller) effectively define it in full. Good related texts also do, for example Reference 7.1.

In SPI one microcontroller is designated the master; it controls all activity on the serial interconnection. The master communicates with one or more slaves. A minimum SPI link uses just one master and one slave, and follows the pattern of Figure 7.2. The master generates and controls the clock, and hence all data transfer. The SDO of one device is connected to the SDI of the other, and vice versa. In every clock cycle one bit is moved from master to slave, and one bit from slave to master; after eight clock cycles a whole byte has been transferred. Thus, data is actually transferred in both directions – in old terminology this is called *full duplex* – and it is up to the devices to decide whether a received byte is intended for it or not. If data transfer in only one direction is wanted, then the data transfer line that is not needed can be omitted.

If more than one slave is needed, then the approach of Figure 7.3 can be used. Only one slave is active at any time, determined by which Slave Select (SS) line the master activates. Note that writing SS indicates that the line is active when low; if it were active high it would simply be SS. The terminology Chip Select ($\overline{\text{CS}}$) is also used for the same role, including in this chapter. Only the slave activated by its $\overline{\text{SS}}$ input responds to the clock signal, and normally only one slave is activated at any one time. The master then communicates with the one active

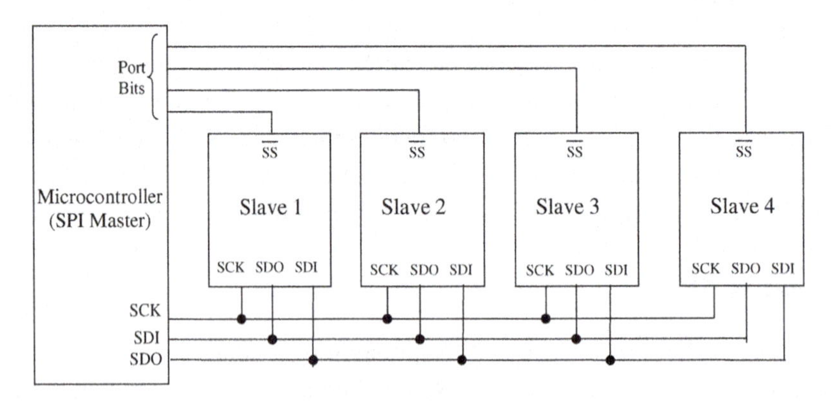

Figure 7.3:
SPI interconnections for multiple slaves

slave just as in Figure 7.2. Notice that for n slaves, the microcontroller needs to commit $(3 + n)$ lines. One of the advantages of serial communication, the small number of interconnections, is beginning to disappear.

7.2.2 SPI on the mbed

As the mbed diagram of Figure 2.1 shows, the mbed has two SPI ports, one appearing on pins 5, 6 and 7, and the other on pins 11, 12 and 13. The application programming interface (API) summary available for SPI master is shown in Table 7.1.

7.2.3 Setting up an mbed SPI Master

Program Example 7.1 shows a very simple setup for a SPI master. The program initializes the SPI port, choosing for it the name **ser_port**, with the pins of one of the possible ports being selected. The **format( )** function requires two variables: the first is the number of bits and the second is the mode. The mode is a feature of SPI which is illustrated in Figure 7.4, with associated codes in Table 7.2. It allows the choice of which clock edge is used to clock data into the shift register (indicated as 'Data strobe' in the diagram), and whether the clock idles high or low. For most applications the default mode, i.e. Mode 0, is acceptable. This program simply applies default values, i.e. 8 bits of data, and Mode 0 format. On the mbed, as with many SPI devices, the same pin is used for SDI if in master mode, or SDO if slave. Hence, this pin gets to be called MISO (master in, slave out). Its partner pin is MOSI.

```
/* Program Example 7.1: Sets up the mbed as SPI master, and continuously sends a single
byte
*/
#include "mbed.h"
SPI ser_port(p11, p12, p13); // mosi, miso, sclk
char switch_word ;           //word we will send

int main() {
  ser_port.format(8,0);         // Setup the SPI for 8 bit data, Mode 0 operation
  ser_port.frequency(1000000); // Clock frequency is 1MHz
  while (1){
    switch_word=0xA1;            //set up word to be transmitted
    ser_port.write(switch_word); //send switch_word
    wait_us(50);
  }
}
```

Program Example 7.1 Minimal SPI master application

Compile, download and run Program Example 7.1 on a single mbed, and observe the data (pin 11) and clock (pin 13) lines simultaneously on an oscilloscope. See how clock and data are active at the same time, and verify the clock data frequency. Check that you can read the transmitted data byte, 0xA1. Is the most significant bit (MSB) or least significant bit (LSB) sent first?

Table 7.1: mbed SPI master API summary

Function	Usage
`SPI`	Create a SPI master connected to the specified pins
`format`	Configure the data transmission mode and data length
`frequency`	Set the SPI bus clock frequency
`write`	Write to the SPI slave and return the response

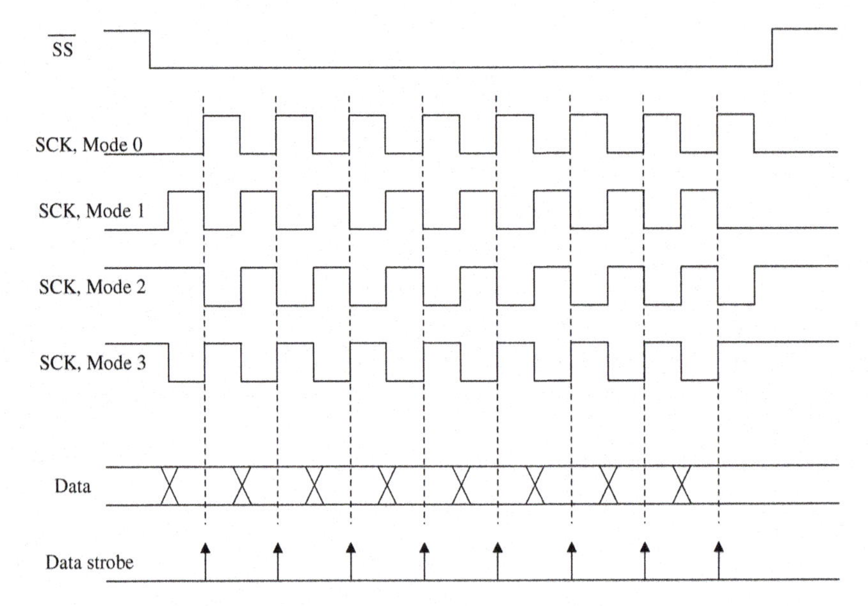

Figure 7.4:
Polarity and phase

Table 7.2: SPI modes

Mode	Polarity	Phase
0	0	0
1	0	1
2	1	0
3	1	1

Exercise 7.1

1. In Program Example 7.1, try invoking each of the SPI modes in turn. Observe both clock and data waveforms on the oscilloscope, and check how they compare to Figure 7.4.
2. Set the SPI format of Program Example 7.1 to 12 and then 16 bits, sending the words 0x8A1 and 0x8AA1, respectively (or to your choice). Check each on an oscilloscope.

7.2.4 Creating a SPI Data Link

We will now develop two programs, one master and one slave, and get two mbeds to communicate. Each will have two switches and two light-emitting diodes (LEDs); the aim will be to get the switches of the master to control the LEDs on the slave, and vice versa.

The program for the master is shown as Program Example 7.2. This is written for the circuit of Figure 7.5. It sets up the SPI port as before, and defines the switch inputs on pins 5 and 6. It declares a variable **switch_word**, the word that will be sent to the slave, and the variable **recd_val**, which is the value received from the slave. For this application the default settings

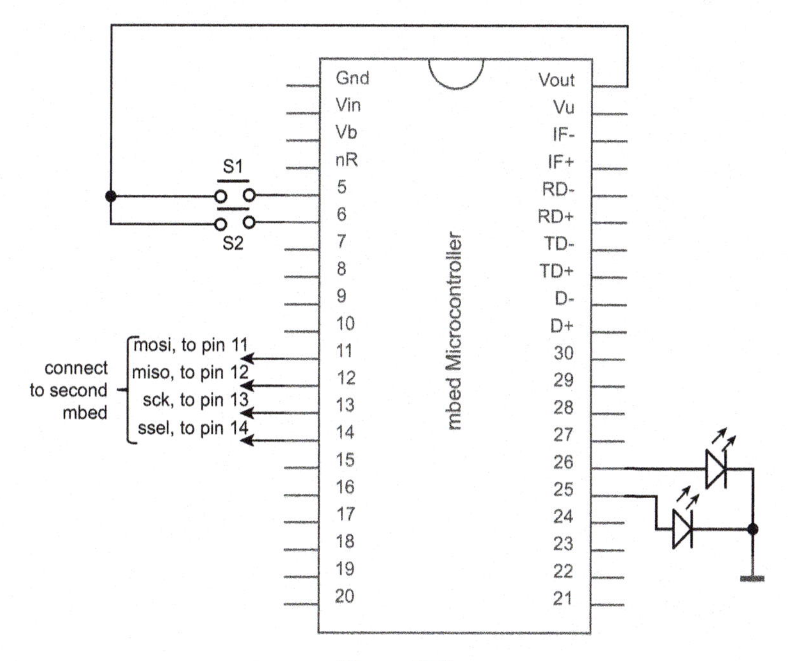

Figure 7.5:
Using SPI to link two mbeds

of the SPI port are chosen, so there is no further initialization in the program. Once in the main loop, the value of **switch_word** is established. To give a pattern to this that will be recognizable on the oscilloscope, the upper four bits are set to hexadecimal A. The two switch inputs are then tested in turn; if they are found to be high, the appropriate bit in **switch_word** is set, by ANDing with 0x01 or 0x02. The **cs** line is set low, and the command to send **switch_word** is made. The return value of this function is the received word, which is read accordingly.

```
/*Program Example 7.2. Sets the mbed up as Master, and exchanges data with a slave,
sending its own switch positions, and displaying those of the slave.
 */
#include "mbed.h"

SPI ser_port(p11, p12, p13); //mosi, miso, sclk
DigitalOut red_led(p25);     //red led
DigitalOut green_led(p26);   //green led
DigitalOut cs(p14);          //this acts as "slave select"
DigitalIn switch_ip1(p5);
DigitalIn switch_ip2(p6);
char switch_word ;     //word we will send
char recd_val;         //value return from slave

int main() {
  while (1){
    //Default settings for SPI Master chosen, no need for further configuration
    //Set up the word to be sent, by testing switch inputs
    switch_word=0xa0;        //set up a recognizable output pattern
    if (switch_ip1==1)
       switch_word=switch_word|0x01;        //OR in lsb
    if (switch_ip2==1)
       switch_word=switch_word|0x02;        //OR in next lsb
    cs = 0;                                 //select slave
    recd_val=ser_port.write(switch_word); //send switch_word and receive data
    cs = 1;
    wait(0.01);

    //set leds according to incoming word from slave
    red_led=0;               //preset both to 0
    green_led=0;
    recd_val=recd_val&0x03; //AND out unwanted bits
    if (recd_val==1)
       red_led=1;
    if (recd_val==2)
       green_led=1;
    if (recd_val==3){
       red_led=1;
       green_led=1;
    }
  }
}
```

Program Example 7.2 The mbed set up as a SPI master, with bidirectional data transfer

Table 7.3: mbed SPI slave API summary

Function	Usage
SPISlave	Create a SPI slave connected to the specified pins
format	Configure the data transmission format
frequency	Set the SPI bus clock frequency
receive	Poll the SPI to see if data has been received
read	Retrieve data from receive buffer as slave
reply	Fill the transmission buffer with the value to be written out as slave on the next received message from the master

The slave program draws upon the mbed functions shown in Table 7.3, and is shown as Program Example 7.3. It is almost the mirror image of the master program, with small but key differences. This emphasizes the very close similarity between the master and slave role in SPI. Let us check the differences. The serial port is initialized with **SPISlave**. Now four pins must be defined, the extra being the slave select input, **ssel**. As the slave will also be generating a word to be sent, and receiving one, it also declares variables **switch_word** and **recd_val**. Change these names if you would rather have something different. The slave program configures its **switch_word** just like the master. Now comes a difference. While the master initiates a transmission when it wishes, the slave must wait. The mbed library does this with the **receive( )** function. This returns 1 if data has been received, and 0 otherwise. Of course, if data has been received from the master, then data has also been sent from slave to master. If there is data, then the slave reads this and sets up the LEDs accordingly. It also sets up the next word to be sent to the master, by transferring its **switch_word** to the transmission buffer, using **reply( )**.

```
/*Program Example 7.3: Sets the mbed up as Slave, and exchanges data with a Master,
sending its own switch positions, and displaying those of the Master. as SPI slave.
*/
#include "mbed.h"
SPISlave ser_port(p11,p12,p13,p14); // mosi, miso, sclk, ssel
DigitalOut red_led(p25);            //red led
DigitalOut green_led(p26);          //green led
DigitalIn switch_ip1(p5);
DigitalIn switch_ip2(p6);
char switch_word ;                  //word we will send
char recd_val;                      //value received from master

int main() {
  //default formatting applied
  while(1) {
    //set up switch_word from switches that are pressed
    switch_word=0xa0;        //set up a recognizable output pattern
    if (switch_ip1==1)
       switch_word=switch_word|0x01;
```

```
    if (switch_ip2==1)
      switch_word=switch_word|0x02;
    if(ser_port.receive()) {   //test if data transfer has occurred
      recd_val = ser_port.read();  // Read byte from master
      ser_port.reply(switch_word); // Make this the next reply
    }
    //now set leds according to received word
    ...
    (continues as in Program Example 7.2)
    ...
  }
}
```

Program Example 7.3 The mbed set up as a SPI slave, with bidirectional data transfer

Now connect two mbeds together, carefully applying the circuit of Figure 7.5. It is simplest if each is powered individually through its own USB cable. Connections to both are identical, i.e. pin 11 goes to pin 11 and so on, so it does not matter which is chosen as master or slave. If you do not want to set up the full circuit straight away, then connect just the switches to the master, and just the LEDs to the slave, or vice versa.

Compile and download Program Example 7.2 into one mbed (which will be the master), and Program Example 7.3 into the other. Once you run the programs, you should find that pressing the switches of the master controls the LEDs of the slave, and vice versa. This is another big step forward; we are communicating data from one microcontroller to another, or from one system to another.

Exercise 7.2

Try the following, and be sure you understand the result. In each case test for data transmission from master to slave, and from slave to master.

1. Remove the pin 11 link, i.e. the data link from master to slave.
2. Remove the pin 12 link, i.e. the data link from slave to master.
3. Remove the pin 13 link, i.e. the clock.
4. Remove the pin 14 link, i.e. the chip select line.

Notice that in some cases if you disconnect a wire, but leave it dangling in the air, you might get odd intermittent behavior which changes if you touch the wire. This is an example of the impact of electromagnetic interference, where interference is being interpreted by an mbed input as a clock or data signal.

7.3 Intelligent Instrumentation and a SPI Accelerometer

7.3.1 Introducing the ADXL345 Accelerometer

Despite (or because of) its age, the SPI standard is wonderfully simple, and hence widely used. It is embedded into all sorts of electronic devices, ICs and gadgets. Given an understanding of how SPI works, we can now communicate with any of these.

With the very high level of integration found in modern ICs, it is common to find a sensor, signal conditioning, an analog-to-digital converter (ADC) and a data interface, all combined onto a single chip. Such devices are part of the new generation of *intelligent instrumentation.* Instead of just having a stand-alone sensor, as we did with the light-dependent resistor in Chapter 5, we can now have a complete measurement subsystem integrated with the sensor, with a convenient serial data output.

The ADXL345 accelerometer, made by Analog Devices, is an example of an integrated 'intelligent' sensor. Its datasheet appears as Reference 7.2. This is also an example of a *microelectromechanical system* (MEMS), in which the accelerometer mechanics are actually fabricated within the IC structure. The accelerometer output is analog in nature, and measures acceleration in three axes. The accelerometer has an internal capacitor mounted in the plane of each axis. Acceleration causes the capacitor plates to move, hence changing the output voltage in proportion to the acceleration or force. The ADXL345 accelerometer converts the analog voltage fluctuations to digital and can output these values over a SPI serial link (or I^2C; see later in this chapter).

Control of the ADXL345 is done by writing to a set of registers through the serial link. Examples of these are shown in Table 7.4. It is clear that the device goes well beyond just making direct measurements. It is possible to calibrate it, change its range and enable it to recognize certain events, for example when it is tapped or in free fall. Measurements are made in terms of *g* (where 1 *g* is the value of acceleration due to earth's gravity, i.e. $9.81\ \mathrm{ms}^{-2}$).

The ADXL345 IC is extremely small and designed for surface mounting on a printed circuit board; therefore, it is ready-mounted on a 'breakout' board, as shown in Figure 7.6. It would otherwise be difficult to handle.

7.3.2 Developing a Simple ADXL345 Program

Program Example 7.4 applies the ADXL345, reading acceleration in three axes, and outputting the data to the host computer screen. The second SPI port (i.e. pins 11, 12 and 13) is used for connecting the accelerometer, applying the connections shown in Table 7.5.

Table 7.4: Selected ADXL345 registers

Address*	Name	Description
0x00	DEVID	Device ID
0x1D	THRESH_TAP	Tap threshold
0x1E/1F/20	OFSX, OFSY, OFSZ	x, y, z-axis offsets
0x21	DUR	Tap duration
0x2D	POWER_CTL	Power-saving features control. Device powers up in standby mode; setting bit 3 causes it to enter Measure mode
0x31	DATA_FORMAT	Data format control Bits:
		7: force a self-test by setting to 1
		6: 1 = 3-wire SPI mode; 0 = 4-wire SPI mode
		5: 0 sets interrupts active high, 1 sets them active low
		4: always 0
		3: 0 = output is 10-bit always; 1 = output depends on range setting
		2: 1 = left align result; 0 = right align result
		1-0: 00 = $\pm$ 2 g; 01 = $\pm$ 4 g; 10 = $\pm$ 8 g; 11 = $\pm$ 2 g
0x33:0x32	DATAX1:DATAX0	x-axis data, formatted according to DATA_FORMAT, in 2's complement.
0x35:0x34	DATAY1:DATAY0	y-axis data, as above
0x37:0x36	DATAZ1:DATAZ0	z-axis data, as above

*In any data transfer the register address is sent first, and formed: bit 7 = $R/\overline{W}$ (1 for read, 0 for write); bit 6:1 for multiple bytes, 0 for single; bits 5-0: the lower 6 bits found in the Address column.

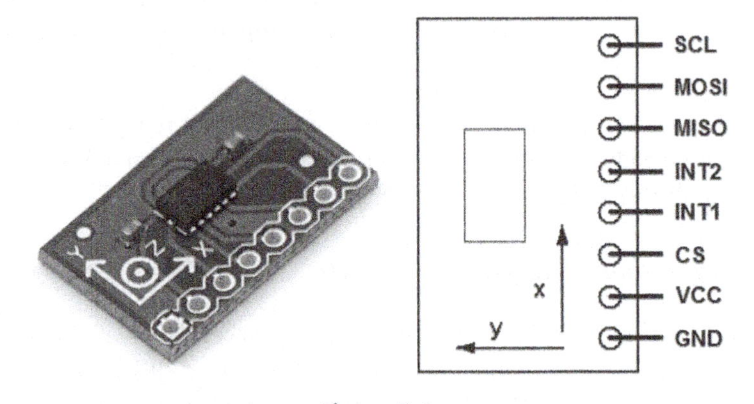

Figure 7.6:
The ADXL345 accelerometer on breakout board. *(Image reproduced with permission of SparkFun Electronics)*

Table 7.5: ADXL345 pin connections to mbed

ADXL345 signal name	mbed pin
Vcc	Vout
Gnd	Gnd
SCL	13
MOSI	11
MISO	12
$\overline{CS}$	14

C code feature

The program initializes a master SPI port, which we have chosen to call **acc**, and sets up the USB link to the host computer. It further declares two arrays; one is a buffer that will hold data read direct from the accelerometer's registers, two for each axis. The second array applies the **int16_t** specifier. This is from the C standard library **stdint**; it tells the compiler that exactly 16-bit (signed) integer type data is being declared. This array will hold the full accelerometer axis values, combined from the two bytes received from the registers.

The main function initializes the SPI port in a manner with which we are familiar. It then loads two of the accelerometer registers, writing the address first, followed by the data byte. It should be possible to work out what is being written by looking at either the datasheet or Table 7.4. A continuous **while** loop is then initiated. Following a pause, a multi-byte read is set up, using an address word formed from information shown in Table 7.4. This fills the **buffer** array. The **data** array is then populated, concatenating (i.e. combining to form a single number) pairs of bytes from the **buffer** array. These values are then scaled to actual *g* values, using the conversion factor from the datasheet, of 0.004 *g* per unit. Results are then displayed on-screen.

```
/*Program Example 7.4: Reads values from accelerometer through SPI, and outputs
continuously to terminal screen.
*/

#include "mbed.h"
SPI acc(p11,p12,p13);           // set up SPI interface on pins 11,12,13
DigitalOut cs(p14);             // use pin 14 as chip select
Serial pc(USBTX, USBRX);        // set up USB interface to host terminal
char buffer[6];                 //raw data array type char
int16_t data[3];                // 16-bit twos-complement integer data
float x, y, z;                  // floating point data, to be displayed on-screen

int main() {
  cs=1;                           //initially ADXL345 is not activated
  acc.format(8,3);                // 8 bit data, Mode 3
```

```
  acc.frequency(2000000);       // 2MHz clock rate
  cs=0;                         //select the device
  acc.write(0x31);              // data format register
  acc.write(0x0B);              // format +/-16g, 0.004g/LSB
  cs=1;                         //end of transmission
  cs=0;                         //start a new transmission
  acc.write(0x2D);              // power ctrl register
  acc.write(0x08);              // measure mode
  cs=1;                         //end of transmission
  while (1) {                   // infinite loop
    wait(0.2);
    cs=0;                       //start a transmission
    acc.write(0x80|0x40|0x32);  // RW bit high, MB bit high, plus address
    for (int i = 0;i<=5;i++) {
      buffer[i]=acc.write(0x00);         // read back 6 data bytes
    }
    cs=1;                                //end of transmission
    data[0] = buffer[1]<<8 | buffer[0];  //combine MSB and LSB
    data[1] = buffer[3]<<8 | buffer[2];
    data[2] = buffer[5]<<8 | buffer[4];
    x=0.004*data[0]; y=0.004*data[1]; z=0.004*data[2]; // convert to float,
                                                       //actual g value
    pc.printf("x = %+1.2fg\t y = %+1.2fg\t z = %+1.2fg\n\r", x, y,z); //print
  }
}
```

Program Example 7.4 Accelerometer continuously outputs three-axis data to terminal screen

Carefully make the connections of Table 7.5, and compile, download and run the code on your mbed. Open a Tera Term screen (as described in Appendix E) on your computer. Accelerometer readings should be displayed to the screen. You will see that when the accelerometer is flat on a table the *z*-axis should read approximately 1 *g*, with the *x*- and *y*-axes approximately 0 *g*. As you rotate and move the device the *g* readings will change. If the accelerometer is shaken or displaced at a quick rate, *g* values in excess of 1 *g* can be observed. Note that there are some inaccuracies in the accelerometer data, which can be reduced in a real application by developing configuration/calibration routines and data averaging functions.

Although we have not used it here, the mbed site provides an ADXL345 library, located in the Cookbook. This simplifies use of the accelerometer, and saves worrying about register addresses and bit values.

Exercise 7.3

Rewrite Program Example 7.4 using the library functions available on the mbed site Cookbook.

7.4 Evaluating SPI

The SPI standard is extremely effective. The electronic hardware is simple and therefore cheap, and data can be transferred rapidly. It does have its disadvantages, however. There is no acknowledgment from the receiver, so in a simple system the master cannot be sure that data has been received. Also there is no addressing. In a system where there are multiple slaves, a separate $\overline{SS}$ line must be run to each slave, as shown in Figure 7.3. Therefore, we begin to lose the advantage that should come from serial communications, i.e. a limited number of interconnect lines. Finally, there is no error checking. Suppose some electromagnetic interference was experienced in a long data link; the data or clock would be corrupted, but the system would have no way of detecting this or correcting for it. You may have experienced this in a small way in Exercise 7.2. Overall, SPI could be graded as: simple, convenient and low cost, but not appropriate for complex or high-reliability systems.

7.5 The Inter-Integrated Circuit Bus

7.5.1 Introducing the I^2C Bus

The inter-integrated circuit (I^2C) standard was developed by Philips to resolve some of the perceived weaknesses of SPI and its equivalents. As its name suggests, it is also intended for interconnection over short distances and generally within a piece of equipment. It uses only

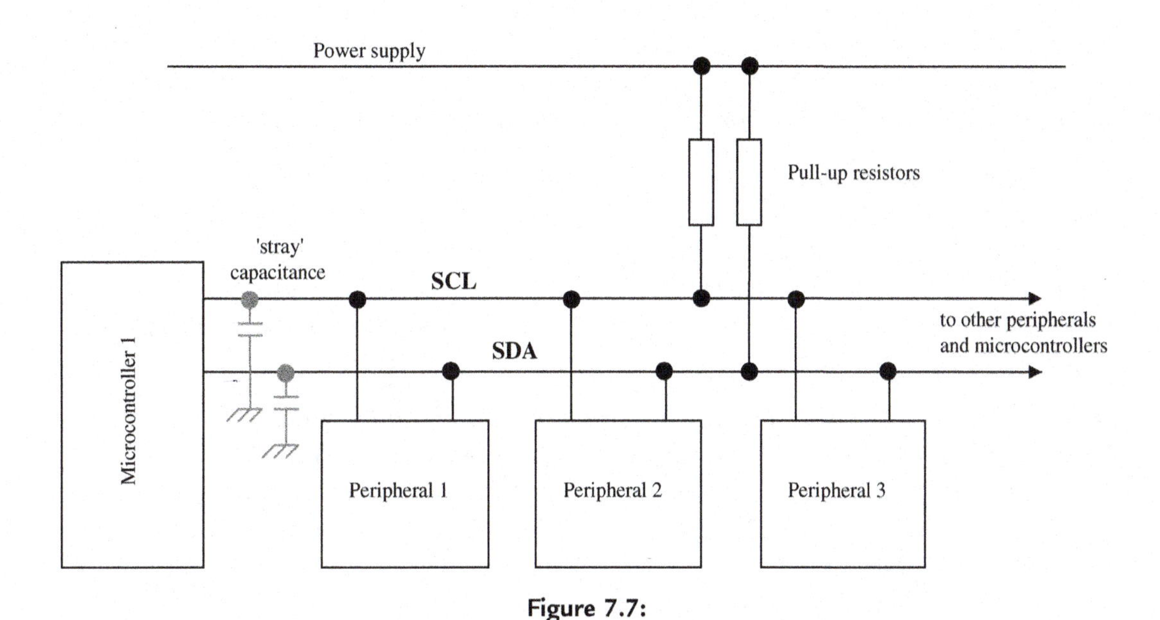

Figure 7.7:
An I^2C-based system

two interconnect wires, but many devices are connected to the bus. These lines are called SCL (serial clock) and SDA (serial data). All devices on the bus are connected to these two lines, as shown in Figure 7.7. The SDA line is bidirectional, so data can travel in either direction, but only one direction at any one time. In the jargon this is called *half duplex*. Like SPI, it is a synchronous serial standard.

One of the interesting features of I^2C, and one that makes it versatile, is that any node connected to it can only pull down the SCL or SDA line to Logic 0; it cannot force the line up to Logic 1. This role is played by a single pull-up resistor connected to each line. When a node pulls a line to Logic 0 and then releases it, it is returned to Logic 1 by the action of the pull-up resistor. There is, however, capacitance associated with the line. Although this is labeled 'stray' capacitance in Figure 7.7, it is in reality mainly unavoidable capacitance that exists in the semiconductor structures connected to the line. This capacitance is thus higher if there are many nodes connected, and lower otherwise. The higher the capacitance and/or pull-up resistance, the longer the rise time of the logic transition from 0 to 1. The I^2C standard requires that the rise time of a signal on SCL or SDA must be less than 1000 ns. Given a known bus setup, it is possible to perform reasonably precise calculations of pull-up resistor required (see Reference 1.1), particularly if you need to minimize power consumption. For simple applications default pull-up resistor values, in the range 2.2−4.7 kΩ, are quite acceptable.

The I^2C protocol has been through several revisions, which have dramatically increased the possible speeds and reflect technological changes, for example in reduced minimum operating voltages. The original version, standard mode, allowed data rates up to 100 kbit/s. Version 1.0, in 1992, increased the maximum data rate to 400 kbit/s. This latter is very well established and still probably accounts for most I^2C implementation. Version 2.0, in 1998, increased the possible bit rate to 3.4 Mbit/s. Version 3 is defined in a surprisingly readable manner in Reference 7.3, and forms the basis of the following description.

Nodes on an I^2C bus can act as master or slave. The master initiates and terminates transfers, and generates the clock. The slave is any device addressed by the master. A system may have more than one master, although only one may be active at any time. Therefore more than one microcontroller could be connected to the bus, and they can claim the master role at different times, when needed. An arbitration process is defined if more than one master attempts to control the bus.

A data transfer is made up of the master signaling a *start condition*, followed by one or two bytes containing address and control information. The start condition (Figure 7.8a) is defined by a high to low transition of SDA when SCL is high. All subsequent data transmission follows the pattern of Figure 7.8b. One clock pulse is generated for each data bit, and data may only change when the clock is low. The byte following the start condition is made up of seven address bits and one data direction bit, as shown in Figure 7.8c. Each slave has a predefined device address; the slaves are therefore responsible for monitoring the bus and

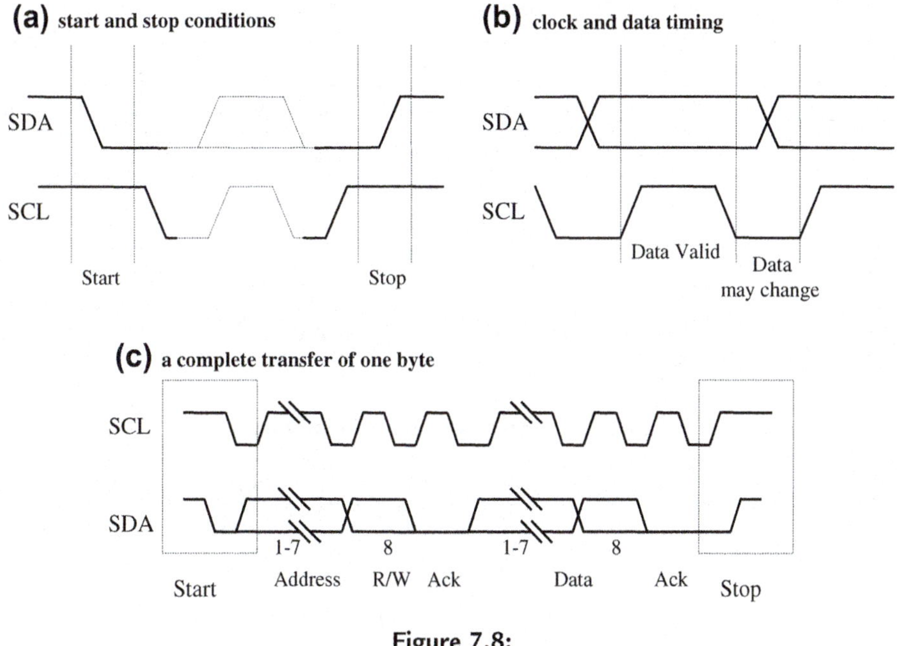

Figure 7.8:
I^2C data transfer

responding only to commands associated with their own address. A slave device that recognizes its address will then be readied either to receive data or to transmit it onto the bus. A 10-bit addressing mode is also available.

All data transferred is in units of one byte, with no limit on the number of bytes transferred in one message. Each byte must be followed by a 1-bit acknowledgment from the receiver, during which time the transmitter relinquishes SDA control. A low to high transition of SDA while SCL is high defines a *stop* condition. Figure 7.8c illustrates the complete transfer of a single byte.

7.5.2 I^2C on the mbed

Figure 2.1 shows us that the mbed offers two I^2C ports, on pins 9 and 10, or 27 and 28. Their use follows the pattern of other mbed peripherals, with available functions shown in Tables 7.6 and 7.7.

7.5.3 Setting up an I^2C Data Link

We will now replicate the action of Section 7.2.4, but using I^2C as the communication link, rather than SPI. We will use the I^2C port on pins 9 and 10. Program Example 7.5, the master,

Table 7.6: mbed I²C master API summary

Function	Usage
I2C	Create an I^2C master interface, connected to the specified pins
frequency	Set the frequency of the I^2C interface
read	Read from an I^2C slave
write	Write to an I^2C slave
start	Creates a start condition on the I^2C bus
stop	Creates a stop condition on the I^2C bus

Table 7.7: mbed I²C slave API summary

Function	Usage
I2CSlave	Create an I^2C slave interface, connected to the specified pins
frequency	Set the frequency of the I^2C interface
receive	Checks to see if this I^2C slave has been addressed
read	Read from an I^2C master
write	Write to an I^2C master
address	Sets the I^2C slave address
stop	Reset the I^2C slave back into the known ready receiving state

follows exactly the pattern of Program Example 7.2, except that SPI-related sections are replaced by those that relate to I^2C. Early in the program an I^2C serial port is configured using the mbed utility **I2C**. The name **i2c_port** is chosen for the port name, and linked to the port on pins 9 and 10. An arbitrary slave address is chosen, 0x52. Following determination of the variable **switch_word**, the I^2C transmission can be seen. This is created from the separate components of a single-byte I^2C transmission, i.e. start – send address – send data – stop, as allowed by the mbed functions. A way of grouping these together is presented later in the chapter. Further down the program we see a request for a byte of data from the slave, with similar message structure. Now the slave address is ORed with 0x01, which sets the $R/\overline{W}$ bit in the address word to indicate Read. The received word is then interpreted in order to set the LEDs, just as we did in the SPI program earlier.

```
/*Program Example 7.5: I2C Master, transfers switch state to second mbed acting as
slave, and displays state of slave's switches on its leds.
*/
#include "mbed.h"
I2C i2c_port(p9, p10);      //Configure a serial port, pins 9 and 10 are sda, scl
DigitalOut red_led(p25);    //red led
```

```
DigitalOut green_led(p26); //green led
DigitalIn switch_ip1(p5);  //input switch
DigitalIn switch_ip2(p6);

char switch_word ;      //word we will send
char recd_val;          //value received from slave
const int addr = 0x52;  //the I2C slave address, an arbitrary even number

int main() {
  while(1) {
    switch_word=0xa0;                //set up a recognizable output pattern
    if (switch_ip1==1)
      switch_word=switch_word|0x01;  //OR in lsb
    if (switch_ip2==1)
      switch_word=switch_word|0x02;  //OR in next lsb
    //send a single byte of data, in correct I2C package
    i2c_port.start();               //force a start condition
    i2c_port.write(addr);           //send the address
    i2c_port.write(switch_word);    //send one byte of data, ie switch_word
    i2c_port.stop();                //force a stop condition
    wait(0.002);
    //receive a single byte of data, in correct I2C package
    i2c_port.start();
    i2c_port.write(addr|0x01);      //send address, with R/W bit set to Read
    recd_val=i2c_port.read(addr);   //Read and save the received byte
    i2c_port.stop();                //force a stop condition
    //set leds according to word received from slave
    red_led=0;                      //preset both to 0
    ...
    (continues as in Program Example 7.2)
    ...
  }
}
```

Program Example 7.5 I^2C data link master

The slave program is shown in Program Example 7.6, and is similar to Program Example 7.3, with SPI features replaced by I^2C. As in SPI, the I^2C slave just responds to calls from the master. The slave port is defined with the mbed utility **I2Cslave**, with **slave** chosen as the port name. Just within the **main** function the slave address is defined, importantly the same 0x52 as we saw in the master program. As before, the **switch_word** value is set up from the state of the switches; this is then saved using the **write** function, in readiness for a request from the master. The **receive( )** function is used to test whether an I^2C transmission has been received. This returns a 0 if the slave has not been addressed, a 1 if it has been addressed to read, and a 3 if addressed to write. If a read has been initiated, then the value already stored is automatically sent. If the value is 3, then the program stores the received value and sets up the LEDs on the master accordingly.

```
/*Program Example 7.6: I2C Slave, when called transfers switch state to mbed acting as
Master, and displays state of Master's switches on its leds.
*/
#include <mbed.h>
I2CSlave slave(p9, p10);      //Configure I2C slave
DigitalOut red_led(p25);      //red led
```

```
DigitalOut green_led(p26);     //green led
DigitalIn switch_ip1(p5);
DigitalIn switch_ip2(p6);
char switch_word ;             //word we will send
char recd_val;                 //value received from master

int main() {
  slave.address(0x52);
  while (1) {
    //set up switch_word from switches that are pressed
    switch_word=0xa0; //set up a recognizable output pattern
    if (switch_ip1==1)
      switch_word=switch_word|0x01;
    if (switch_ip2==1)
      switch_word=switch_word|0x02;
    slave.write(switch_word); //load up word to send
    //test for I2C, and act accordingly
    int i = slave.receive();
    if (i == 3){       //slave is addressed, Master will write
      recd_val= slave.read();
          //now set leds according to received word
    ...
   (continues as in Program Example 7.2)
    ...
  }
}
```

Program Example 7.6 I^{2}C data link slave

Connect two mbeds together with an I^{2}C link, applying the circuit diagram of Figure 7.9. This is very similar to Figure 7.5, except that the SPI connection is removed, and replaced by the I^{2}C connection. It is essential to include the pull-up resistors; values of 4.7 kΩ are shown, but they can be anywhere in the range 2.2–4.7 kΩ. Note that each mbed should have two switches and two LEDs connected, but there should just be one pair of pull-up resistors between them. Compile and download Program Example 7.5 into either mbed, and Example 7.6 into the other. You should find that the switches of one mbed control the LEDs of the other, and vice versa.

Monitor SCL and SDA lines on an oscilloscope. This may require careful oscilloscope triggering. With appropriate setting of the oscilloscope time base you should be able to see the two messages being sent between the mbeds. Identify as many features of I^{2}C as you can, including the idle high condition, the start and stop conditions, and the address byte with the R/$\overline{W}$ bit embedded.

Exercise 7.4

In Program Example 7.6, replace the line `if (i == 3)` with `if (i == 4)`; in other words, the condition cannot be satisfied. Compile, download and run. Notice that the slave still

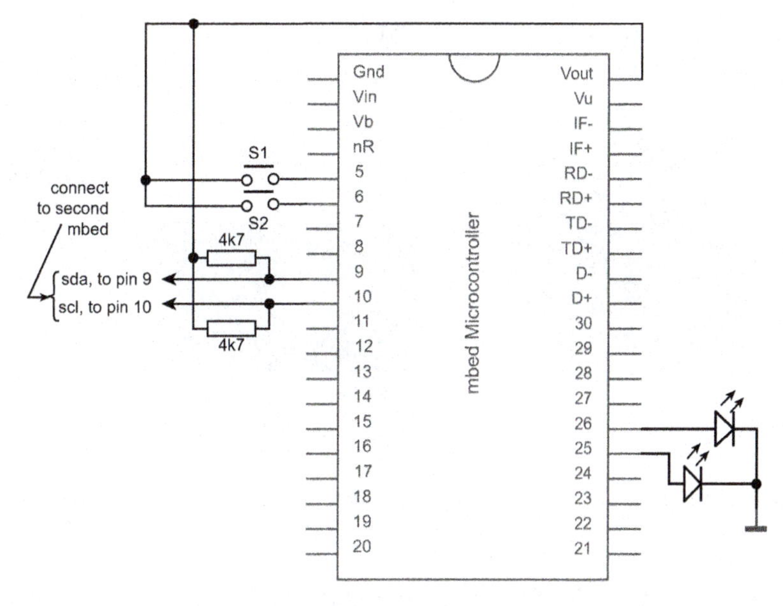

Figure 7.9:
Linking two mbeds with I^2C

continues to write to the master, but is now unable to respond to the master's message. Why is this?

■

7.6 Communicating with an I^2C Temperature Sensor

Just as we did with the SPI port and the accelerometer, we can use the mbed I^2C port to communicate with a very wide range of peripheral devices, including many intelligent sensors. The Texas Instruments TMP102 temperature sensor (Reference 7.4) has an I^2C data link. This is similar to the accelerometer introduced in Section 7.3, in that an analog sensing device is integrated with an ADC and a serial port, producing an ideal and easy-to-use system element. Note from the datasheet that the TMP102 actually makes use of the system management bus (SMbus). This was defined by Intel in 1995, and is based on I^2C. In simple applications the two standards can be mixed; for more advanced applications it is worth checking the small differences between them.

The TMP102 is a tiny device, as required of a temperature sensor. Like the accelerometer, it is used mounted on a small breakout board, seen in Figure 7.10a. It has six possible connections, shown in Table 7.8. Connections on connector JP1 are the essential ones of power supply and I^2C. On connector JP2 the sensor has an address pin, ADD0; this is used to select the address

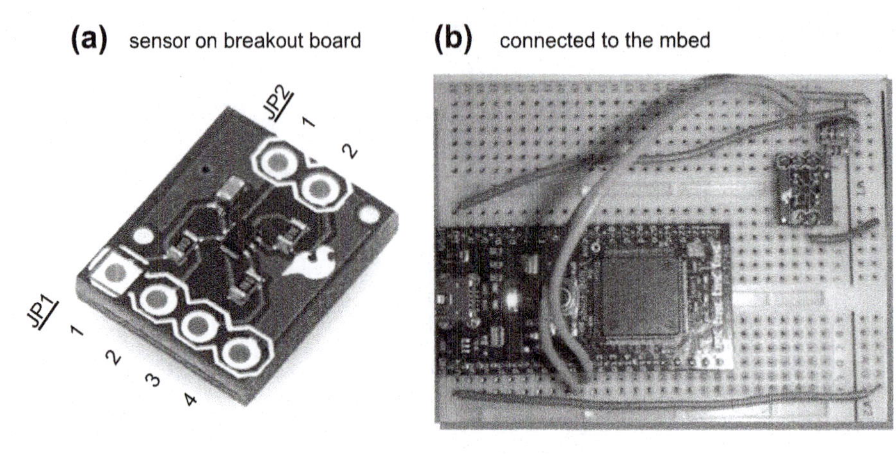

Figure 7.10:
The TMP102 temperature sensor *(Image reproduced with permission of SparkFun Electronics)*

of the device, as seen in the table. This allows four different address options, hence four of the same sensor can be used on the same I^2C bus.

Program Example 7.7 can be applied to link the mbed to the sensor. It defines an I^2C port on pins 9 and 10, and names this **tempsensor**. The serial link to be used to communicate with the PC is set up. As sensor pin ADD0 is tied to ground, Table 7.8 shows that the sensor address will be 0x90; this is defined in the program, with name **addr**. Two small arrays are also defined, one to hold the sensor configuration data, and the other to hold the raw data read from the sensor. A further variable **temp** will hold the scaled decimal equivalent of the reading.

Table 7.8: Connecting the TMP102 sensor to the mbed

Signal	TMP102 pin	mbed pin	Notes	
Vcc (3.3 V)	JP1: 1	40		
SDA	JP1: 2	9	2.2 kΩ pull-up to 3.3 V	
SCL	JP1: 3	10	2.2 kΩ pull-up to 3.3 V	
Gnd (0 V)	JP1: 4	1		
Alert	JP2: 1	1		
			Connect to	Slave address
ADD0	JP2: 2	1	0V	0x90
			Vcc	0x91
			SDA	0x92
			SCL	0x93

The configuration options can be found from the TMP102 datasheet. To set the configuration register we first need to send a data byte of 0x01 to specify that the pointer register is set to 'Configuration Register'. This is followed by two configuration bytes, 0x60 and 0xA0. These select a simple configuration setting, initializing the sensor to normal mode operation. These values are sent at the start of the **main** function. Note the format of the write command used here, which is able to send multi-byte messages. This is different from the approach used in Program Example 7.5, where only a single byte was sent in any one message. Using the I^2C **write( )** function, we need to specify the device address and the array of data, followed by the number of bytes to send.

The sensor will now operate and acquire temperature data, so we simply need to read the data register. To do this we need first to set the pointer register value to 0x00. In this command we have only sent one data byte to set the pointer register. The program now starts an infinite loop to read continuously the 2-byte temperature data. This data is then converted from a 16-bit reading to an actual temperature value. The conversion required (as specified by the datasheet) is to shift the data right by 4 bits (it is actually only 12-bit data held in two 8-bit registers) and to multiply by the specified conversion factor, 0.0625 degrees C per LSB. The value is then displayed on the PC screen.

```
/*Program Example 7.7: Mbed communicates with TMP102 temperature sensor, and scales and
displays readings to screen.
*/
#include "mbed.h"
I2C tempsensor(p9, p10);    //sda, scl
Serial pc(USBTX, USBRX);    //tx, rx
const int addr = 0x90;
char config_t[3];
char temp_read[2];
float temp;

int main() {
  config_t[0] = 0x01;                       //set pointer reg to 'config register'
  config_t[1] = 0x60;                       // config data byte1
  config_t[2] = 0xA0;                       // config data byte2
  tempsensor.write(addr, config_t, 3);
  config_t[0] = 0x00;                       //set pointer reg to 'data register'
  tempsensor.write(addr, config_t, 1);      //send to pointer 'read temp'
  while(1) {
    wait(1);
    tempsensor.read(addr, temp_read, 2);    //read the two-byte temp data
    temp = 0.0625 * (((temp_read[0] << 8) + temp_read[1]) >> 4); //convert data
    pc.printf("Temp = %.2f degC\n\r", temp);
  }
}
```

Program Example 7.7 Communicating by I^2C with the TMP102 temperature sensor

Connect the sensor according to the information in Table 7.8, leading to a circuit that looks something like Figure 7.10b. Once again, the SDA and SCL lines each need to be pulled up to 3.3 V through a resistor, of value between 2.2 and 4.7 kΩ. Compile, download and run the code of Program Example 7.7. Test that the displayed temperature increases when you press your finger against the sensor. You can try placing the sensor on something warm, for example a radiator, in order to check that it responds to temperature changes. If you have a calibrated temperature sensor, try to compare readings from both sources.

Exercise 7.5

Rewrite Program Example 7.5 using the I²C master **read( )** and **write( )** functions, as seen in Program Example 7.7. Compile, download and test that it works as expected.

7.7 Using the SRF08 Ultrasonic Range Finder

The SRF08 ultrasonic range finder, as shown in Figure 7.11, can be used to measure the distance between the sensor and an acoustically reflective surface or object in front of it. It makes the measurement by transmitting a pulse of ultrasound from one of its transducers, and then measuring the time for an echo to return to the other. If there is no echo it times out. The distance to the reflecting object is proportional to the time taken for the echo to return. Given knowledge of the speed of sound in air, the actual distance can be calculated. The SRF08 has

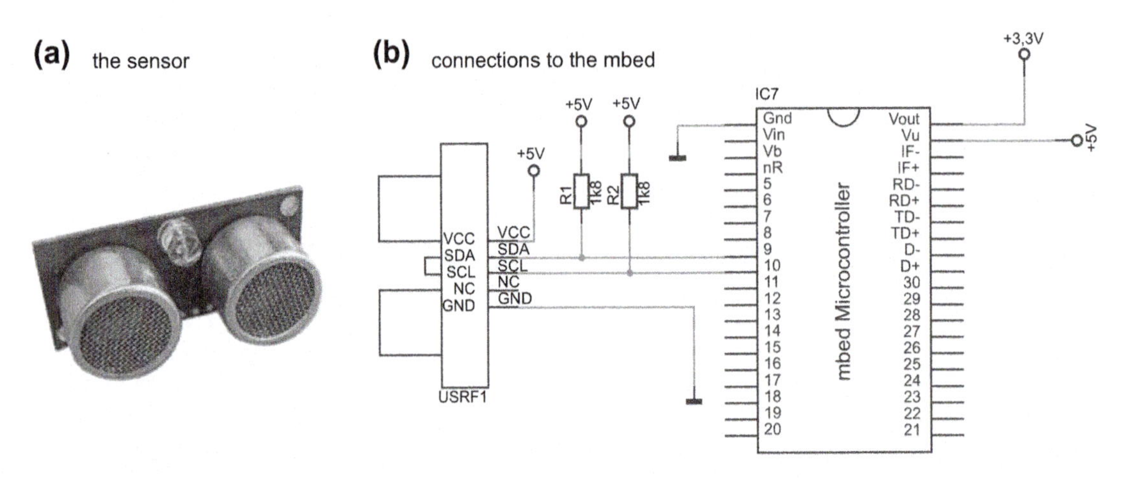

Figure 7.11:
The SRF08 ultrasonic range finder. *(Image reproduced with permission of SparkFun Electronics)*

an I^2C interface. Data is readily available for it, for example Reference 7.5, although at the time of writing not as a formalized document.

The SRF08 can be connected to the mbed as shown in Figure 7.11b. It must be powered from 5 V, with I^2C pull-up resistors connected to that voltage. The mbed remains powered from 3.3 V, and is able to tolerate the higher voltage being presented at its I^2C pins.

Having worked through the preceding programs, it should be easy to grasp how Program Example 7.8 works. Note the following information, taken from the device data:

- The SRF08 I^2C address is 0xE0.
- The pointer value for the command register is 0x00.
- A data value of 0x51 to the command register initializes the range finder to operate and return data in cm.
- A pointer value of 0x02 prepares for 16-bit data (i.e. two bytes) to be read.

```
/*Program Example 7.8: Configures and takes readings from the SRF08 ultrasonic range
finder, and displays them on screen.
*/
#include "mbed.h"
I2C rangefinder(p9, p10); //sda, scl
Serial pc(USBTX, USBRX); //tx, rx
const int addr = 0xE0;
char config_r[2];
char range_read[2];
float range;

int main() {
  while (1) {
    config_r[0] = 0x00;                         //set pointer reg to 'cmd register'
    config_r[1] = 0x51;                         //initialize, result in cm
    rangefinder.write(addr, config_r, 2);
    wait(0.07);
    config_r[0] = 0x02;                         //set pointer reg to 'data register'
    rangefinder.write(addr, config_r, 1);       //send to pointer 'read range'
    rangefinder.read(addr, range_read, 2);      //read the two-byte range data
    range = ((range_read[0] << 8) + range_read[1]);
    pc.printf("Range = %.2f cm\n\r", range); //print range on screen
    wait(0.05);
  }
}
```

Program Example 7.8 Communicating by I^2C with the SRF08 range finder

Connect the circuit of Figure 7.11b. Compile Program Example 7.8 and download to the mbed. Verify correct operation by placing the range finder a known distance from a hard, flat surface. Then explore its ability to detect irregular surfaces, narrow objects (e.g. a broom handle) and distant objects.

7.8 Evaluating I^2C

As we have seen, the I^2C protocol is well established and versatile. Like SPI, it is widely applied to short distance data communication. However, it goes well beyond SPI in its ability to set up more complex networks, and to add and subtract nodes with comparative ease. Although we have not really explored it here, it provides for a much more reliable system. If an addressed device does not send an acknowledgment, the master can act upon that fault. Does this mean that I^2C is going to meet all our needs for serial communication, in any application? The answer is a clear No, for at least two reasons. One is that the bandwidth is comparatively limited, even in the faster versions of I^2C. The second is the security of the data. While fine in, say, a domestic appliance, I^2C is still susceptible to interference and does not check for errors. Therefore, one would be unlikely to consider using it in a medical, motor vehicle or other high-reliability application.

7.9 Asynchronous Serial Data Communication

The synchronous serial communication protocols described so far this chapter, in the form of SPI and I^2C, are extremely useful ways of moving data around. The question remains, however: do we really need to send that clock signal wherever the data goes? Although it allows an easy way of synchronizing the data, it does have these disadvantages:

- An extra (clock) line needs to go to every data node.
- The bandwidth needed for the clock is always twice the bandwidth needed for the data; therefore, it is the demands of the clock that limit the overall data rate.
- Over long distances, clock and data themselves could lose synchronization.

7.9.1 Introducing Asynchronous Serial Data

For the reasons just stated, several serial standards have been developed that do not require a clock signal to be sent with the data. This is generally called *asynchronous* serial communication. It is now up to the receiver to extract all timing information directly from the signal itself. This has the effect of laying new and different demands on the signal, and making transmitter and receiver nodes somewhat more complex than comparable synchronous nodes.

A common approach to achieving asynchronous communication is based on this:

- Data rate is predetermined – both transmitter and receiver are preset to recognize the same data rate. Hence, each node needs an accurate and stable clock source, from which the data rate can be generated. Small variations from the theoretical value can, however, be accommodated.

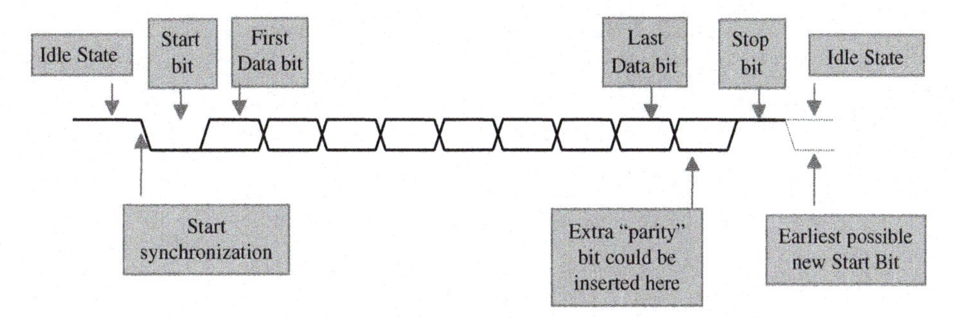

Figure 7.12:
A common asynchronous serial data format

- Each byte or word is *framed* with a Start and Stop bit. These allow synchronization to be initiated before the data starts to flow.

An asynchronous data format, of the sort used by such standards as RS-232, is shown in Figure 7.12. There is now only one data line. This idles in a predetermined state, in this example at Logic 1. The start of a data word is initiated by a *Start* bit, which has polarity opposite to that of the idle state. The leading edge of the Start bit is used for synchronization. Eight data bits are then clocked in. A ninth bit, for parity checking, is also sometimes used. The line then returns to the idle state, which forms a Stop bit. A new word of data can be sent immediately, following the completion of a single Stop bit, or the line may remain in the idle state until it is needed again.

An asynchronous serial port integrated into a microcontroller or peripheral device is generally called a UART, standing for *universal asynchronous receiver/transmitter.* In its simplest form, a UART has one connection for transmitted data, usually called TX, and another for received data, called RX. The port should sense when a start bit has been initiated, and automatically clock in and store the new word. It can initiate a transmission at any time. The data rate that receiver and transmitter will operate at must be predetermined; this is specified by its *baud rate.* For the present purposes baud rate can be viewed as being equivalent to bit rate; for more advanced applications one should check the distinctions between these two terms.

7.9.2 Applying Asynchronous Communication on the mbed

If we look back at Figure 2.3. we see that the LPC1768 has four UARTs. Three of these appear on the pinout of the mbed, simply labeled 'Serial', on pins 9 and 10, 13 and 14, and 27 and 28. Their not insignificant API summary is given in Table 7.9.

We will now repeat what we have already done with SPI and I^{2}C, which is to connect two mbeds to demonstrate a serial link, but this time an asynchronous one. You can view Figure 7.13, the build we will use, as a variation on Figures 7.5 and 7.9; notice that it is slightly simpler than either of these. We will apply Program Example 7.9. It will be the same

Table 7.9: Serial (asynchronous) API summary

Function	Usage
`Serial`	Create a serial port, connected to the specified transmit and receive pins
`baud`	Set the baud rate of the serial port
`format`	Set the transmission format used by the serial port
`putc`	Write a character
`getc`	Read a character
`printf`	Write a formatted string
`scanf`	Read a formatted string
`readable`	Determine if there is a character available to read
`writeable`	Determine if there is space available to write a character
`attach`	Attach a function to call whenever a serial interrupt is generated

program we load into both mbeds. It follows a similar pattern to Program Example 7.2, but replaces all SPI code with UART code. The code itself applies a number of functions from Table 7.9, and – reading the comments – should not be too difficult to follow.

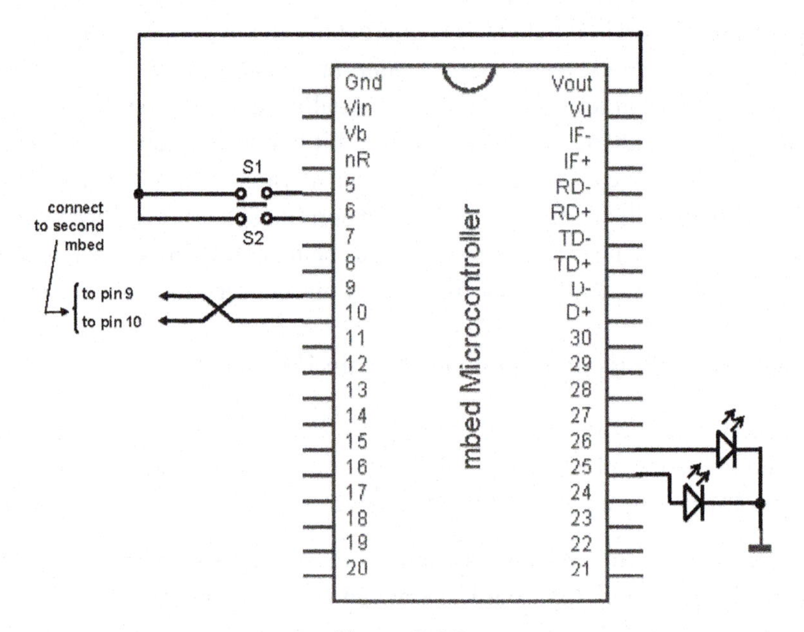

Figure 7.13:
Linking two mbed UARTs

```
/*Program Example 7.9: Sets the mbed up for async communication, and exchanges data with
a similar node, sending its own switch positions, and displaying those of the other.
 */
#include "mbed.h"
Serial async_port(p9, p10);          //set up TX and RX on pins 9 and 10
DigitalOut red_led(p25);             //red led
DigitalOut green_led(p26);           //green led
DigitalOut strobe(p7);               //a strobe to trigger the scope
DigitalIn switch_ip1(p5);
DigitalIn switch_ip2(p6);
char switch_word ;                   //the word we will send
char recd_val;                       //the received value

int main() {
  async_port.baud(9600);             //set baud rate to 9600 (ie default)
  //accept default format, of 8 bits, no parity
  while (1){
    //Set up the word to be sent, by testing switch inputs
    switch_word=0xa0;                //set up a recognizable output pattern
    if (switch_ip1==1)
      switch_word=switch_word|0x01;  //OR in lsb
    if (switch_ip2==1)
      switch_word=switch_word|0x02;  //OR in next lsb
    strobe =1;                       //short strobe pulse
    wait_us(10);
    strobe=0;
    async_port.putc(switch_word);    //transmit switch_word
    if (async_port.readable()==1)    //is there a character to be read?
      recd_val=async_port.getc();    //if yes, then read it
    ...
    (continues as in Program Example 7.2)
    ...
  }
}
```

Program Example 7.9 Bidirectional data transfer between two mbed UARTs

Connect two mbeds as shown in Figure 7.13, and compile and download Program Example 7.9 into each. You should find that the switches from one mbed control the LEDs from the other, and vice versa.

Exercise 7.6

On an oscilloscope, observe the data waveform of one of the TX lines. It helps to trigger the oscilloscope from the strobe pulse (pin 7). See how the pattern changes as the switches are pressed.

1. What is the time duration of each data bit? How does this relate to the baud rate? Change the baud rate and measure the new duration.

2. Is the data byte transmitted MSB first or LSB? How does this compare to the serial protocols seen earlier in this chapter?
3. What is the effect of removing one of the data links in Figure 7.13?

■

7.9.3 Applying Synchronous Communication with the Host Computer

While the mbed has three UARTs connecting to the external pins, the LPC1768 has a fourth. This is reserved for communication back to the USB link, and can be seen in the mbed block diagram in Figure 2.2. This UART acts just like any of the others, in terms of its use of the API (Table 7.8). We have been using it already, for the first time in Program Example 5.4. As we saw there, the mbed compiler will recognize **pc**, **USBTX** and **USBRX** as identifiers to set up this connection, as in the line:

```
Serial pc(USBTX, USBRX);
```

With **pc** thus created, the API member functions can be exploited. The requirement to set up the host computer correctly in order to communicate with this link was outlined in Section 5.3.

7.10 Mini-Project: Multi-Node I^2C Bus

Design a simple circuit which has a temperature sensor, range finder and mbed connected to the same I^2C bus. Merge Program Examples 7.8 and 7.9, so that measurements from each sensor are displayed in turn on the screen. Verify that the I^2C protocol works as expected.

Chapter Review

- Serial data links provide a ready means of communication between microcontroller and peripherals, and/or between microcontrollers.
- SPI is a simple synchronous standard, which is still very widely applied. The mbed has two SPI ports and a supporting library.
- While a very useful standard, SPI has certain very clear limitations, relating to a lack of flexibility and robustness.
- The I^2C protocol is a more sophisticated serial alternative to SPI; it runs on a two-wire bus, and includes addressing and acknowledgment.
- I^2C is a flexible and versatile standard. Devices can be readily added to or removed from an existing bus, multi-master configurations are possible, and a master can detect if a slave fails to respond, and can take appropriate action. Nevertheless, I^2C has limitations which mean that it cannot be used for high-reliability applications.
- A very wide range of peripheral devices is available, including intelligent sensors, which communicate through SPI and I^2C.

- A useful asynchronous alternative to I^2C and SPI is provided by the UART. The mbed has four of these, one of which provides a communication link back to the host computer.

Quiz

1. What do the abbreviations SPI, I^2C and UART stand for?
2. Draw up a table comparing the advantages and disadvantages of using SPI versus I^2C for serial communications.
3. What are the limitations for the number of devices that can be connected to a single SPI, I^2C or UART bus?
4. A SPI link is running with a 500 kHz clock. How long does it take for a single message containing one data byte to be transmitted?
5. An mbed configured as SPI master is to be connected to three other mbeds, each configured as slave. Sketch a circuit showing how this interconnection could be made. Explain your sketch.
6. An mbed is to be set up as SPI master, using pins 11, 12 and 13, running at a frequency of 4 MHz, with 12-bit word length. The clock should idle at Logic 1, and data should be latched on its negative edge. Write the necessary code to set this up.
7. Repeat Question 4, but for I^2C, ensuring that you calculate time for the complete message.
8. Repeat Question 5, but for I^2C. Identify carefully the advantages and disadvantages of each connection.
9. You need to set up a serial network, which will have one master and four slaves. Either SPI or I^2C can be used. Every second, data has to be distributed, such that one byte is sent to Slave 1, four to Slave 2, three to Slave 3 and four to Slave 4. If the complete data transfer must take not more than 200 μs, estimate the minimum clock frequency that is allowable for SPI and I^2C. Assume there are no other timing overheads.
10. Repeat Question 4, but for asynchronous communication through a UART, assuming a baud rate of 500 kHz. Ensure that you calculate time for the complete message.

References

7.1. Spasov, P. (1996). Microcontroller Technology, The 68HC11. 2nd edition. Prentice Hall.
7.2. ADXL345 Datasheet, Rev. C. www.analog.com
7.3. NXP Semiconductors. The I2C Bus Specification and User Manual. Rev. 03. 2007. Document number UM10204.
7.4. Texas Instruments. TMP102. Low Power Digital Temperature Sensor. August 2007. Rev. October 2008. Document number SBOS397B.
7.5. SRF08 data. http://www.robot-electronics.co.uk/htm/srf08tech.shtml. Accessed 30 July 2011.

CHAPTER 8

Liquid Crystal Displays

Chapter Outline

8.1 Display Technologies

Light-emitting diodes (LEDs), particularly individual LEDs and seven-segment displays, have already been encountered in previous chapters. On their own, LEDs can only inform us of a few different states, for example to indicate whether something is on or off, or whether a variable is cleared or set. It is possible to be innovative in the way LEDs are used, for example, different flashing speeds can be used to represent different states, but clearly there is

Fast and Effective Embedded Systems Design. DOI: 10.1016/B978-0-08-097768-3.00008-8

a limit to how much information a single LED can communicate. With seven-segment displays it is possible to display numbers and a few characters from the alphabet, but there is a problem with seven-segment displays in that in simple use they require a microcontroller output for each LED segment. The multiplexing technique described in Section 3.6.3 can be used, but a limitation is still seen. Equally, LEDs are relatively power hungry, which is an issue for power-conscious designs. Embedded systems designers need to be clever in the way they use the available input and output capabilities of a microcontroller, but obviously the use of LEDs as a display has its limitations. If text-based messages with a number of characters are required to be communicated with a user, then a more advanced display type is needed, such as the liquid crystal displays (LCDs) that are commonly used in consumer electronics.

8.1.1 Introducing Liquid Crystal Technology

Liquid crystal displays are of great importance these days, both in the electronic world in general, and in the embedded system environment. Their main advantages are their extremely low power requirements, light weight and high flexibility of application. They have been one of the enabling technologies for battery-powered products such as the digital watch, the laptop computer and the mobile telephone, and are available in a huge range of indicators and displays. However, LCDs do have some disadvantages; these include limited viewing angle and contrast in some implementations, sensitivity to temperature extremes, and very high cost for the more sophisticated graphical displays.

Although LCDs do not emit light, they can reflect incident light, or transmit or block backlight. The principle of an LCD is illustrated in Figure 8.1. The liquid crystal is an organic compound which responds to an applied electric field by changing the alignment of its molecules, and hence the light polarization that it introduces. A small quantity of liquid crystal is contained between two parallel glass plates. A suitable field can be applied if transparent electrodes are located on the glass surface. In conjunction with the external polarizing light filters, light is either blocked or transmitted by the display cell.

The electrodes can be made in any pattern desired. These may include single digits or symbols, or the standard patterns of bargraph, seven-segment, dot matrix, starburst and so on. Alternatively, they may be extended to complex graphical displays, with addressable pixels.

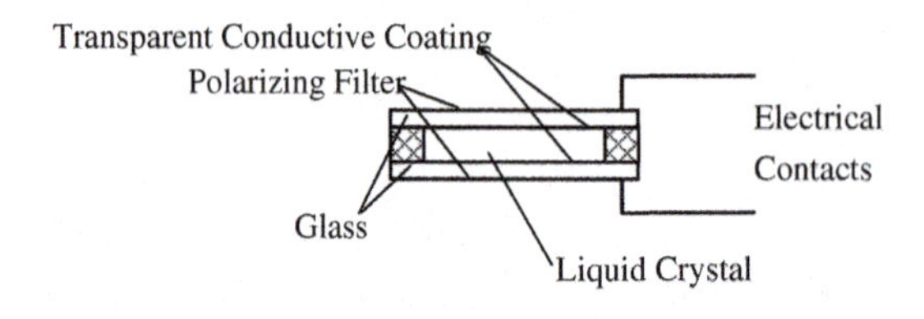

Figure 8.1:
A simple liquid crystal structure

8.1.2 Liquid Crystal Character Displays

A popular, indeed ubiquitous, form of LCD is the character display, as seen in Figure 8.2. These are widely available from one line of characters to four or more, and are commonly seen on many domestic and office items, such as photocopiers, burglar alarms and DVD players. Driving this complex array of tiny LCD dots is far from simple, so such displays always contain a hidden microcontroller, customized to drive the display. The first such controller to gain widespread acceptance was the Hitachi HD44780. While this has been superseded by others, they have kept the interface and internal structure of the Hitachi device. It is important to know its main features, in order to design with it.

The HD44780 contains an 80-byte random access memory (RAM) to hold the display data, and a read only memory (ROM) for generating the characters. It has a simple instruction set, including instructions for initialization, cursor control (moving, blanking, blinking) and clearing the display. Communication with the controller is made via an 8-bit data bus, three control lines and an enable/strobe line (E). These are itemized in Figure 8.3a.

Data written to the controller is interpreted either as instruction or as display data (Figure 8.3b), depending on the state of the RS (Register Select) line. An important use of reading data back from the LCD is to check the controller status via the Busy flag. As some instructions take a finite time to implement (e.g. a minimum of 40 μs is required to receive one character code), it is sometimes useful to be able to read the Busy flag and wait until the LCD controller is ready to receive further data.

The controller can be set up to operate in 8-bit or 4-bit mode. In the latter mode only the four most significant bits of the bus are used, and two write cycles are required to send a single byte. In both cases the most significant bit doubles as the Busy flag when a Read is undertaken.

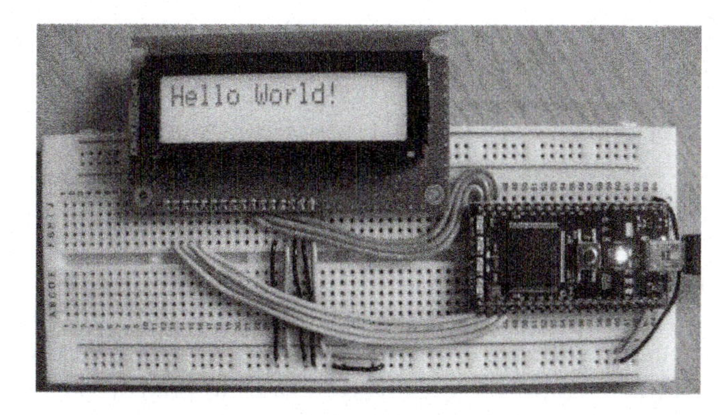

Figure 8.2:
The mbed driving an LCD

(a)

RS	Register Select:	0 = Instruction register	1 = Data Register
R/$\overline{W}$	Selects read or write		
E	Synchronizes read and write operations		
DB4 - DB7	Higher order bits of data bus; DB7 also used as Busy flag		
DB0 - DB3	Lower order bits of data bus; not used for 4-bit operation		

(b)

RS	R/$\overline{W}$	E	Action
0	0	↓	Write instruction code
0	1	⎍	Read busy flag and address counter
1	0	↓	Write data
1	1	⎍	Read data

(c)

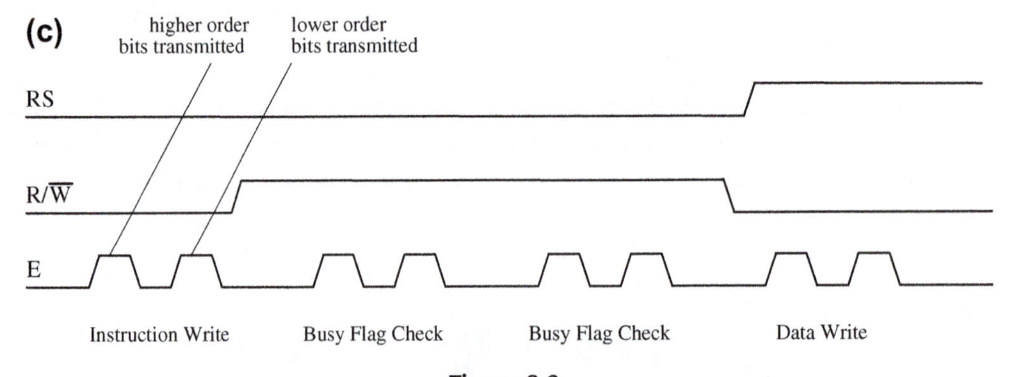

Figure 8.3:
(a) HD44780 user interface lines; (b) HD44780 data and instruction transfers; (c) HD44780 timing for 4-bit interface

8.2 Using the PC1602F LCD

The mbed processor can be interfaced to an external LCD, in order to display messages on the screen. Interfacing an LCD requires a few involved steps to prepare the device and achieve the desired display. The following tasks must be considered in order to successfully interface the LCD:

- Hardware integration: the LCD will need to be connected to the correct mbed pins.
- Modular coding: as there are many processes that need to be completed, it makes sense to define LCD functions in modular files.
- Initializing the LCD: a specific sequence of control signals must be sent to the LCD to initialize it.
- Outputting data: we will need to understand how the LCD converts control data into legible display data.

Here, we will use the 2 × 16 character Powertip PC1602F LCD, although a number of similar LCDs can be found with the same hardware configuration and functionality.

8.2.1 Introducing the PC1602F Display

The PC1602F display is a 2 × 16 character display with an onboard data controller chip and an integrated backlight, as seen in Figure 8.2. The LCD has 16 connections, defined in Table 8.1.

In this example we will use the LCD in 4-bit mode. This means that only the upper 4 bits of the data bus (DB4–DB7) are connected. The two halves of any byte are sent in turn on these

Table 8.1: PC1602F pin descriptions

Pin number	Pin name	Function
1	V_{SS}	Power supply (GND)
2	V_{DD}	Power supply (5 V)
3	V_0	Contrast adjust
4	RS	Register select signal
5	$R/\overline{W}$	Data read / write
6	E	Enable signal
7	DB0	Data bus line bit 0
8	DB1	Data bus line bit 1
9	DB2	Data bus line bit 2
10	DB3	Data bus line bit 3
11	DB4	Data bus line bit 4
12	DB5	Data bus line bit 5
13	DB6	Data bus line bit 6
14	DB7	Data bus line bit 7
15	A	Power supply for LED backlight (5 V)
16	K	Power supply for LED backlight (GND)

lines. As a result, the LCD can be controlled with only seven lines, rather than the 11 lines required for 8-bit mode. Every time a nibble (a 4-bit word is sometimes called a *nibble*) is sent, the E line must be pulsed, as seen in Figure 8.3. The display is initialized by sending control instructions to the configuration registers in the LCD. This is done by setting RS and $R/\overline{W}$ low; once the LCD has been initialized, display data can be sent by setting the RS bit high. As before, the E bit must be pulsed for every nibble of display data sent.

8.2.2 Connecting the PC1602F to the mbed

An mbed pin should be attached to each of the LCD data pins that are used, configured as digital output. Four outputs are required to send the 4-bit instruction and display data and two outputs are required to manipulate the RS and E control lines.

The suggested interface configuration for connecting the mbed to the PC1602F is as shown in Table 8.2. Note that, in simple applications, the LCD can be used only in write mode, so $R/\overline{W}$ can be tied permanently to ground (mbed pin 1). If rapid control of the LCD is required, i.e. if it needs updating faster than approximately every 1 ms, then the $R/\overline{W}$ input can be activated to enable reading from the LCD's Busy flag. The Busy flag changes state when a display request has been completed, so by testing it the LCD can be programmed to operate as quickly as possible. If the $R/\overline{W}$ line is tied to ground, then a 1 ms delay between data transfers is adequate to ensure that all internal processes can complete before the next transfer.

Table 8.2: Wiring table for connecting the PC1602F to the mbed

mbed pin number	LCD pin number	LCD pin name	Power connection
1	1	V_{SS}	0 V
39	2	V_{DD}	5 V
1	3	V_0	0 V
19	4	RS	
1	5	$R/\overline{W}$	0 V
20	6	E	
21	11	DB4	
22	12	DB5	
23	13	DB6	
24	14	DB7	
39	15	A	5 V
1	16	K	0 V

A: anode; K: cathode of the backlight LED.

■	■	■	■	■	■	■	■	■	■	■	■	■	■	■	■
14	13	12	11	10	9	8	7	6	5	4	3	2	1	16	15
DB7	DB6	DB5	DB4	DB3	DB2	DB1	DB0	E	R/$\overline{W}$	RS	V_0	V_{DD}	V_{SS}	A	K

Figure 8.4:
PC1602F physical pin layout

The LCD pins DB0, DB1, DB2 and DB3 are left unconnected as these are not required when operating in 4-bit data mode. The PC1602F has the pin layout shown in Figure 8.4.

8.2.3 Using Modular Coding to Interface the LCD

Initializing and interfacing a peripheral device, such as an LCD, can be done effectively by using modular files to define various parts of the code. We will therefore use three files for this application. The files required are:

- a main code file (**main.cpp**), which can call functions defined in the LCD feature file
- an LCD definition file (**LCD.cpp**), which will include all the functions for initializing and sending data to the LCD
- an LCD header file (**LCD.h**), which will be used to declare data and function prototypes.

We will declare the following functions in the LCD header file:

- **toggle_enable( )**: a function to toggle/pulse the enable bit
- **LCD_init( )**: a function to initialize the LCD
- **display_to_LCD( )**: a function to display characters on the LCD.

The **LCD.h** header file should therefore define the function prototypes as shown in Program Example 8.1.

```
/* Program Example 8.1: LCD.h header file
*/
#ifndef LCD_H
#define LCD_H
#include "mbed.h"
void toggle_enable(void);        //function to toggle/pulse the enable bit
void LCD_init(void);             //function to initialize the LCD
void display_to_LCD(char value); //function to display characters
#endif
```

Program Example 8.1 LCD header file

In the LCD definition file (**LCD.cpp**), we need to define the mbed objects required to control the LCD. Here, one digital output each can be used for RS and E, and an mbed **BusOut** object for the 4-bit data. The suggested mbed object definitions need to conform to the pin connections listed in Table 8.2.

To control and initialize the LCD, we send a number of control bytes, each sent as two 4-bit nibbles. As each nibble is asserted, the LCD requires the E bit to be pulsed. This is done by the **toggle_enable( )** function, detailed in Program Example 8.2.

8.2.4 Initializing the Display

A specific initialization procedure must be programmed in order for the PC1602F display to operate correctly. Full details are provided in the Powertip PC1602F datasheet (Reference 8.1) and also in Reference 8.2.

Reading this data, we see that we first need to wait a short period (approximately 20 ms), then set the RS and E lines to zero, and then send a number of configuration messages to set up the LCD. We then need to send configuration data to the Function Mode, Display Mode and Clear Display registers in order to initialize the display. These are now introduced.

Function Mode

To set the LCD function mode, the RS, R/$\overline{\text{W}}$ and DB0-7 bits should be set as shown in Figure 8.5. Data bus values are sent as two nibbles.

If, for example, we send a binary value of 00101000 (0x28 hex) to the LCD data pins, this defines 4-bit mode, 2 line display and 5x7 dot characters. In the given example we would therefore send the value 0x2, pulse E, then send 0x8, then pulse E again. The C/C++ code Function Mode register commands are shown in the **LCD_init( )** function detailed in Program Example 8.2.

Display Mode

The Display Mode control register must also be set up during initialization. Here, we need to send a command to switch the display on, and to determine the cursor function. The Display Mode register is defined as shown in Figure 8.6.

RS	R/$\overline{\text{W}}$
0	0

DB7	DB6	DB5	DB4	DB3	DB2	DB1	DB0
0	0	1	BW	N	F	X	X

BW = 0 → 4 bit mode   N = 0 → 1 line mode   F = 0 → 5×7 pixels

BW = 1 → 8 bit mode   N = 1 → 2 line mode   F = 1 → 5×10 pixels

X = Don't care bits (can be 0 or 1)

Figure 8.5:
Function Mode control register

RS	R/$\overline{W}$
0	0

DB7	DB6	DB5	DB4	DB3	DB2	DB1	DB0
0	0	0	0	1	P	C	B

P = 0 → display off   C = 0 → cursor off   B = 0 → cursor no blink

P = 1 → display on   C = 1 → cursor on   B = 1 → cursor blinking

Figure 8.6:
Display Mode control register

In order to switch the display on with a blinking cursor, the value 0x0F (in two 4-bit nibbles) is required. The C/C++ code Display Mode register commands are shown in the **LCD_init( )** function detailed in Program Example 8.2.

Clear Display

Before data can be written to the display, the display must be cleared, and the cursor reset to the first character in the first row (or any other location that you wish to write data to). The Clear Display command is shown in Figure 8.7.

RS	R/$\overline{W}$
0	0

DB7	DB6	DB5	DB4	DB3	DB2	DB1	DB0
0	0	0	0	0	0	0	1

Figure 8.7:
Clear Display command

8.2.5 Sending Display Data to the LCD

A function called (in this example) **display_to_LCD( )** displays characters on the LCD screen. Characters are displayed by setting the RS flag to 1 (data setting), then sending a data byte describing the ASCII character to be displayed.

The term ASCII (American Standard Code for Information Interchange) has been introduced previously in Chapter 6; it is a method for defining alphanumeric characters as 8-bit values. When communicating with displays, by sending a single ASCII byte, the display is informed which particular character should be shown. The complete ASCII table is included with the LCD datasheet, but for interest some common ASCII values for display on the LCD are shown in Table 8.3.

The **display_to_LCD( )** function needs to accept an 8-bit value as a data input, so that it can display the desired character on the LCD screen. In C/C++ the **char** data type can be used to define an 8-bit number. It can be seen in the **LCD.h** header file (Program Example 8.1) that

Table 8.3: Common ASCII values

		Less significant bits (lower nibble)															
		0x0	0x1	0x2	0x3	0x4	0x5	0x6	0x7	0x8	0x9	0xA	0xB	0xC	0xD	0xE	0xF
More significant bits (upper nibble)	0x0																
	0x1																
	0x2		!	"	#	$	%	&	'	(	)	*	+	,	-	.	/
	0x3	0	1	2	3	4	5	6	7	8	9	:	;	<	=	>	?
	0x4	@	A	B	C	D	E	F	G	H	I	J	K	L	M	N	O
	0x5	P	Q	R	S	T	U	V	W	X	Y	Z	[	\	]	^	_
	0x6	‘	a	b	C	d	e	f	g	h	i	j	k	l	m	n	o
	0x7	p	q	r	S	t	u	v	w	x	y	z	{	|	}	~	

display_to_LCD( ) has already been defined as a function with a **char** input. Referring to Table 8.3, it can be seen, for example, that if we send the data value 0x48 to the display, the character 'H' will be displayed.

The **display_to_LCD( )** function is shown in Program Example 8.2. Note that, as we are using 4-bit mode, the most significant bits of the ASCII byte must be shifted right in order to be output on the 4-bit bus created through **BusOut**. The lower 4 bits can then be output directly.

8.2.6 The Complete LCD.cpp *Definition*

The LCD definition file (**LCD.cpp**) contains the three C functions – **toggle_enable( )**, **LCD_init( )** and **display_to_LCD( )** – as described above. The complete listing of **LCD.cpp** is shown in Program Example 8.2. Note that the **toggle_enable( )** function has two 1 ms delays, which remove the need to monitor the Busy flag; the downside to this is that we have introduced a timing delay into our program, the consequences of which will be discussed in more detail in Chapter 9.

```
/* Program Example 8.2: Declaration of objects and functions in LCD.cpp file
*/
#include "LCD.h"
DigitalOut RS(p19);
DigitalOut E(p20);
BusOut data(p21, p22, p23, p24);

void toggle_enable(void){
  E=1;
  wait(0.001);
  E=0;
```

```
  wait(0.001);
}
//initialize LCD function
void LCD_init(void){
  wait(0.02);               // pause for 20 ms
  RS=0;                     // set low to write control data
  E=0;                      // set low
  //function mode
  data=0x2;                 // 4 bit mode (data packet 1, DB4-DB7)
  toggle_enable();
  data=0x8;                 // 2-line, 7 dot char (data packet 2, DB0-DB3)
  toggle_enable();
  //display mode
  data=0x0;                 // 4 bit mode (data packet 1, DB4-DB7)
  toggle_enable();
  data=0xF;                 // display on, cursor on, blink on
  toggle_enable();
  //clear display
  data=0x0;                 //
  toggle_enable();
  data=0x1;                 // clear
  toggle_enable();
}
//display function
void display_to_LCD(char value){
  RS=1;                 // set high to write character data
  data=value>>4;        // value shifted right 4 = upper nibble
  toggle_enable();
  data=value;           // value bitmask with 0x0F = lower nibble
  toggle_enable();
}
```

Program Example 8.2 Declaration of objects and functions in LCD.cpp

8.2.7 Utilizing the LCD Functions

We can now develop a main control file (**main.cpp**) to utilize the LCD functions described above. Program Example 8.3 initializes the LCD, displays the word 'HELLO' and then displays the numerical characters from 0 to 9.

```
/* Program Example 8.3 Utilizing LCD functions in the main.cpp file
*/
#include "LCD.h"
int main() {
  LCD_init();                 // call the initialize function
  display_to_LCD(0x48);       // 'H'
  display_to_LCD(0x45);       // 'E'
  display_to_LCD(0x4C);       // 'L'
  display_to_LCD(0x4C);       // 'L'
  display_to_LCD(0x4F);       // 'O'
```

```
  for(char x=0x30;x<=0x39;x++){
  display_to_LCD(x);              // display numbers 0-9
  }
}
```

Program Example 8.3 File main.cpp utilizing LCD functions

Exercise 8.1

Applying Table 8.2, connect a Powertip PC1602F LCD to an mbed and construct a new program with the files **main.cpp**, **LCD.cpp** and **LCD.h** as described in the program examples above. Compile and run the program.

1. Verify that the word 'HELLO' and the numerical characters 0–9 are correctly displayed with a flashing cursor.
2. Change the program so that your name appears on the display, after the word 'HELLO', instead of the numerical characters 0–9.

Exercise 8.2

If you look at the **toggle_enable( )** function in Program Example 8.2, you will notice that it sets the E line high and low, waiting 1 ms in each case. This saves us testing the Busy flag, as seen in Figure 8.3c, as the display will have exited from its Busy state during the 1 ms delay. For many embedded applications, however, this loss of 2 ms would be quite unacceptable.

Try applying the information so far supplied to write a revised program, so that the delays are removed, and the Busy flag is tested. You will now need to activate $R/\overline{W}$ and set it to 1 in order to read the Busy flag, which is indicated by data bit DB7. Once Busy is clear, the program should proceed. Estimate how much time is saved by your new program. You can test this with an oscilloscope by making your program write a digit continuously to the display, and measuring on the oscilloscope the time between the E pulses.

8.2.8 Adding Data to a Specified Location

The display is mapped out with memory address locations so that each display unit has a unique address. The display pointer can therefore be set before outputting data, and the data will then appear in the position specified. The display address layout is shown in Figure 8.8.

If the program counter is set to address 0x40, data will be displayed at the first position on the second line. To change the pointer address, the desired 6-bit address value must be sent in a control byte with bit 7 also set, as shown in Figure 8.9.

Display Position		1	2	3	4	5	6	7	8	9	10	11	12	13	14	15	16
Display Pointer Address	1st Line	00	01	02	03	04	05	06	07	08	09	0A	0B	0C	0D	0E	0F
	2nd Line	40	41	42	43	44	45	46	47	48	49	4A	4B	4C	4D	4E	4F

Figure 8.8:
Screen display pointer address values

A new function can be created to set the location of the display pointer prior to writing. We will call this function **set_location( )**, as shown in Program Example 8.4.

```
/* Program Example 8.4 function to set the display location. Parameter "location" holds
address of display unit to be selected
*/
void set_location(char location){
  RS=0;
  data=(location|0x80)>>4;                    // upper nibble
  toggle_enable();  data=location&0x0F;       // lower nibble
  toggle_enable();
}
```

Program Example 8.4 Function to change the display pointer position

Notice that bit DB7 is set by ORing the location value with 0x80.

Exercise 8.3

Add the **set_location( )** function shown in Program Example 8.4 to the LCD.cpp definition. You will also need to declare the function prototype in LCD.h. Now add **set_location( )** function calls to **main.cpp** so that the word 'HELLO' appears in the center of the first line and the numerical characters 0–9 appear in the center of the second line of the display.

Exercise 8.4

Modify the **LCD_init( )** function to disable the flashing cursor. Here, you will need to modify the value sent to the Display Mode register.

RS	R/$\overline{W}$
0	0

DB7	DB6	DB5	DB4	DB3	DB2	DB1	DB0
1	AC6	AC5	AC4	AC3	AC2	AC1	AC0

AC0-AC6 describe the 6-bit display pointer address

Figure 8.9:
Display pointer control

8.3 Using the mbed TextLCD Library

There is an mbed library which makes the use of an alphanumeric LCD much simpler and quicker to program. The mbed **TextLCD** library is also more advanced than the simple functions we have created; in particular, the **TextLCD** library performs the laborious LCD setup routine for us. The **TextLCD** definition also tells the LCD object which pins are used for which functions.

The object definition is defined in the following manner:

```
TextLCD lcd(int rs, int e, int d0, int d1, int d2, int d3);
```

We need to ensure that our pins are defined in the same order. For our particular hardware setup (described in Table 8.2) this will be:

```
TextLCD lcd(p19, p20, p21, p22, p23, p24);
```

Simple **printf( )** statements are used to display characters on the LCD screen; this statement is also discussed in Section B.9 of Appendix B. The **printf( )** function allows formatted print statements to be used. This means that text strings and formatted data can be sent to a display, meaning that a function call is not required for each individual character (as was the case in Program Example 8.3).

When using the **printf( )** function with the mbed **TextLCD** library, the display object's name is also required, so if the **TextLCD** object is defined as in the examples above (with the object name **lcd**), then a 'Hello World' string can be written as follows:

```
lcd.printf("Hello World!");
```

When using predefined mbed libraries, such as **TextLCD**, the library file needs to be imported to the mbed program. The mbed **TextLCD** library can (at the time of writing) be accessed from the following location:

http://mbed.org/users/simon/libraries/TextLCD/livod0

The library header file must also be included with the **#include** statement in the **main.cpp** file or relevant project header files. Program Example 8.5 is a simple Hello World example using the **TextLCD** library.

```
/*Program Example 8.5: TextLCD library example
*/
#include "mbed.h"
#include "TextLCD.h"

TextLCD lcd(p19, p20, p21, p22, p23, p24); //rs,e,d0,d1,d2,d3
int main() {
  lcd.printf("Hello World!");
}
```

Program Example 8.5 TextLCD Hello World

The cursor can be moved to a chosen position to allow you to choose where to display data. This uses the **locate( )** function as follows. The display is laid out as two rows (0–1) of 16 columns (0–15). The locate function defines the column first followed by the row. For example, add the following statement to move the cursor to the second line and fourth column:

```
lcd.locate(3,1);
```

Any **printf( )** statements after this will be printed at the new cursor location.

Exercise 8.5

1. Create a new program and import the **TextLCD** library file (right click on the project and select 'import library').
2. Add Program Example 8.5 to the main.cpp file and compile and run. Verify that the program correctly displays the Hello World characters.
3. Experiment further with the locate function and verify that the Hello World string can be positioned to a desired location on the display.

The screen can also be cleared with the following command:

```
lcd.cls();
```

Program Example 8.6 displays a count variable on the LCD. The count variable increments every second.

```
/* Program Example 8.6: LCD Counter example
*/
#include "mbed.h"
#include "TextLCD.h"
TextLCD lcd(p19, p20, p21, p22, p23, p24); // rs, e, d0, d1, d2, d3
int x=0;
int main() {
  lcd.printf("LCD Counter");
  while (1) {
    lcd.locate(5,1);
    lcd.printf("%i",x);
    wait(1);
    x++;
  }
}
```

Program Example 8.6 LCD counter

Exercise 8.6

1. Implement Program Example 8.6 as a new program. Do not forget to import the TextLCD library!
2. Increase the speed of the counter and investigate how the cursor position changes as the count value increases.

8.4 Displaying Analog Input Data on the LCD

An analog input can be defined – before the **main( )** function – on pin 18 as follows:

```
AnalogIn Ain(p18);
```

We can now connect a potentiometer between 3.3 V (mbed pin 40) and 0 V (mbed pin 1) with the wiper connected to pin 18. The analog input variable has a floating point value between 0 and 1, where 0 is 0 V and 1 represents 3.3 V. We will multiply the analog input value by 100 to display a percentage between 0 and 100%, as shown in Program Example 8.7. An infinite loop can be used so that the screen updates automatically. To do this it is necessary to clear the screen and add a delay to set the update frequency.

```
/*Program Example 8.7: Display analog input data
*/
#include "mbed.h"
#include "TextLCD.h"
TextLCD lcd(p19, p20, p21, p22, p23, p24); //rs,e,d0, d1,d2,d3
AnalogIn Ain(p17);
float percentage;

int main() {
  while(1){
    percentage=Ain*100;
    lcd.printf("%1.2f",percentage);
    wait(0.002);
    lcd.cls();
  }
}
```

Program Example 8.7 Display analog input data

Exercise 8.7

1. Implement Program Example 8.7 and verify that the potentiometer modifies readings between 0 and 100%.
2. Modify the **wait( )** statement to be a larger value and evaluate the change in performance. It is interesting to make a mental note that a certain range of update rates appears irritating to view (too fast), while others may be perceived as too slow.

Exercise 8.8

Create a program to make the mbed and display act like a standard voltmeter, as shown in Figure 8.10. Potential difference should be measured between 0 and 3.3 V and displayed to the screen. Note the following:

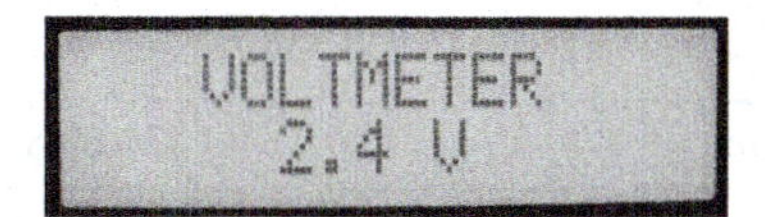

Figure 8.10:
Voltmeter display

- You will need to convert the 0.0–1.0 analog input value to a value that represents 0 –3.3 V.
- An infinite loop is required to allow the voltage value to update continuously as the potentiometer position changes.
- Check the display with the reading from an actual voltmeter: is it accurate?
- Increase the number of decimal places that the voltmeter reads. Evaluate the noise and accuracy of the voltmeter readings with respect to the mbed's ADC resolution.

■

8.5 More Advanced LCDs

8.5.1 Color LCDs

Nowadays, LCD technology is used in advanced displays for mobile phones, PC monitors and televisions. For color displays, each pixel is made up of three subpixels for red, green and blue. Each subpixel can be set to 256 different shades of its color, so it is therefore possible for a single LCD pixel to display 256 * 256 * 256 = 16.8 million different colors. The pixel color is therefore usually referred to by a 24-bit value, where the highest 8 bits define the red shade, the middle 8 bits the green shade and the lower 8 bits the blue shade, as shown in Table 8.4.

Given that each pixel needs to be assigned a 24-bit value and a 1280 × 1024 LCD computer display has over 1 million pixels (1280 * 1024 = 1 310 720), a lot of data needs to be sent to a color LCD. Standard color LCDs are set to refresh the display at a frequency of 60 Hz, so the digital input requirements are much greater than those associated with the alphanumeric LCD discussed previously in this chapter.

8.5.2 Controlling a SPI LCD Mobile Phone Display

A mobile phone screen, like the Nokia 6610, represents state-of-the-art, low-cost, high-volume display technology. Yet many have an interface based on the good old serial peripheral interface (SPI) standard, introduced in Chapter 7. It is hard to find formal published

Table 8.4: 24-bit color values

Color	24-bit value
Red	0xFF0000
Green	0x00FF00
Blue	0x0000FF
Yellow	0xFFFF00
Orange	0xFF8000
Purple	0x800080
Black	0x000000
White	0xFFFFFF

data on the Nokia display; Reference 8.3, however, gives some useful background. We can use the Nokia 6610 display with the mbed. Here, we will see how to connect it and how to type data on the screen from the PC keyboard. Instead of using the basic SPI library, we will use the predesigned mbed MobileLCD library. You will need to import this library from:

http://mbed.co.uk/projects/cookbook/svn/MobileLCD/tests/MobileLCD

The mobile screen has 10 pins, as shown in Figure 8.11. It can be connected to the mbed using the connections shown in Table 8.5.

It is possible to clear the screen, set the background color, fill blocks, draw pixel images, move the pointer and perform many other operations. Program Example 8.8 writes characters to the display from the computer keyboard, and uses the # key to clear the screen.

```
/*Program Example 8.8: Program which reads character from computer screen, and displays
on Nokia LCD display.
*/
#include "mbed.h"
#include "MobileLCD.h"                    // include the Nokia display library
MobileLCD lcd(p11, p12, p13, p15, p16); // mosi,miso,clk,cs,rst
Serial pc(USBTX, USBRX);                  // host terminal comms setup
char c;                                   // char variable for keyboard input
void screen_setup(void);                  // function prototype
int main() {
  pc.printf("\n\rType something to be displayed:\n\r");
  screen_setup();                         // call the screen setup function
  while(1){
    c = pc.getc();                        // c = character input from computer keyboard
    wait(0.001);
    if (c=='#'){                          // perform the following if "#" is pressed
      screen_setup();                     // call the screen setup function
```

```
        lcd.locate(0,0);                     // move the cursor back to row 0 column 0
      }
      else{
        lcd.printf("%c",c);                  // print character on the LCD screen
        pc.printf("%c",c);                   // print character on the terminal screen
      }
    }
  }
//function definition for screen_setup
void screen_setup(void) {
  lcd.background(0x0000FF);                  // set the background color
  lcd.cls();                                 // clear the screen
}
```

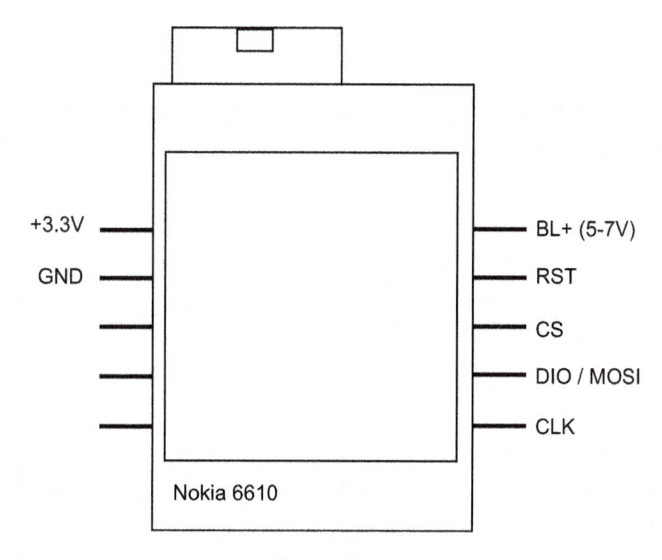

Figure 8.11:
Nokia 6610 display connections

Table 8.5: Connections between Nokia display and mbed

Nokia 6610 pin	mbed pin
+3.3 V	40
GND	1
BL+	39
RST	9
CS	8
MOSI	5
CLK	7

Program Example 8.8 Writing characters to the Nokia display

Having connected the circuit, create a new project, and enter and compile the example code. Once it is running on the mbed, and with the terminal application open, you should be able to type to the LCD screen.

Exercise 8.9

Add the following fill functions to the **screen_setup( )** function (after the lcd.cls command) in order to fill some areas of the LCD:

```
lcd.fill(2, 51, 128, 10, 0x00FF00);
lcd.fill(50, 1, 10, 128, 0xFF0000);
```

It is also possible to draw on the screen pixel by pixel. The following loop will create the function for a sine wave and print this to the screen. Add this to the setup function.

```
for(int i=0; i<130; i++) {
  lcd.pixel(i, 80 + sin((float)i / 5.0)*10, 0x000000);
}
```

We have seen that LCDs allow greater flexibility than simple LED-based displays, and that simple alphanumeric and advanced color LCD devices can be utilized with microcontrollers and embedded systems.

8.6 Mini-Project: Digital Spirit Level

Design, build and test a digital spirit level based on the mbed. Use an ADXL345 (Section 7.3) accelerometer to measure the angle of orientation in two planes, a digital push-to-make switch to allow calibration and zeroing of the orientation, and a color LCD to output the measured orientation data, in degrees.

To help you proceed, consider the following:

1. Design your display to show a pixel or an image moving around the LCD screen with respect to the orientation of the accelerometer.
2. Add the digital switch to allow simple calibration and zeroing of the data.
3. Improve your display output to give accurate measurements of the two-plane orientation angles in degrees from the horizontal (i.e. horizontal = 0°).

Chapter Review

- Liquid crystal displays (LCDs) use an organic crystal which can polarize and block light when an electric field is introduced.
- Many types of LCDs are available and, when interfaced with a microcontroller, they allow digital control of alphanumeric character displays and high-resolution color displays.
- The PC1602F is a 16-column by 2-row character display which can be controlled by the mbed.
- Data can be sent to the LCD registers to initialize the device and to display character messages.
- Character data is defined using the 8-bit ASCII table.
- The mbed **TextLCD** library can be used to simplify working with LCDs and allow the display of formatted data using the **printf( )** function.
- Color LCDs use a serial interface to display data, with each pixel given a 24-bit color setting.
- The mbed **MobileLCD** library can be used to interface a Nokia 6610 mobile phone-type LCD.

Quiz

1. What are the advantages and disadvantages associated with using an alphanumeric LCD in an embedded system?
2. What types of cursor control are commonly available on alphanumeric LCDs?
3. How does the mbed **BusOut** object help to simplify interfacing an alphanumeric display?
4. What is the function of the E input on an alphanumeric display such as the PC1602F?
5. What does ASCII stand for?
6. What are the ASCII values associated with the numerical characters from 0 to 9?
7. What is the correct **printf( )** C/C++ code required to display the value of a floating point variable called 'ratio' to two decimal places ?
8. List five practical examples of a color LCD used in an embedded system.
9. If a color LCD is filled to a single background color, what colors will the following 24-bit codes give:
   (a) 0x00FFFF
   (b) 0x00007F
   (c) 0x7F7F7F?

References

8.1. Rapid Electronics. PC1602F datasheet. http://www.rapidonline.com/pdf/57-0913.pdf
8.2. HD44780 LCDisplays. http://www.a-netz.de/lcd.en.php
8.3. Nokia 6100 LCD Display Driver. http://www.sparkfun.com/tutorial/Nokia%206100%20LCD%20Display%20Driver.pdf

CHAPTER 9

Interrupts, Timers and Tasks

Chapter Outline

Fast and Effective Embedded Systems Design. DOI: 10.1016/B978-0-08-097768-3.00009-X

9.1 Time and Tasks in Embedded Systems

9.1.1 Timers and Interrupts

The very first diagram in this book, Figure 1.1, shows the key features of an embedded system. Among these is *time*. Embedded systems have to respond in a timely manner to events as they happen. Usually, this means they have to be able to do the following:

- measure time durations
- generate time-based events, which may be single or repetitive
- respond with appropriate speed to external events, which may occur at unpredictable times.

In doing all of these the system may find that it has a conflict of interest, with two actions needing attention at the same time. For example, an external event may demand attention just when a periodic event needs to take place. Therefore, the system may need to distinguish between events that have a high level of urgency and those that do not, and take action accordingly.

It follows that we need a set of tools and techniques to allow effective time-based activity to occur. Key features of this toolkit are interrupts and timers, the subject of this chapter. In brief, a timer is just what its name implies, a digital circuit that allows us to measure time, and hence make things happen when a certain time has elapsed. An interrupt is a mechanism whereby a running program can be interrupted, with the central processing unit (CPU) then being required to jump to some other activity. In our study of interrupts and timers, we make major steps forward in our understanding of how programs can be structured, and how we can move towards more sophisticated program design. That is where the third topic in the chapter title comes in – the concept of program *tasks*.

9.1.2 Tasks

In almost all embedded programs, the program has to undertake a number of different activities. Entering briefly the world of small-scale farming, consider the requirements for a temperature controller for a mushroom-growing shed, working in a cold environment. These friendly little fungi grow best under tightly controlled conditions of temperature and humidity. Their growth goes through separate phases, at which time different preferred temperatures may apply. The system will need to control, display and log the temperature, keep track of time, respond to changes in setting by the user, and control the heater and fan. Each is a fairly distinct activity, and each will require a block of code. In programming terminology we call these distinct activities *tasks*. Our ability to partition any program into tasks becomes an important skill in more advanced program design. Once a program has more than one task, we enter the domain of *multi-tasking*. As tasks increase in number, the challenge of responding to the needs of all tasks becomes greater, and many techniques are developed to do this.

9.1.3 Event-Triggered and Time-Triggered Tasks

Tasks performed by embedded systems tend to fall into two categories: event triggered and time triggered. Tasks that are event triggered occur when a particular external event happens, at a time which is usually not predictable. Tasks that are time triggered happen periodically, at a time determined by the microcontroller. To continue with our example of the mushroom shed, Table 9.1 lists some possible tasks and suggests whether each is event or time triggered. For those that are time triggered, it becomes necessary to indicate how frequently the tasks occur; suggestions for this are also shown.

9.2 Responding to Event-Triggered Events

9.2.1 Polling

A simple example of an event-triggered activity is when a user pushes a button. This can happen at any time, without warning, but when it does the user expects a response. One way of programming for this is to continuously test that external input. This is illustrated in Figure 9.1, where a program is structured as a continuous loop. Within this, it tests the state of two input buttons, and responds to them if activated. This way of checking external events is called *polling*; the program ensures that it periodically checks input states and responds if there is a need. This is the sort of approach we have used so far in this book, whenever needed. It works well for simple systems; however, it is not adequate for more complex programs.

Suppose the program of Figure 9.1 is extended, so that a microcontroller has 20 input signals to test in each loop. On most loop iterations, the input data may not even change, so we are running the polling for no apparent benefit. Worse, the program might spend time checking the value of unimportant inputs, while not recognizing very quickly when a major fault condition has arisen.

There are two main problems with polling:

- The processor cannot perform any other operations during a polling routine.
- All inputs are treated as equal; the urgent change has to wait its turn before it is recognized by the computer.

Table 9.1: Example tasks: temperature controller for a mushroom shed

Task	Event or time triggered
Measure temperature	Time (every minute)
Compute and implement heater and fan settings	Time (every minute)
Respond to user control	Event
Record and display temperature	Time (every minute)
Orderly switch to battery backup in case of power loss	Event

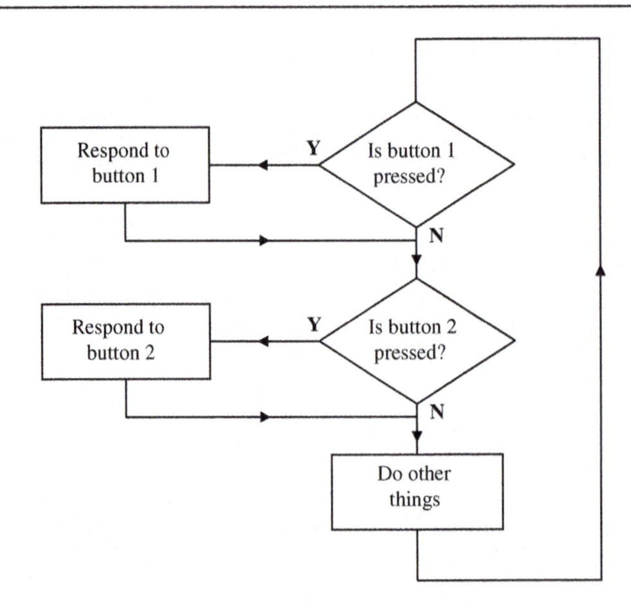

Figure 9.1:
A simple program using polling

A better solution is for input changes to announce themselves; time is not wasted finding out that there is no change. The difficulty lies in knowing when the input value has changed. This is the purpose of the interrupt system.

9.2.2 Introducing Interrupts

The interrupt represents a radical alternative to the polling approach just described. With an interrupt, the hardware is designed so that the external variable can stop the CPU in its tracks, and demand attention. Suppose you lived in a house, and were worried that a thief might come in during the night. You *could* arrange an alarm clock to wake you up every half hour to check there was no thief, but you would not get much sleep. In this case you would be *polling* the possible thief 'event'. Alternatively, you could fit a burglar alarm. You would then sleep peacefully, *unless* the alarm went off, interrupted your sleep, and you would jump up and chase the burglar. In very simple terms, this is the basis of the computer interrupt.

Interrupts have become a hugely important part of the structure of any microprocessor or microcontroller, allowing external events and devices to force a change in CPU activity. In early processors interrupts were mainly used to respond to really major external events; designs allowed for just one, or a small number of interrupt sources. The interrupt concept was, however, found to be so useful that more and more possible interrupt sources were introduced, sometimes for dealing with rather routine matters.

In responding to interrupts, most microprocessors follow the general pattern of the flow diagram shown in Figure 9.2. The CPU completes the current instruction it is executing. It is

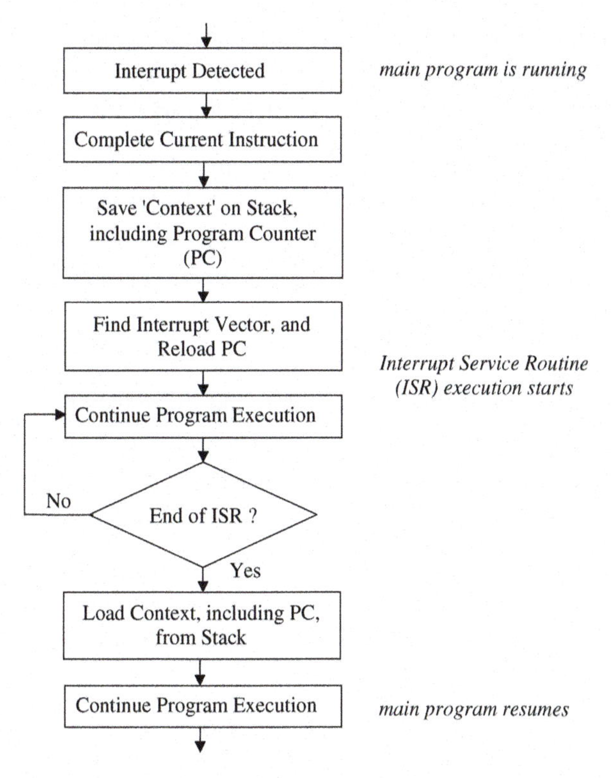

Figure 9.2:
A typical microprocessor interrupt response

about to go off and find a completely different piece of code to execute, so it must save key information about what it has just been doing; this is called the *context*. The context includes at least the value of the Program Counter (this tells the CPU where it should come back to when the interrupt has completed) and generally a set of key registers, for example those holding current data values. All this is saved on a small block of memory local to the CPU, called the *Stack*. The CPU then runs a section of code called an Interrupt Service Routine (ISR); this has been specifically written to respond to the interrupt that has occurred. The address of the ISR is found through a memory location called the *Interrupt Vector*. On completing the ISR, the CPU returns to the point in the main code immediately after the interrupt occurred, finding this by retrieving the Program Counter from the Stack where it left it. It then continues program execution as if nothing happened. Later, in Section 9.4, we add to this explanation, and set it more in the context of the mbed.

9.3 Simple Interrupts on the mbed

The mbed application programming interface (API) exploits only a small subset of the interrupt capability of the LPC1768 microcontroller, mainly focusing on the external interrupts. Their use is very flexible, as any of pins 5–30 can be used as an interrupt input, excepting only pins 19 and 20. The available API functions are shown in Table 9.2. Using these we can create an interrupt input, write the corresponding ISR and link the ISR with the interrupt input.

Program Example 9.1 is a very simple interrupt program, adapted from the mbed website. It is made up of a continuous loop, which switches on and off LED4 (labeled **flash**). The interrupt input is specified as pin 5, and labeled button. A tiny ISR is written for it, called ISR1, which is structured exactly as a function. The address of this function is attached to the rising edge of the interrupt input, in the line

```
button.rise(&ISR1);
```

When the interrupt is activated, by this rising edge, the ISR executes, and LED1 is toggled. This can occur at any time in program execution. The program has effectively one time-triggered task, the switching of LED4, and one event-triggered task, the switching of LED1.

```
/* Program Example 9.1: Simple interrupt example. External input causes interrupt,
while led flashes
*/
#include "mbed.h"
InterruptIn button(p5);   //define and name the interrupt input
DigitalOut led(LED1);
DigitalOut flash(LED4);

void ISR1() {             //this is the response to interrupt, i.e. the ISR
  led = !led;
}

int main() {
  button.rise(&ISR1);     // attach the address of the ISR function to the
                          // interrupt rising edge
  while(1) {              // continuous loop, ready to be interrupted
    flash = !flash;
    wait(0.25);
  }
}
```

Program Example 9.1 Introductory use of an interrupt

Table 9.2: API interrupts summary

Function	Usage
InterruptIn	Create an InterruptIn connected to the specified pin
rise	Attach a function to call when a rising edge occurs on the input
rise	Attach a member function to call when a rising edge occurs on the input
fall	Attach a function to call when a falling edge occurs on the input
fall	Attach a member function to call when a falling edge occurs on the input
mode	Set the input pin mode

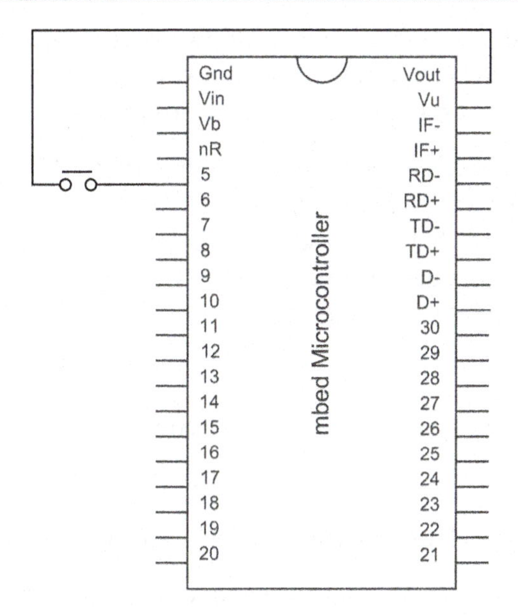

Figure 9.3:
Circuit build for Program Example 9.1

Compile and run this program, applying the very simple build shown in Figure 9.3. Notice that when you push the button the interrupt is taken high, and LED1 changes state; LED4 meanwhile continues its flashing, almost unperturbed. The program depends on the internal pull-down resistor, which is enabled by default during setup. *If* you experience erratic behavior with this program, you may be experiencing switch bounce. In this case, fast forward to Section 9.10 for an introduction to this important topic.

Exercise 9.1

Change Program Example 9.1 so that:

1. The interrupt is triggered by a falling edge on the input.
2. There are two ISRs, from the same push-button input. One toggles LED1 on a rising interrupt edge, and the other toggles LED2 on a falling edge.

■

9.4 Getting Deeper into Interrupts

We can take our first generalized understanding of interrupts further, and try to understand a little more what goes on inside the mbed. It should be pointed out straight away that this is not essential as far as using the mbed API is concerned. In fact, although the LPC1768 microcontroller, inside the mbed, has a very sophisticated interrupt structure, the mbed only uses a small part of this, and that only in a modest way. Therefore, go straight to the next section if you do not want to get into any deeper interrupt detail.

Okay, so you're still there! Let's extend our ideas of interrupts. While we said earlier that an interrupt was possibly like a thief coming in the night, imagine now a different scenario. Suppose you are a teacher of a big class made up of enthusiastic but poorly behaved kids; you have set them a task to do but they need your help. Tom calls you over, but while you are helping him Jane starts clamoring for attention.

Do you:

- tell Jane to be quiet and wait until you've finished with Tom?

OR

- tell Tom you'll come back to him, and go over to sort Jane out?

To make matters worse, your school principal has asked you to let the members of the school band out of class half an hour early, but you really want them to finish their work before they go. This influences the above decision. Suppose Jane is in the band, and Tom is not. Therefore, in this situation you decide you must leave Tom to help Jane. This school classroom situation is reflected in almost any embedded system. There could be a number of interrupt sources, all possibly needing attention. Some will be of great importance, others much less. Therefore, most processors contain four important mechanisms:

- Interrupts can be *prioritized,* in other words some are defined as more important than others. If two occur at the same time, then the higher priority one executes first.
- Interrupts can be *masked*, i.e. switched off, if they are not needed, or are likely to get in the way of more important activity. This masking could be just for a short period, for example while a critical program section completes.
- Interrupts can be *nested.* This means that a higher priority interrupt can interrupt one of lower priority, just like the teacher leaving Tom to help Jane. Working with nested interrupts increases the demands on the programmer and is strictly for advanced players only. Not all processors permit nested interrupts, and some allow you to switch nesting on or off.
- The location of the ISR in memory can be selected, to suit the memory map and programmer's wishes.

Let us take on just a couple more important interrupt concepts, these ones from the point of view of the interrupt source. Go to the moment in the above scenario when Jane suddenly realizes she needs help, and puts her hand up. Some short time later the teacher comes over. The delay between her putting up her hand and the teacher actually arriving is called the interrupt *latency.* Latency may be due to a number of things; in this case the teacher has to notice Jane's hand in the air, may need to finish with another pupil, and then has to walk over. Once the teacher arrives, Jane puts her hand down. While Jane is waiting with her hand in the air, patiently we hope, her interrupt is said to be *pending*.

These concepts and capabilities hint at some of the deep magic that can be achieved with advanced interrupt structures.

We can put all of this into more technical terms, and hence refine our understanding of interrupt action. This was first illustrated in the flow diagram of Figure 9.2. Further detail on part of this figure is now shown in Figure 9.4. The interrupt being asserted is like Jane putting up her hand. In a microprocessor, the interrupt input will be a logic signal; depending on its input configuration it may be active high or low, or triggered by a rising or falling edge. This input will cause an internal 'flag' to be set. This is normally just a single bit in a register, which records the fact that an interrupt has occurred. This does not necessarily mean that the interrupt automatically gets the attention it seeks. If it is not enabled (i.e. it is masked), then there will be no response. The flag is left high, however, as the program may later enable that interrupt or the program may just poll the interrupt flag. Back to the flow diagram: if another ISR is already running then again the incoming interrupt may not get a response, at least not immediately. If it is higher priority and nested interrupts are allowed, then it will be allowed to run. If it is lower priority, it will have to wait for the other ISR to complete. The subsequent actions in the flow diagram, as already

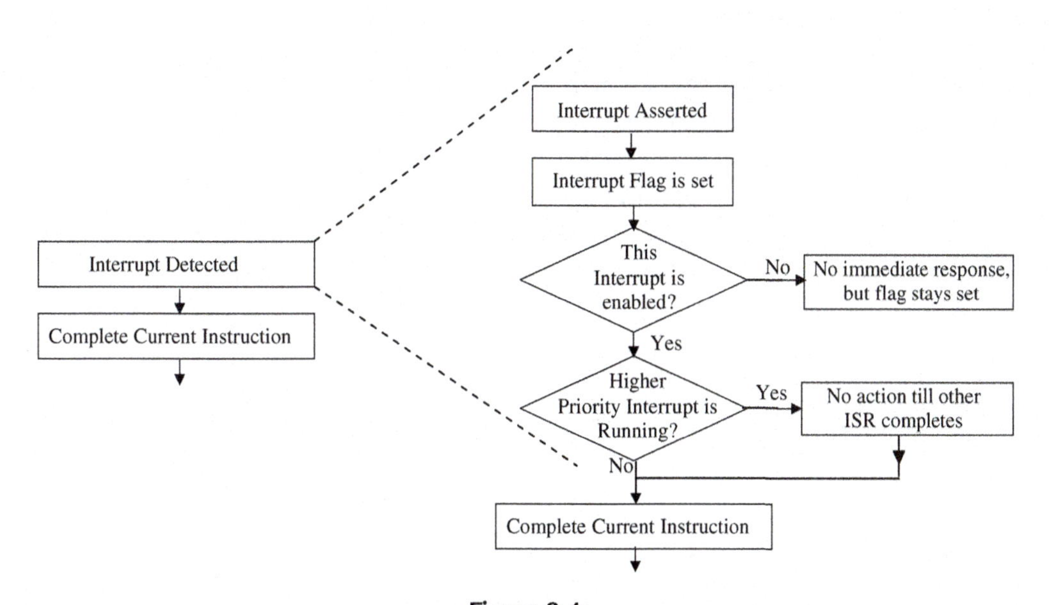

Figure 9.4:
A typical microprocessor interrupt response — some greater detail

seen in Figure 9.2, then follow. Note that the figure is potentially misleading, as it implies these actions happen in turn. To get low latency, they should happen as rapidly as possible; a good interrupt management system will allow some of the actions to take place in parallel. For example, the interrupt vector could be accessed while the current instruction is completing.

9.4.1 Interrupts on the LPC1768

Now let's get back to microprocessor hardware. Recall that the mbed contains the LPC1768 microprocessor, and that the LPC1768 contains the ARM Cortex core. It is to the Cortex core that we must turn first, as it provides the interrupt structure used by the microcontroller. Back in Figure 2.3, we saw the Cortex core, set within the LPC1768 microprocessor. Management of all interrupts in the Cortex is undertaken by the formidable-sounding *Nested Vectored Interrupt Controller* (NVIC). You could think of this as managing the processes overviewed in Figures 9.2 and 9.4. The NVIC is also a bit like a digital electronic control box with a lot of unconnected wires hanging out. When the Cortex is embedded into a microcontroller, such as the LPC1768, the chip designer assigns and configures those features (through the 'loose wires') of the NVIC that are needed for that application. For example, the Cortex allows for 240 possible interrupts, both external and from the peripherals, and 256 possible priority levels. The LPC1768, however, has 'only' 33 interrupt sources, with 32 possible programmable priority levels.

9.4.2 Testing Interrupt Latency

Now that we have met the concept of interrupt latency, we can test it in the mbed. Program Example 9.2 adapts Program Example 9.1, but the interrupt is now generated by an external square wave, instead of an external button push. This makes it easier to see on an oscilloscope. When an interrupt occurs, the external LED is pulsed high for a fixed duration.

```
/* Program Example 9.2: Tests interrupt latency. External input causes interrupt, which
pulses external LED while LED4 flashes continuously.
*/
#include "mbed.h"
InterruptIn squarewave(p5);    //Connect input square wave here
DigitalOut led(p6);
DigitalOut flash(LED4);

void pulse() {                 //ISR sets external led high for fixed duration
   led = 1;
  wait(0.01);
   led = 0;
}
int main() {
```

```
  squarewave.rise(&pulse);       // attach the address of the pulse function to
                                 // the rising edge
  while(1) {                     // interrupt will occur within this endless loop
  flash = !flash;
  wait(0.25);
  }
}
```

Program Example 9.2 Testing interrupt latency

Create a new project from Program Example 9.2. Connect an external LED between pin 6 and ground, ensuring correct polarity. Connect to pin 5 a signal generator set to logic-compatible square wave output (probably labeled 'TTL compatible'), running initially at around 10 Hz. Once connected, with the program running, the external LED should flash at this rate, i.e. around 10 times a second.

Exercise 9.2

Connect two inputs of an oscilloscope to the interrupt input, and the LED output, triggering from the interrupt. Increase the input frequency to around 50 Hz. Set the oscilloscope time base to 5 μs per division. You should be able to see the rising edge of the interrupt input, and a few microseconds later the LED output rising. The time delay between the two is an indication of latency. The rise of the LED will be flickering a little, as the delay will depend on what the CPU is doing at the moment the interrupt occurs. It is important to note that the latency as measured here depends on both hardware and software factors.

9.4.3 Disabling Interrupts

Interrupts are an essential tool in embedded design. But because they can occur at any time, they can have unexpected or undesirable side effects. In some situations it is essential to disable (mask) the interrupt. These include when you are undertaking a time-sensitive activity, or a complex calculation that must be completed in one go. As any incoming interrupt will have left its flag set, it can be responded to once interrupts are enabled again. Of course, a delay in response has been introduced and the latency greatly compromised.

You can simply disable interrupts, as shown here:

```
__disable_irq();                   //disable interrupts
//activity which can't be interrupted
__enable_irq();                    //enable interrupts
```

Note that each line starts with *two* underscores.

Exercise 9.3

Using Program Example 9.2, disable the interrupt for the duration of the `wait(0.25)` delay. Connect the oscilloscope as in Exercise 9.2 and observe the output again. Comment on the outcome.

Experiment with different values for this wait function. Try then splitting it into two waits, running immediately after each other, with one having interrupts disabled and the other enabled.

9.4.4 Interrupts from Analog Inputs

Aside from digital inputs, it is useful to generate interrupts when analog signals change, for example if an analog temperature sensor exceeds a certain threshold. One way to do this is by applying a comparator. A comparator is just that, it compares two input voltages. If one input is higher than the other then the output switches to a high state; if it is lower, the output switches low. A comparator can easily be configured from an operational amplifier (op amp), as shown in Figure 9.5. Here, an input voltage, labeled V_{in}, is compared to a threshold voltage derived from a potential divider, made from the two resistors R_1 and R_2. These are connected to the supply voltage, labeled V_{sup}. For the right choice of op amp or comparator, and with suitable supply voltages, the output is a Logic 1 when the input is above the threshold value, and Logic 0 otherwise. The threshold voltage just mentioned, labeled V_- in the diagram, is calculated using Equation 9.1:

$$V_- = V_{sup}R_2/(R_1 + R_2) \tag{9.1}$$

As an example, we might want to generate an interrupt from a temperature input, using an LM35 temperature sensor (Figure 5.8), with the interrupt triggered if the temperature exceeds 30°C. As we know, this sensor has an output of 10 mV/°C, which would lead to an output

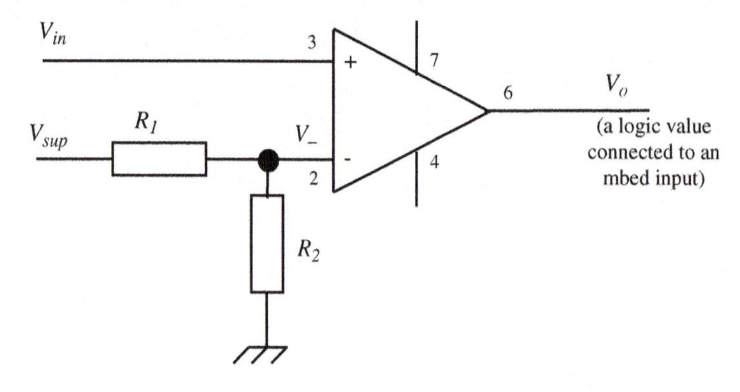

Figure 9.5:
A comparator circuit

voltage of 300 mV at the trigger point proposed. To set V_{-} to 300 mV in Figure 9.5, we apply Equation 9.1, with a V_{sup} value of 3.3 V; we find that $R_1 = 0.1R_2$. Values of $R_1 = 10k$, and $R_2 = 1k$ could therefore be chosen.

Exercise 9.4

Unlike many op amps, the ICL7611 can be run from very low supply voltages, even the 3.3 V of the mbed. Using an LM35 IC temperature sensor, and an ICL7611 op amp connected as a comparator, design and build a circuit that causes an interrupt when the temperature exceeds 30°C. Write a program that lights an LED when the interrupt occurs. The pin connections shown in Figure 9.5 can be applied. Pin 7 is the positive supply, and can be connected to the mbed 3.3 V; pin 4 is the negative supply, and is connected to 0 V. For this op amp, also connect pin 8 to the positive supply rail. (This pin controls the output drive capability; check the datasheet for more information.)

9.4.5 Conclusion on Interrupts

This section has provided a good introduction to interrupts and the main concepts associated with them. They become an essential part of the toolkit of any embedded designer. We have so far limited ourselves to single-interrupt examples. Where multiple interrupts are used, the design challenges become considerably greater – interrupts can have a very destructive effect if not used well. The design of advanced multiple interrupt programs is, however, beyond the scope of this book.

9.5 An Introduction to Timers

A simple program, for example the very first Program Example 2.1, uses **wait( )** functions to perform timing operations, for example to introduce a delay of 200 ms. This is easy and very convenient, but during this delay loop the microcontroller cannot perform any other activity; the time spent waiting is just wasted time. As we try to write more demanding programs, this simple timing technique becomes inadequate. We need a way of letting timing activity go on in the background, while the program continues to do useful things. We turn to the digital hardware for this.

9.5.1 The Digital Counter

It is an easy task in digital electronics to make electronic counters; you simply connect together a series of bistables or flip-flops. Each one holds one bit of information, and all those bits together form a digital word. This is illustrated in very simple form in Figure 9.6, with

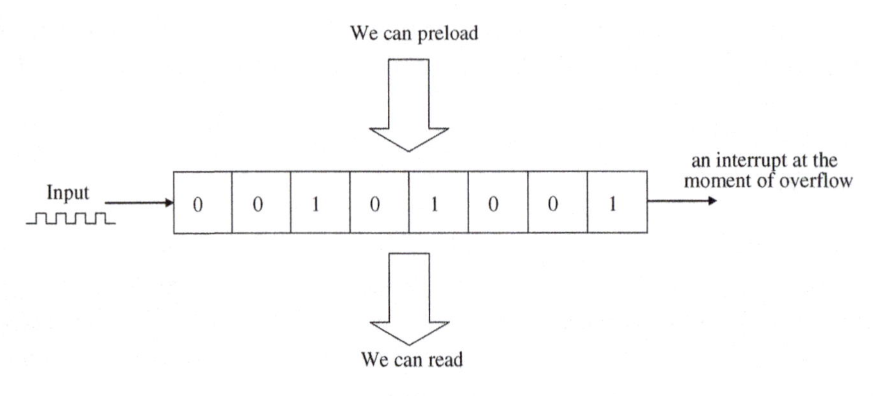

Figure 9.6:
A simple 8-bit counter

each little block representing one flip-flop, each holding one bit of the overall number. If the input of this arrangement is connected to a clock signal then the counter will count, in binary, the number of clock pulses applied to it. It is easy to read the overall digital number held in the counter, and it is not difficult to arrange the necessary logic to preload it with a certain number, or to clear it to zero.

The number that a counter can count up to is determined by the number of bits in the counter. In general, an n-bit counter can count from 0 to $(2^n - 1)$. For example, an 8-bit counter can count from 0000 0000 to 1111 1111, or 0 to 255 in decimal. Similarly, a 16-bit counter can count from 0 to 65 535. If a counter reaches its maximum value, and the input clock pulses keep on coming, then it overflows back to zero, and starts counting up all over again. All is not lost if this happens – in fact we have to be ready to deal with it. Many microcontroller counters cause an interrupt as the counter overflows; this interrupt can be used to record the overflow, and the count can continue in a useful way.

9.5.2 Using the Counter as a Timer

The input signal to a counter can be a series of pulses coming from an external source, for example counting people going through a door. Alternatively, it can be a fixed-frequency logic signal, such as the clock source within a microcontroller. Very importantly, if that clock source is a known and stable frequency, then the counter becomes a timer. As an example, if the clock frequency is 1.000 MHz (hence with period of 1 μs), then the count will update every microsecond. If the counter is cleared to zero, and then starts counting, the value held in the counter will give the elapsed time since the counting started, with a resolution of 1 μs. This can be used to measure time, or trigger an event when a certain time has elapsed. It can also be used to control time-based activity, for example serial data communication, or a pulse width modulation (PWM) stream.

Alternatively, if the counter is just free-running with a continuous clock signal, then the 'interrupt on overflow' occurs repeatedly. This becomes very useful where a periodic interrupt is needed. For example, if an 8-bit counter is clocked with a clock frequency of 1 MHz, it will reach its maximum value and overflow back to zero in 256 μs (it is the 256th pulse that causes the overflow from 255 to 0). If it is left running continuously, then this train of interrupt pulses can be used to synchronize timed activity, for example it could define the baud rate of a serial communication link.

Timers based on these principles are an incredibly important feature of any microcontroller. Indeed, most microcontrollers have many more than one timer, applied to a variety of different tasks. These include generating timing in PWM or serial links, measuring the duration of external events and producing other timed activity.

9.5.3 Timers on the mbed

To find out what hardware timers the mbed has, we turn back to Figure 2.3 and Reference 2.4, the LPC1768 user manual. We find that the microcontroller has four general-purpose timers, a *Repetitive Interrupt Timer* and a *System Tick Timer.* All are based on the principles just described. The mbed makes use of these in three distinct applications, described in the following sections. These are the Timer, used for simple timing applications, Timeout, which calls a function after a predetermined delay, and Ticker, which repeatedly calls a function at a predetermined rate. It also applies a *Real Time Clock* to keep track of time of day and the date.

9.6 Using the mbed Timer

The mbed Timer allows basic timing activities to take place, for comparatively short durations. A Timer can be created, started, stopped and read. There is no limit on the number of Timers that can be set up. The API summary is shown in Table 9.3. The mbed site notes that

Table 9.3: API summary for Timer

Function	Usage
`start`	Start the timer
`stop`	Stop the timer
`reset`	Reset the timer to 0
`read`	Get the time passed in seconds
`read_ms`	Get the time passed in milliseconds
`read_us`	Get the time passed in microseconds

the Timer is based on the 32-bit counters, and can time up to a maximum of $(2^{31} - 1)$ microseconds, i.e. something over 30 minutes. Based on the theory above, one might expect the Timer to count up to $(2^{32} - 1)$. However, one bit is reserved in the API object as a sign bit, so only 31 bits are available for counting.

Program Example 9.3 gives a simple but interesting timing example, and is taken from the mbed site. It measures the time taken to write a message to the screen, and displays that message. Compile and run the program, with Tera Term activated (Appendix E). Then carry out Exercise 9.5 to make some further measurements and calculations.

```
/* Program Example 9.3: A simple Timer example, from mbed website.
Activate Tera Term terminal to test.
                                                                        */
#include "mbed.h"
Timer t;                       // define Timer with name "t"
Serial pc(USBTX, USBRX);

int main() {
  t.start();                   //start the timer
  pc.printf("Hello World!\n");
  t.stop();                    //stop the timer
  pc.printf("The time taken was %f seconds\n", t.read()); //print to pc
}
```

Program Example 9.3 A simple Timer application

Exercise 9.5

Run Program Example 9.3, and note from the computer screen readout the time taken for the message to be written. Then write some other messages, of differing lengths, and record in each case the number of characters and the time taken. Can you relate the times taken to the baud rate used? Can you make any other deductions? If necessary, check Section 7.9 to recall some of the timing issues relating to the asynchronous serial data link used.

9.6.1 Using Multiple mbed Timers

We are now going to apply the Timer in a different way, to run one function at one rate and another function at another rate. Two LEDs will be used to show this; you will quickly realize that the principle is powerful, and can be extended to more tasks and more activities. Program Example 9.4 shows the program listing. The program creates two Timers, named **timer_fast** and **timer_slow**. The main program starts these running, and tests when each exceeds a certain number. When the time value is exceeded, a function is called, which flips the associated LED.

```
/*Program Example 9.4: Program which runs two time-based tasks
*/
#include "mbed.h"
Timer timer_fast;                // define Timer with name "timer_fast"
Timer timer_slow;                // define Timer with name "timer_slow"
DigitalOut ledA(LED1);
DigitalOut ledB(LED4);

void task_fast(void);            //function prototypes
void task_slow(void);

int main() {
  timer_fast.start();            //start the Timers
  timer_slow.start();
  while (1){
    if (timer_fast.read()>0.2){ //test Timer value
       task_fast();              //call the task if trigger time is reached
       timer_fast.reset();       //and reset the Timer
    }
    if (timer_slow.read()>1){   //test Timer value
       task_slow();
       timer_slow.reset();
    }
  }
}

void task_fast(void){            //"Fast" Task
   ledA = !ledA;
}

void task_slow(void){            //"Slow" Task
  ledB = !ledB;
}
```

Program Example 9.4 Running two timed tasks

Create a project around Program Example 9.4, and run it on the mbed alone. Check the timing with a stopwatch or an oscilloscope.

Exercise 9.6

Experiment with different repetition rates in Program Example 9.4, including ones that are not multiples of each other. Add a third and then a fourth timer to it, flashing all mbed LEDs at different rates.

9.6.2 Testing the Timer Duration

We quoted above the maximum value that the Timer can reach. As with many microcontroller features, it is very important to understand the operating limits and to be certain to remain within them. Program Example 9.5 measures the Timer limit in a simple way. It clears the Timer and then sets it running, displaying a time update to the Tera Term screen every second.

In doing this it keeps its own record of seconds elapsed and compares this with the elapsed time value given by the Timer. From the program structure we expect the Timer value always to be just ahead of the recorded time value. At some point, however, the Timer overflows back to zero and this condition is no longer satisfied. The program detects this and sends a message to the screen. The highest value reached by the Timer is recorded on the screen.

```
/* Program Example 9.5: Tests Timer duration, displaying current time values to terminal
*/
#include "mbed.h"
Timer t;
float s=0;                          //seconds cumulative count
float m=0;                          //minutes cumulative count
DigitalOut diag (LED1);
Serial pc(USBTX, USBRX);
int main() {
 pc.printf("\r\nTimer Duration Test\n\r");
 pc.printf("-------------------\n\n\r");
 t.reset();                         //reset Timer
 t.start();                         // start Timer
 while(1){
   if (t.read()>=(s+1)){ //has Timer passed next whole second?
     diag = 1;                      //If yes, flash LED and print a message
     wait (0.05);
     diag = 0;
     s++ ;
     //print the number of seconds exceeding whole minutes
     pc.printf("%1.0f seconds\r\n",(s-60*(m-1)));
   }
   if (t.read()>=60*m){
     printf("%1.0f minutes \n\r",m);
     m++ ;
   }
   if (t.read()<s){                 //test for overflow
     pc.printf("\r\nTimer has overflowed!\n\r");
     for(;;){}                      //lock into an endless loop doing nothing
   }
 }                                 //end of while
}
```

Program Example 9.5 Testing Timer duration

Exercise 9.7

Create a project around Program Example 9.5 and run it on an mbed, enabling also a connection to Tera Term (see Appendix E). No further hardware is needed. Calculate precisely the time duration you expect from the Timer, i.e. $(2^{31} - 1)$ microseconds. You can leave the program running while doing other things. Come back, however, just after half an hour, and watch intently to see whether it overflows! Do your predicted and measured times agree?

In some embedded programs there comes a moment when we just want the program to stop; there is simply nothing more for it to do. Program Example 9.3 is one such. Yet we have no instruction available to us that just says stop; the CPU is designed to go on and on running until it is switched off. Notice at the end of this program example how the stop is implemented, by trapping program execution in an endless loop that does nothing.

9.7 Using the mbed Timeout

Program Example 9.4 showed the mbed Timer being used to trigger time-based events in an effective way. However, we needed to poll the Timer value to know when the event should be triggered. The Timeout allows an event to be triggered by an interrupt, with no polling needed. Timeout sets up an interrupt to call a function after a specified delay. There is no limit on the number of Timeouts created. The API summary is shown in Table 9.4.

9.7.1 A Simple Timeout Application

A simple first example of Timeout is shown in Program Example 9.6. This causes an action to be triggered a fixed period after an external event. This simple program is made up of the **main( )** function and a **blink( )** function. A **Timeout** is created, named **Response**, along with some familiar digital input and output. Looking in the **main( )** function, we see an **if** declaration, which tests if the button is pressed. If it is, the **blink( )** function gets attached to the **Response** Timeout. We can expect that two seconds after this attachment is made, the **blink( )** function will be called. To aid in the diagnostics, the button also switches on LED3. As a continuous task, the state of LED1 is reversed every 0.2 s. This program is thus a microcosm of many embedded systems programs. A time-triggered task needs to keep going, while an event-triggered task needs to take place at unpredictable times.

Table 9.4: API summary for Timeout

Function	Usage
`attach`	Attach a function to be called by the Timeout, specifying the delay in seconds
`attach`	Attach a member function to be called by the Timeout, specifying the delay in seconds
`attach_us`	Attach a function to be called by the Timeout, specifying the delay in microseconds
`attach_us`	Attach a member function to be called by the Timeout, specifying the delay in microseconds
`detach`	Detach the function

```
/*Program Example 9.6: Demonstrates Timeout, by triggering an event a fixed duration
after a button press.
*/
#include "mbed.h"
Timeout Response;       //create a Timeout, and name it "Response"
DigitalIn button (p5);
DigitalOut led1(LED1);
DigitalOut led2(LED2);
DigitalOut led3(LED3);

void blink() {          //this function is called at the end of the Timeout
  led2 = 1;
  wait(0.5);
  led2=0;
}

int main() {
  while(1) {
    if(button==1){
      Response.attach(&blink,2.0); // attach blink function to Response Timeout,
                                   //to occur after 2 seconds
      led3=1;                      //shows button has been pressed
    }
    else {
      led3=0;
    }
    led1=!led1;
    wait(0.2);
  }
}
```

Program Example 9.6 Simple Timeout application

Compile Program Example 9.6 and download to an mbed, with the build of Figure 9.3. Observe the response as the push-button is pressed.

Exercise 9.8

With Program Example 9.6 running, answer the following questions:

1. Is the 2 s Timeout timed from when the button is pressed or when it is released? Why is this?
2. When the event-triggered task occurs (i.e. the blinking of LED2), what impact does it have on the time-triggered task (i.e. the flashing of LED1)?
3. If you tap the button very quickly, you will see that it is possible for the program to miss it entirely (even though electrically we can prove that the button has been pressed). Why is this?

9.7.2 Further Use of Timeout

Program Example 9.6 has demonstrated use of the Timeout nicely, but the questions in Exercise 9.8 throw up some of the classic problems of task timing, notably that execution of the event-triggered task can interfere with the timing of the time-triggered task.

Program Example 9.7 does the same thing as the previous example, but in a much better way. Glancing through it, we see two Timeouts created, and an interrupt. The latter connects to the push-button, and replaces the digital input which previously took that role. There are now three functions, in addition to **main( )**. This has become extremely short and simple, and is concerned primarily with keeping the time-triggered task going. Now response to the push-button is by interrupt, and it is within the interrupt function that the first Timeout is set up. When it is spent, the **blink( )** function is called. This sets the LED output, but then enables the second Timeout, which will trigger the end of the LED blink.

```
/*Program Example 9.7: Demonstrates the use of Timeout and interrupts, to allow response
to an event-driven task while a time-driven task continues.
*/
#include "mbed.h"
void blink_end (void);
void blink (void);
void ISR1 (void);
DigitalOut led1(LED1);
DigitalOut led2(LED2);
DigitalOut led3(LED3);
Timeout Response;              //create a Timeout, and name it Response
Timeout Response_duration;     //create a Timeout, and name it Response_duration
InterruptIn button(p5);        //create an interrupt input, named button
void blink() {                 //This function is called when Timeout is complete
   led2=1;
   // set the duration of the led blink, with another timeout, duration 0.1 s
   Response_duration.attach(&blink_end, 1);
}
void blink_end() {             //A function called at the end of Timeout Response_duration
   led2=0;
}
void ISR1(){
   led3=1;    //shows button is pressed; diagnostic and not central to program
   //attach blink1 function to Response Timeout, to occur after 2 seconds
   Response.attach(&blink, 2.0);
}
int main() {
   button.rise(&ISR1);         //attach the address of ISR1 function to the rising edge
   while(1) {
     led3=0;                   //clear LED3
     led1=!led1;
     wait(0.2);
  }
}
```

Program Example 9.7 Improved use of Timeout

Compile and run the code of Program Example 9.7, again using the build of Figure 9.3.

Exercise 9.9

Repeat all the questions of Exercise 9.8 for the most recent program example, noting and explaining the differences you find.

9.7.3 Timeout used to Test Reaction Time

C code feature

Program Example 9.8 shows an interesting recreational application of Timeout, in which the timeout duration is itself a variable. It tests reaction time by blinking an LED, and timing how long it takes for the player to hit a switch in response. To add challenge, a 'random' delay is generated before the LED is lit. This uses the C library function **rand( )**. The program should be understandable from the comments it contains.

The circuit build is the same as seen in Figure 9.3. Now the push-button is the switch the player must hit to show a reaction. A Tera Term terminal should be enabled.

```
/*Program Example 9.8: Tests reaction time, and demos use of Timer and Timeout functions
*/
#include "mbed.h"
#include <stdio.h>
#include <stdlib.h>          //contains rand() function
void measure ();
Serial pc(USBTX, USBRX);
DigitalOut led1(LED1);
DigitalOut led4(LED4);
DigitalIn responseinput(p5); //the player hits the switch connected here to respond
Timer t;    //used to measure the response time
Timeout action;        //the Timeout used to initiate the response speed test

int main (){
  pc.printf("Reaction Time Test\n\r");
  pc.printf("------------------\n\r");
  while (1) {
    int r_delay;   //this will be the "random" delay before the led is blinked
    pc.printf("New Test\n\r");
    led4=1;                  //warn that test will start
    wait(0.2);
    led4=0;
    r_delay = rand() % 10 + 1; // generates a pseudorandom number range 1-10
    pc.printf("random number is %i\n\r", r_delay); // allows test randomness;
                             //removed for normal play
    action.attach(&measure,r_delay); // set up Timeout to call measure()
                           // after random time
    wait(10);   //test will start within this time, and we then return to it
```

```
  }
}
void measure (){   // called when the led blinks, and measures response time
  if (responseinput ==1){              //detect cheating!
    pc.printf("Don't hold button down!");
  }
  else{
    t.start();        //start the timer
    led1=1;           //blink the led
    wait(0.05);
    led1=0;
    while (responseinput==0) {
      //wait here for response
    }
    t.stop();            //stop the timer once response detected
    pc.printf("Your reaction time was %f seconds\n\r", t.read());
    t.reset();
  }
}
```

Program Example 9.8 Reaction time test: applying Timer and Timeout

Exercise 9.10

Run Program Example 9.8 for a period of time and note the sequence of 'random' numbers. Run it again. Do you recognize a pattern in the sequence of 'random' numbers? In fact, it is difficult for a computer to generate true random numbers, although various tricks and algorithms are used to create *pseudorandom* sequences of numbers.

9.8 Using the mbed Ticker

The mbed Ticker feature sets up a recurring interrupt, which can be used to call a function periodically, at a rate decided by the programmer. There is no limit on the number of Tickers created. The API summary is shown in Table 9.5.

Table 9.5: API summary for Ticker

Function	Usage
`attach`	Attach a function to be called by the Ticker, specifying the interval in seconds
`attach`	Attach a member function to be called by the Ticker, specifying the interval in seconds
`attach_us`	Attach a function to be called by the Ticker, specifying the interval in microseconds
`attach_us`	Attach a member function to be called by the Ticker, specifying the interval in microseconds
`detach`	Detach the function

We can demonstrate Ticker by returning to the very first program example in this book, number 2.1. This simply flashes an LED every 200 ms. Creating a periodic event is one of the most natural and common requirements in an embedded system, so it is not surprising that it appeared in our first program. We created the 200 ms period by using a delay function; these are useful in a limited way, but when they are running they tie up the CPU, so that it can do nothing else productive.

Program Example 9.9 simply replaces the Delay functions with the Ticker.

```
/* Program Example 9.9: Simple demo of "Ticker". Replicates behavior of first
led flashing program.
*/
#include "mbed.h"
void led_switch(void);
Ticker time_up;                  //define a Ticker, with name "time_up"
DigitalOut myled(LED1);

void led_switch(){               //the function that Ticker will call
  myled=!myled;
}

int main(){
  time_up.attach(&led_switch, 0.2);        //initializes the ticker
  while(1){     //sit in a loop doing nothing, waiting for Ticker interrupt
   }
}
```

Program Example 9.9 Applying Ticker to our very first program

It should be easy to follow what is going on in this program. The major step forward is that the CPU is now freed to do anything that is needed, while the task of measuring the time between LED changes is handed over to the Timer hardware, running in the background.

We have already called functions periodically with the Timer feature, so at first Ticker does not seem to add anything really new. Remember, however, that we have to poll the Timer value in Program Example 9.4 to test its value, and instigate the related function. Using Ticker, it is an interrupt that calls the associated function when time is up. As already discussed, this is a more effective approach to programming.

9.8.1 Using Ticker for a Metronome

Program Example 9.10 uses the mbed to create a metronome, using the Ticker facility. If you have not met one before, a metronome is an aid to musicians, setting a steady beat, against which they can play their music. The musician generally selects a beat rate in the range 40–208 beats per second. Old metronomes were based on elegant clockwork mechanisms, with a swinging pendulum arm. Most are made these days electronically. Normally the indication given to the musician is a loud audible 'tick', sometimes accompanied by an LED flash. Here we just restrict ourselves to the LED.

The main program **while** loop checks the up and down buttons, adjusts the beat rate accordingly, and displays the current rate to screen. This loops continuously, but lurking in the background is the Ticker. This has been initialized before the **while** loop, in the line

```
beat_rate.attach(&beat, period); //initializes the beat rate
```

Once the time indicated by **period** has elapsed, the **beat** function is called. At this moment the Ticker is updated, possibly with a new value of **period**, and the LED is flashed to indicate a beat. Program execution then returns to the main **while** loop, until the next Ticker occurrence.

```
/*Program Example 9.10: Metronome. Uses Ticker to set beat rate
*/
#include "mbed.h"
#include <stdio.h>
Serial pc(USBTX, USBRX);
DigitalIn up_button(p5);
DigitalIn down_button(p6);
DigitalOut redled(p19);             //displays the metronome beat
Ticker beat_rate;                   //define a Ticker, with name "beat_rate"
void beat(void);
float period (0.5);                 //metronome period in seconds, initial value 0.5
int rate (120);                     //metronome rate, initial value 120

int main() {
  pc.printf("\r\n");
  pc.printf("mbed metronome!\r\n");
  pc.printf("_______________\r\n");
  period = 1;
  redled = 1;                       //diagnostic
  wait(.1);
  redled = 0;
  beat_rate.attach(&beat, period);  //initializes the beat rate
  //main loop checks buttons, updates rates and displays
  while(1){
    if (up_button ==1)              //increase rate by 4
      rate = rate + 4;
    if (down_button ==1)            //decrease rate by 4
      rate = rate - 4;
    if (rate > 208)                 //limit the maximum beat rate to 208
      rate = 208;
    if (rate < 40)                  //limit the minimum beat rate to 40
      rate = 40;
    period = 60/rate;               //calculate the beat period
    pc.printf("metronome rate is %i\r", rate);
    //pc.printf("metronome period is %f\r\n", period); //optional check
    wait (0.5);
  }
}

  void beat() {                     //this is the metronome beat
    beat_rate.attach(&beat, period);  //update beat rate at this moment
```

```
    redled = 1;
    wait(.1);
    redled = 0;
  }
```

Program Example 9.10 Metronome, applying Ticker

The hardware build for the metronome is simple, and is shown in Figure 9.7. The Tera Term terminal should be enabled. Build the hardware, and compile and download the program. With a stopwatch or another timepiece check that the beat rates are accurate.

Exercise 9.11

We have left a bit of polling in Program Example 9.10. Rewrite it so that response to the external pins is done with interrupts.

9.8.2 Reflecting on Multi-Tasking in the Metronome Program

Having picked up the idea of program tasks at the beginning of this chapter, we have seen a sequence of program examples that could be described as multi-tasking. This started with

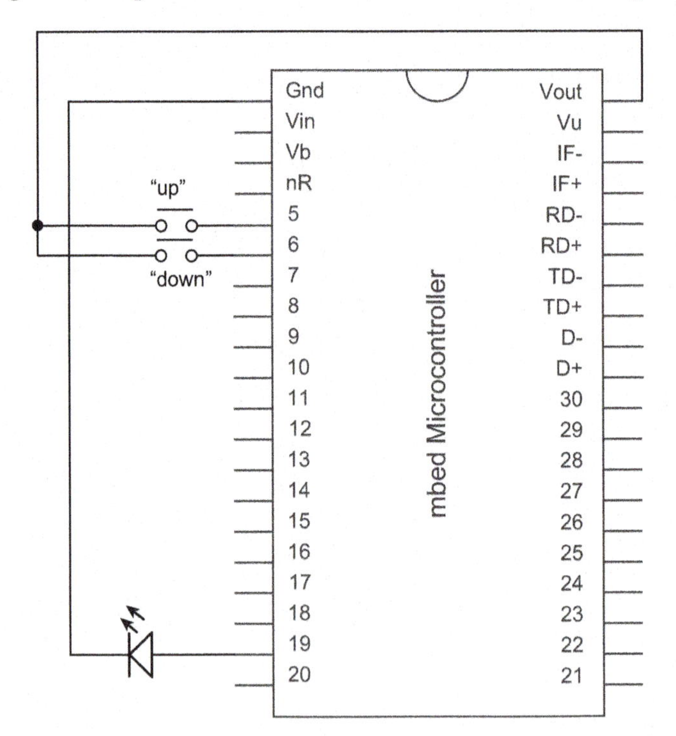

Figure 9.7:
Metronome build

Program Example 9.1, which runs one time-triggered and one event-triggered task. Many of the subsequent programs have clear multi-tasking features. In the metronome example, the program has to keep a regular beat going. While doing this, it must also respond to inputs from the user, calculate beat rates and write to the display. There is, therefore, one time-triggered task, the beat, and at least one event-triggered task, the user input. Sometimes it is difficult to decide which program activities belong together in one task. In the metronome example the user response seems to contain several activities, but they all relate closely to each other, so it is sensible to view them as the same task.

We have developed here a useful program structure, whereby time-triggered tasks can be linked to a Timer or Ticker function, and event-triggered tasks to an external interrupt. This will be useful as long as there are not too many tasks, and as long as they are not too demanding of CPU time. When that happens, we will need to look at more sophisticated solutions, either with a real-time operating system or with a multi-processor hardware. Both of these are beyond the scope of this book, but topics to look out for as your knowledge and confidence increase.

9.9 The Real Time Clock

The Real Time Clock (RTC) is an ultra-low-power peripheral on the LPC1768, which is implemented by the mbed. The RTC is a timing/counting system that maintains a calendar and time-of-day clock, with registers for seconds, minutes, hours, day, month, year, day of month and day of year. It can also generate an alarm for a specific date and time. It runs from its own 32 kHz crystal oscillator, and can have its own independent battery power supply. It can thus be powered, and continue in operation, even if the rest of the microcontroller is powered down. The mbed API does not create any C++ objects, but just implements standard functions from the standard C library, as shown in Table 9.6. Simple examples for use can be found on the mbed website.

Table 9.6: API summary for Real Time Clock

Function	Usage
`Time`	Get the current time
`set_time`	Set the current time
`mktime`	Converts a tm structure to a timestamp
`localtime`	Converts a timestamp to a tm structure
`ctime`	Converts a timestamp to a human-readable string
`strftime`	Converts a tm structure to a custom format human-readable string

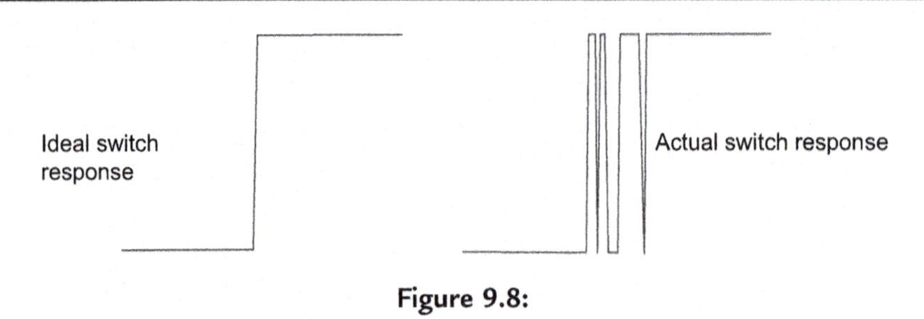

Figure 9.8:
Demonstrating switch bounce

9.10 Switch Debouncing

With the introduction of interrupts, we now have some choices to make when writing a program to a particular design specification. For example, Program Example 3.3 uses a digital input to determine which of two LEDs to flash. The digital input value is continuously polled within an infinite loop. However, we could equally have designed this program with an event-driven approach, to flip a control variable every time the digital input changes. Importantly, there are some inherent timing constraints within Program Example 3.3 which have not previously been discussed. One is that the frequency of polling is actually quite slow, because once the switch input has been tested a 0.4 s flash sequence is activated. This means that the system has a response time of at worst 0.4 s, because it only tests the switch input once for every program loop. When the switch changes position it could take up to 0.4 s for the LED to change, which is very slow in terms of embedded systems.

Interrupt-driven systems can have much quicker response rates to switch presses, because the response to the digital input can take place while other tasks are running. However, when a system can respond very rapidly to a switch change, a new issue arises which needs addressing, called *switch bounce*. This is due to the fact that the mechanical contacts of a switch literally bounce together as the switch closes. This can cause a digital input to swing wildly between Logic 0 and Logic 1 for a short time after a switch closes, as illustrated in Figure 9.8. The solution to switch bounce is a technique called *switch debouncing*.

First, the problem with switch bounce can be identified by evaluating a simple event-driven program. Program Example 9.11 attaches a function to a digital interrupt on pin 5, so the circuit build shown in Figure 9.3 can again be used. The function simply toggles (flips) the state of the mbed's onboard LED1 for every raising edge on pin 18.

```
/* Program Example 9.11: Toggles LED1 every time p18 goes high. Uses hardware build
shown in Figure 9.3.
*/
#include "mbed.h"
InterruptIn button(p5);    // Interrupt on digital pushbutton input p18
DigitalOut led1(LED1);     // mbed LED1
void toggle(void);         // function prototype

int main() {
  button.rise(&toggle);     // attach the address of the toggle
}                          // function to the rising edge

void toggle() {
   led1=!led1;
}
```

Program Example 9.11 Toggles LED1 every time mbed pin 5 goes high

Implement program 18 with a push-button or SPDT (single-pole, double-throw) type switch on pin 18. Depending a little on the type of switch you use, you will see that the program does not work very well. It can become unresponsive or the button presses can become out of synch with the LED. This demonstrates the problem with switch bounce.

From Figure 9.8 it is easy to see how a single button press or change of switch position can cause multiple interrupts and hence the LED can get out of synch with the button. The switch can be 'debounced' with a timer feature. The debounce feature needs to ensure that once the raising edge has been seen, no further raising edge interrupts should be implemented until a calibrated time period has elapsed. In reality, some switches move positions more cleanly than others, so the exact timing required needs some tuning. To assist, switch manufacturers often provide data on switch bounce duration. The downside to including debouncing of this type is that the implemented timing period also reduces the response of the switch, although not as much as that discussed with reference to polling.

Switch debouncing can be implemented in a number of ways. In hardware, there are little configurations of logic gates that can be used (see Reference 5.1). In software, timers, bespoke counters or other programming methods can be used. Program Example 9.12 solves the switch bounce issue by starting a timer on a switch event and ensuring that 10 ms has elapsed before allowing a second event to be processed.

```
/* Program Example 9.12: Event driven LED switching with switch debounce
*/
#include "mbed.h"
InterruptIn button(p18);      // Interrupt on digital pushbutton input p18
DigitalOut led1(LED1);        // digital out to LED1
Timer debounce;               // define debounce timer
void toggle(void);            // function prototype
int main() {
  debounce.start();
  button.rise(&toggle);        // attach the address of the toggle
```

```
}                            // function to the rising edge
void toggle() {
if (debounce.read_ms()>10)   // only allow toggle if debounce timer
  led1=!led1;                 // has passed 10 ms
  debounce.reset();           // restart timer when the toggle is performed
}
```

Program Example 9.12 Event-driven LED switching with switch debounce

Exercise 9.12

1. Experiment with modifying the debounce time to be shorter or greater values. There comes a point where the timer does not effectively solve the debouncing, and at the other end of the scale responsiveness can be reduced too. What is the best debounce time for the switch you are using?
2. Implement Program Example 9.12 using an mbed Timeout object instead of the Timer object. Are there any advantages or disadvantages to each approach?
3. Rewrite Program Example 3.3 to do the same thing but with an event-driven approach (i.e. using interrupts). How much improvement can be made to the system's responsiveness to switch changes?

9.11 Mini-Projects

9.11.1 A Self-Contained Metronome

The metronome described in Section 9.8.1 is interesting, but it does not result in something that a musician would really want to use. So try revising the program, and its associated build, to make a self-contained battery-powered unit, using an LCD instead of the computer screen to display beat rate. Experiment also with getting a loudspeaker to 'tick' along with the LED. If you succeed in this, then try including the facility to play 'concert A' (440 Hz), or another pitch, to allow the musicians to tune their instruments.

9.11.2 Accelerometer Threshold Interrupt

We met the ADXL345 accelerometer, with its SPI serial interface, in Section 7.3. Although we did not use them at the time, it is interesting to note that the device has two interrupt outputs, as seen in Figure 7.6. These can be connected to an mbed digital input to run an interrupt routine whenever an acceleration threshold is exceeded. For example, the accelerometer might be a crash detection sensor in a vehicle which, when a specified acceleration value is exceeded, activates an airbag.

Use an accelerometer on a cantilever arm to provide the acceleration data. Figure 9.9 shows the general construction. A plastic 30 cm or 1 foot ruler, clamped at one end to a table, can be

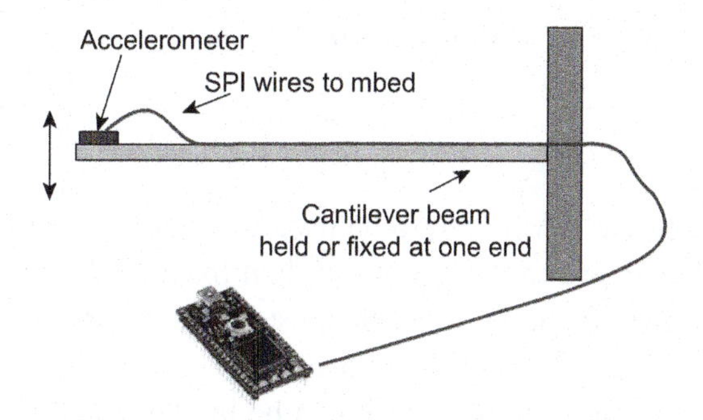

Figure 9.9:
Cantilever beam held or fixed one end

used. Set the accelerometer to generate an interrupt whenever a threshold in the z-axis is exceeded. Connect this as an interrupt input to the mbed, and program it so a warning LED lights for one second whenever the threshold is exceeded. You can experiment with the actual threshold value to alter the sensitivity of the detection system.

Chapter Review

- Signal inputs can be repeatedly tested in a loop, a process known as polling.
- An interrupt allows an external signal to interrupt the action of the CPU, and start code execution from somewhere else in the program.
- Interrupts are a powerful addition to the structure of the microprocessor. Generally, multiple interrupt inputs are possible, which adds considerably to the complexity of both hardware and software.
- It is easy to make a digital counter circuit, which counts the number of logic pulses presented at its input. Such a counter can be integrated into a microcontroller structure.
- Given a clock signal of known and reliable frequency, a counter can readily be used as a timer.
- Timers can be structured in different ways so that interrupts can be generated from their output, for example to give a continuous sequence of interrupt pulses.
- Switch debounce is required in many cases to avoid multiple responses being triggered by a single switch press.

Quiz

1. Explain the differences between using polling and event-driven techniques for testing the state of one of the digital input pins on a microcontroller.
2. List the most significant actions that a CPU takes when it responds to an enabled interrupt.

3. Explain the following terms with respect to interrupts:
   (a) priority
   (b) latency
   (c) nesting.
4. A comparator circuit and LM35 are to be used to create an interrupt source, using the circuit of Figure 9.5. The comparator is supplied from 5.0 V and the temperature threshold is to be approximately 38°C. Suggest values for R_1 and R_2. Resistor values of 470, 680, 820, 1k, 1k2, 1k5 and 10k are available.
5. Describe in overview how a timer circuit can be implemented in hardware as part of a microprocessor's architecture.
6. What is the maximum value, in decimal, that a 12-bit and a 24-bit counter can count up to?
7. A 4.0 MHz clock signal is connected to the inputs of a 12-bit and a 16-bit counter. Each starts counting from zero. How long does it take before each reaches its maximum value?
8. A 10-bit counter, clocked with an input frequency of 512 kHz, runs continuously. Every time it overflows, it generates an interrupt. What is the frequency of that interrupt stream?
9. What is the purpose of the mbed's Real Time Clock? Give an example of when it might be used.
10. Describe the issue of switch bounce and explain how timers can be used to overcome this.

CHAPTER 10

Memory and Data Management

Chapter Outline

10.1 A Memory Review

10.1.1 Memory Function Types

Broadly speaking, a microprocessor needs memory for two reasons: to hold its program, and to hold the data that it is working with; these are often called *program memory* and *data memory*.

To meet these needs a number of different semiconductor memory technologies is available, which can be embedded on the microcontroller chip. Memory technology is divided broadly into two types: volatile and non-volatile. Non-volatile memory retains its data when power is removed, but tends to be more complex to write to in the first place. For historical reasons it is still often called ROM (read only memory). Non-volatile memory is generally

Fast and Effective Embedded Systems Design. DOI: 10.1016/B978-0-08-097768-3.00010-6

required for program memory, so that the program data is there and ready when the processor is powered up. Volatile memory loses all data when power is removed, but is easy to write to. Volatile memory is traditionally used for data memory; it is essential to be able to write to memory easily, and there is little expectation for data to be retained when the product is switched off. For historical reasons it is often called RAM (random access memory), although this terminology tells us little that is useful. These categorizations of memory, however, give an over-simplified picture. It can be useful to change the contents of program memory, and there are times when we want to save data long term. Moreover, new memory technologies now provide non-volatile memory that is easy to write to.

10.1.2 Essential Electronic Memory Types

In any electronic memory we want to be able to store all the 1s and 0s that make up our data. There are several ways that this can be done, and a few essential ones are outlined here.

A simple one-bit memory is a coin. It is stable in two positions, with either 'heads' facing up, or 'tails'. We can try to balance the coin on its edge, but it would pretty soon fall over. The coin is stable in two states, and we call this *bistable*. It could be said that 'heads' represents Logic 1 and 'tails' Logic 0. With eight coins, an 8-bit number can be represented and stored. If we had 10 million coins, we could store the data that makes up one photograph of good resolution, but that would take up a lot of space indeed!

There are various electronic alternatives to the coin, which take up much less space. One is to use an electronic bistable (or 'flip-flop') circuit, as shown in Figure 10.1. The two circuits of

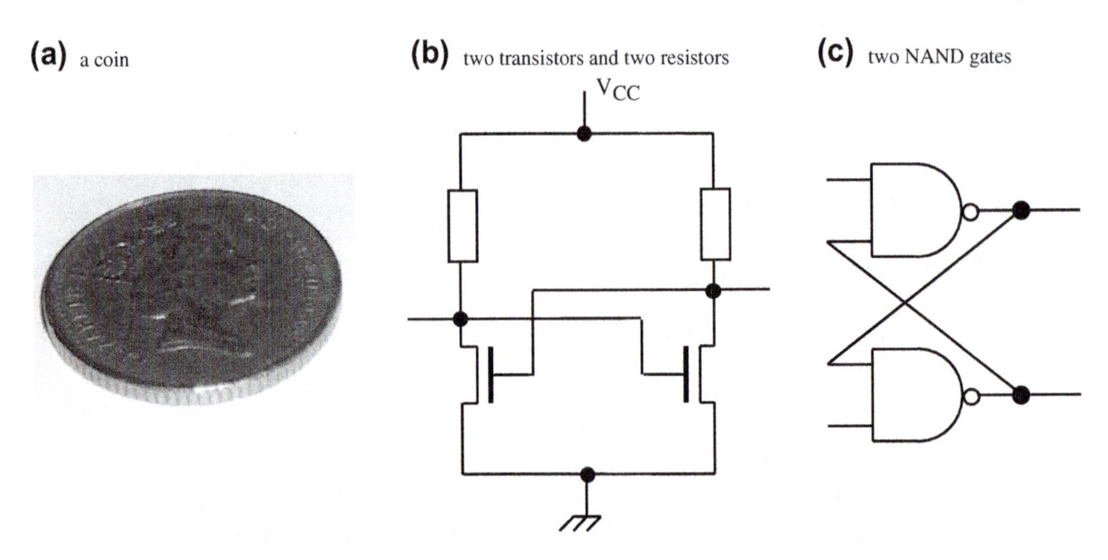

Figure 10.1:
Three ways of implementing a one-bit memory unit

Figure 10.1b and c are stable in only two states, and each can be used to store one bit of data. Circuits like these have been the bedrock of volatile memory.

Looking at the bigger picture, there are several different types of volatile and non-volatile memory, as shown in Figure 10.2 and described in some detail in References 1.1 and 10.1. Static random access memory (SRAM) consists of a vast array of memory cells based on the circuit of Figure 10.1b. These have to be addressable, so that just the right group of cells is written to, or read from, at any one time. To make this possible, two extra transistors are added to the two outputs. To reduce the power consumption, the two resistors are usually replaced by two transistors also. That means six transistors per memory cell. Each transistor takes up a certain area on the integrated circuit (IC), so when the circuit is replicated thousands or millions of times, it can be seen that this memory technology is not actually very space efficient. Despite that, it is of great importance; it is low power, can be written to and read from with ease, can be embedded onto a microcontroller, and hence forms the standard way of implementing data memory in most embedded systems. All data is lost when power is removed.

Dynamic random access memory (DRAM) is intended to do the same thing as SRAM with a reduced silicon area. Instead of using a number of transistors, one bit of information is stored in a tiny capacitor, like a small rechargeable battery. Such capacitors can be fabricated in large numbers on an IC. In order to select the capacitor for reading or writing, a simple transistor switch is required. Unfortunately, owing to the small capacitors and leakage currents on the chip, the memory loses its charge over a short period of time (around 10 to 100 ms). So the DRAM needs to be accessed every few milliseconds to refresh the charges, otherwise the information is lost. DRAM has about four times larger storage capacity than SRAM at about the same cost and chip size, with a compromise of the extra work involved in regular refreshing. It is, moreover, power hungry, so inappropriate for any

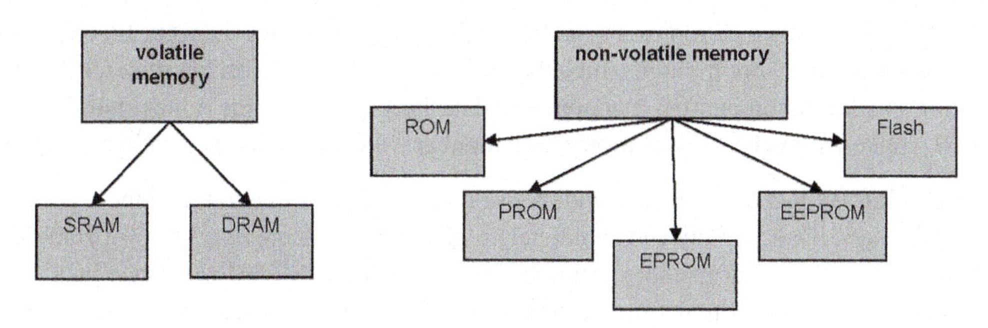

Figure 10.2:
Electronic memory types. SRAM: static random access memory; DRAM: dynamic random access memory; ROM: read only memory; PROM: programmable ROM; EPROM: electrically programmable ROM; EEPROM: electrically erasable and programmable ROM

battery-powered device. DRAM has found wide application as data memory in mains-powered computers, such as the PC.

The original ROMs and programmable read only memories (PROMs) could only ever be programmed once, and have now completely disappeared from normal usage. The first type of non-volatile reprogrammable semiconductor memory, electrically programmable read only memory (EPROM), represented a huge step forward – a non-volatile memory could now be reprogrammed. With a process called *hot electron injection* (HEI), electrons can be forced through a very thin layer of insulator, onto a tiny conductor embedded within the insulator, and can be trapped there almost indefinitely. This conductor is placed so that it interferes with the action of a field effect transistor (FET). When it is charged/discharged, the action of the FET is disabled/enabled. This modified FET effectively becomes a single memory cell, far more dense than the SRAM discussed previously. Moreover, the memory effect is non-volatile; trapped charge is trapped charge! This programming does require a comparatively high voltage (around 25 V), so generally needs a specialist piece of equipment. The memory is erased by exposure to intense ultraviolet light; EPROMs can always be recognized by the quartz window on the IC, which allows this to happen.

The next step beyond HEI was *Nordheim Fowler Tunneling*. This requires even finer memory cell dimensions, and gives a mechanism for the trapped charge to be retrieved electrically, which had not previously been possible. With electrically erasable and programmable read only memory (EEPROM), words of data are individually writeable, readable and erasable, and are non-volatile. The downside of this is that more transistors are needed to select each word. In many cases this flexibility is not required. A revised internal memory structure led to *flash* memory; in this, the ability to erase individual words is not available. Whole blocks have to be erased at any one time, 'in a flash'. This compromise leads to a huge advantage: flash memory is very high density indeed, more or less the highest available. This memory type has been a key feature of many recent products that we have now become used to, like digital cameras, memory sticks, solid-state hard drives and so on. A curious feature of flash and EEPROM, unlike most electronics, is that they exhibit a wear-out mechanism. Electrons can get trapped in the insulator through which they are forced when a write operation takes place. Therefore this limitation is often mentioned in datasheets, for example a maximum of 100 000 write/erase cycles. This is a very high number, and is not experienced in normal use.

Although EPROM had become very widely used, and had been integrated onto microcontrollers, it was rapidly and completely replaced by flash memory. Now in embedded systems, the two dominant memory technologies are flash and SRAM. A glance back at Figures 2.2 and 2.3 shows how important they are to the mbed. Program memory on the LPC1768 is flash, and data memory is SRAM. On the mbed card, the 'USB disk' is a flash IC. The other technology that we are likely to meet at times is EEPROM; this is still used where the ability to rewrite single words of data remains essential.

10.2 Using Data Files with the mbed

Armed with a little knowledge about memory technologies, let's explore how to access and use the mbed memory. The **LocalFileSystem** library allows us to set up a local file system for accessing the mbed flash universal serial bus (USB) disk drive. This allows programs to read and write files on the same disk drive that is used to hold the mbed programs, and which are accessed from the host computer. Once the system has been set up, the standard C/C++ file access functions can be used to open, read and write files.

10.2.1 Reviewing some Useful C/C++ Library Functions

In C/C++ we can open files, read and write data and also scan through files to specific locations, even searching for particular types of data. The functions for input and output operations are all defined by the C Standard Input and Output Library (**stdio.h**), introduced in Section B.9 of Appendix B. Here, we will use the functions summarized in Table 10.1 (this is effectively Table B.6, repeated here for convenience).

Data can be stored in files (as chars) or as words and strings (as character arrays). The mbed provides a mechanism to allow storage in its USB disk, to save data that can be recalled at a later date.

10.2.2 Defining the mbed Local File System

The compiler needs to know where to store and retrieve files; this is done using the mbed **LocalFileSystem** declaration. This sets up the mbed as an accessible flash memory storage unit and defines a directory for storing local files. To implement, simply add the line:

Table 10.1: Useful stdio library functions

Function	Format	Summary action
fopen	`FILE * fopen ( const char * filename, const char * mode);`	Opens the file of name filename
fclose	`int fclose ( FILE * stream );`	Closes a file
fgetc	`int fgetc ( FILE * stream );`	Gets a character from a stream
fgets	`char * fgets ( char * str, int num, FILE * stream );`	Gets a string from a stream
fputc	`int fputc ( int character, FILE * stream );`	Writes a character to a stream
fputs	`int fputs ( const char * str, FILE * stream );`	Writes a string to a stream
fseek	`int fseek ( FILE * stream, long int offset, int origin );`	Moves file pointer to specified location

str: An array containing the null-terminated sequence of characters to be written.
stream: Pointer to a FILE object that identifies the stream where the string is to be written.

```
LocalFileSystem local("local"); //Create local file system named "local"
```

to the declarations section of a program.

10.2.3 Opening and Closing Files

A file stored on the mbed (in this example called '**datafile.txt**') can therefore be opened with the following command:

```
FILE* pFile = fopen("/local/datafile.txt","w");
```

The **fopen( )** function call uses the * operator to assign a pointer with name **pFile** to the file at the specific location given. From here onwards we can access the file by referring to its pointer (**pFile**) rather than having to use the specific filename.

We also need to specify whether we want read or write access to the file. This is done by the 'w' specifier, which is referred to as an *access mode* (in this case denoting *write* access). If the file does not already exist the **fopen( )** function will automatically create it in the specified location. Several other file open access modes, and their specific meanings, are shown in Table B.7 and are elaborated further in Reference 10.2. The three most common access modes are given also in Table 10.2. The *append* access mode (denoted by 'a') is useful for opening an existing file and writing additional data to the end of that file.

When you have finished using a file for reading or writing it is essential to close it, for example using:

```
fclose(pFile);
```

If you fail to do this you might lose all access to the mbed (see Section 10.3.1)!

10.2.4 Writing and Reading File Data

If the intention is to store numerical data, this can be done in a simple way by storing individual 8-bit data values. The **fputc** function allows this, as follows:

```
char write_var=0x0F;
fputc(write_var, pFile);
```

Table 10.2: Common access modes for fopen

Access mode	Meaning	Action
'r'	Read	Opens an existing file for reading
'w'	Write	Creates a new empty file for writing. If a file of the same name already exists it will be deleted and replaced with a blank file
'a'	Append	Appends to a file. Write operations result in data being appended to the end of the file. If the file does not exist a new blank file will be created

This stores the 8-bit variable **write_var** to the data file. The data can also be read from a file to a variable as follows:

```
read_var = fgetc(pFile);
```

Using the **stdio.h** functions, it is also possible to read and write words and strings and search or move through files looking for particular data elements. The C/C++ **fseek( )** function can be used to search through text files.

10.3 Example mbed Data File Access

10.3.1 File Access

Program Example 10.1 creates a data file and writes the arbitrary value 0x23 to that file. The file is saved on the mbed USB disk. The program then opens and reads back the data value and displays it to the screen in a host terminal application.

```
/* Program Example 10.1: read and write char data bytes
*/
#include "mbed.h"
Serial pc(USBTX,USBRX);              // setup terminal link
LocalFileSystem local("local");      // define local file system
int write_var;
int read_var;                        // create data variables

int main () {
  FILE* File1 = fopen("/local/datafile.txt","w");    // open file
    write_var=0x23;                                  // example data
    fputc(write_var, File1);                         // put char (data value) into file
    fclose(File1);                                   // close file

  FILE* File2 = fopen ("/local/datafile.txt","r");   // open file for reading
    read_var = fgetc(File2);                         // read first data value
    fclose(File2);                                   // close file
    pc.printf("input value = %i \n",read_var);       // display read data value
}
```

Program Example 10.1 Saving data to a file

Create a new project and add the code in Program Example 10.1. Run the program, and verify that the data file is created on the mbed and read back correctly. If you navigate to and open the file **datafile.txt** in a standard text editor program (such as Microsoft Wordpad), you should see a hash character (#) in the top left corner. This is because the ASCII character for 0x23 is the hash character (recall Table 8.3).

Exercise 10.1

Change Program Example 10.1 to experiment with other data values; check these against their associated ASCII codes. Write the numbers 1–10 and view them on the screen.

Note that when the microcontroller program opens a file on the local drive, the mbed will be marked as 'removed' on a host PC. This means the PC will often display a message such as 'insert a disk into drive' if you try to access the mbed at this time; this is normal, and stops both the mbed and the PC trying to access the USB disk at the same time. Note also that the USB drive will only reappear when all file pointers are closed in your program, or the microcontroller program exits. If a running program on the mbed does not correctly close an open file, you will no longer be able to see the USB drive when you plug the mbed into your PC. It is therefore important for a programmer to take care when using files to ensure that all files are closed when they are not being used.

If a running program on the mbed does not exit correctly, to allow you to see the drive again (and load a new program), use the following procedure:

1. Unplug the mbed.
2. Hold the mbed reset button down.
3. While still holding the button, plug in the mbed. The mbed USB drive should appear on the host computer screen.
4. Keep holding the button until the new program is saved onto the USB drive.

10.3.2 String File Access

Program Example 10.2 creates a file and writes text data to that file. The file is saved on the mbed. The program then opens and reads back the text data and displays it to the screen in a host terminal application.

```
/* Program Example 10.2: Read and write text string data
*/
#include "mbed.h"
Serial pc(USBTX,USBRX);              // setup terminal link
LocalFileSystem local("local");      // define local file system
char write_string[64];               // character array up to 64 characters
char read_string[64];                // create character arrays (strings)

int main () {
  FILE* File1 = fopen("/local/textfile.txt","w");      // open file access
  fputs("lots and lots of words and letters", File1);  // put text into file
  fclose(File1);                                       // close file

  FILE* File2 = fopen ("/local/textfile.txt","r");     // open file for reading
  fgets(read_string,256,File2);                        // read first data value
  fclose(File2);                                       // close file
  pc.printf("text data: %s \n",read_string);           // display read data string
}
```

Program Example 10.2 Saving a string to a file

Compile and run the program, and verify that the text file is created and read back correctly – if you open the file **textfile.txt**, found on the mbed, the correct text data should be found within.

When reading data from a file, the file pointer can be moved with the **fseek( )** function. For example, the following command will reposition the file pointer to the 8th byte in the text file:

```
fseek (File2 , 8 , SEEK_SET ); // move file pointer to byte 8 from the start
```

The **fseek( )** function needs three input terms; first, the name of the file pointer; secondly, the value to offset the file pointer to; and thirdly, an 'origin' term which tells the function where exactly to apply the offset. The term **SEEK_SET** is a predefined origin term (defined in the **stdio** library) which ensures that the 8-byte offset is applied from the start of the file.

Exercise 10.2

Add the following **fseek( )** statement to Program Example 10.2 just prior to the data being read back:

```
fseek (File2 , 8 , SEEK_SET ); // move file pointer to byte 8
```

Verify that Tera Term only displays the data after byte 8 of the data file. Remember byte values increment from zero, so it will actually be after the 9th character in the file.

10.3.3 Using Formatted Data

It is possible to format data stored in a file. This can be done with the **fprintf( )** function, which has very similar syntax to **printf( )**, except that the filename pointer is also required. We may want, for example, to log specific events to a data file and include variable data values such as time, sensor input data and output control settings. Program Example 10.3 shows use of the **fprintf( )** function in an interrupt controlled push-button project. Each time the push-button is pressed, the light-emitting diode (LED) toggles and changes state. Also on each button press, the file **log.txt** is updated to include the time elapsed since the previous button press, and the current LED state. Program Example 10.3 also implements a simple debounce timer (as described in Section 9.10) to avoid multiple interrupts and file write operations.

```
/* Program Example 10.3: Interrupt toggle switch with formatted data logging to text
file
*/

#include "mbed.h"
InterruptIn button(p30);          // Interrupt on digital input p30
DigitalOut led1(LED1);            // digital out to onboard LED1
Timer debounce;                   // define debounce timer
LocalFileSystem local("local");   // define local file system
void toggle(void);                // function prototype

int main() {
```

```
 debounce.start();                 // start debounce timer
 button.rise(&toggle);             // attach the toggle function to the rising edge
}
void toggle() {                    // perform toggle if debounce time has elapsed
    if (debounce.read_ms()>200)
    led1=!led1;                                        // toggle LED
    FILE* Logfile = fopen ("/local/log.txt","a"); // open file for appending
    fprintf(Logfile,"time=%.3fs: setting led=%d\n\r",debounce.read(),led1.read());
    fclose(Logfile);                                   // close file
    debounce.reset();                                  // reset debounce timer
  }
  }
```

Program Example 10.3 Push-button LED toggle with formatted data logging

Note that the text file **log.txt** may not display full formatting in a simple text viewer such as Microsoft Notepad. If line breaks are not displaying correctly then try a more advanced text file viewer such as Microsoft Wordpad.

Exercise 10.3

Create a program which prompts the user to type some text data into a terminal application. When the user presses return the text is captured and stored in a file on the mbed.

Ensure that the data is correctly written to the data file by opening it with a standard text viewer program.

10.4 Using External Memory with the mbed

A flash SD (secure digital) card can be used with the mbed via the serial peripheral interface (SPI) protocol, as described in Reference 10.3. Using a micro SD card with a card holder cradle (as shown in Figure 10.3), it is possible to access the SD card as an external memory.

Figure 10.3:
An SD card with holder. *(Image reproduced with permission of SparkFun Electronics)*

Table 10.3: Connections for SPI access to the SD card

MicroSD breakout	mbed pin
CS	8 (DigitalOut)
DI	5 (SPI mosi)
Vcc	40 (Vout)
SCK	7 (SPI sclk)
GND	1 (GND)
DO	6 (SPI miso)
CD	No connection

Table 10.4: Library files and import paths for implementing the SD card interface

Library	Import path
SDFileSystem	http://mbed.org/users/simon/programs/SDFileSystem/5yj8f
FATFileSystem	http://mbed.org/projects/libraries/svn/FATFileSystem/trunk?rev=29

The SD card can be configured into SPI communication mode, which requires the serial connections described in Table 10.3. This example uses the mbed SPI port on pins 5, 6 and 7 and an arbitrary digital output on pin 8 to act as the SPI chip select signal.

To implement the SD card interface, it is necessary to import the mbed libraries shown in Table 10.4.

Having imported the described files and libraries, and connected the SD card as suggested, Program Example 10.4 writes a test text file to the card.

```
/* Program Example 10.4: writing data to an SD card
*/
#include "mbed.h"
#include "SDFileSystem.h"
SDFileSystem sd(p5, p6, p7, p8, "sd");            // MOSI, MISO, SCLK, CS
Serial pc(USBTX, USBRX);

int main() {
  FILE *File = fopen("/sd/sdfile.txt", "w");      // open file
  if(File == NULL) {                              // check for file pointer
    pc.printf("Could not open file for write\n");  // error if no pointer
  }
  else{
    pc.printf("SD card file successfully opened\n"); // if pointer ok
  }
  fprintf(File, "Here's some sample text on the SD card");  // write data
  fclose(File);                                             // close file
}
```

Program Example 10.4 Writing data to an SD card

Compile Program Example 10.4 and verify that the SD card is correctly accessed by viewing the created text file **sdfile.txt** in a standard text editor program.

Program Example 10.4 requires an enhanced definition statement for the **SDFileSystem** object, which in this case is named **sd**. Within the interface definition, `SDFileSystem sd(p5, p6, p7, p8, "sd")`, we see that not only are the SPI interface connection pins defined, but the name of the **SDFileSystem** object is also defined as an input variable in quotation marks. Notice that the line:

```
if(File == NULL) {
```

effectively performs an error check to ensure that the file was opened correctly by the previous **fopen( )** call. If the file pointer **File** has a **NULL** value then it means it has not been created and the **fopen( )** call was not successfully implemented.

Exercise 10.4

Create a program which records 5 seconds of analog data to the screen and to a text file on an SD card. Use a potentiometer to generate the analog input data with sample period of 100 ms. Ensure that your data file records the elapsed time and voltage data.

You can then open your data file from within a standard spreadsheet application, such as Microsoft Excel. This will enable you to plot a chart and to visualize the analog data.

10.5 Introducing Pointers

Pointers are used in C/C++ to indicate where a particular element or block of data is stored in memory. Pointers are also discussed in Section B.8.2. We will look here in a little more detail at a specific mbed example using functions and pointers.

When a pointer is defined it can be set to a particular memory address and C/C++ syntax allows the data at that address to be accessed. Pointers are required for a number of reasons; one is because the C/C++ standard does not allow us to pass arrays of data to and from functions, so we must use pointers instead. In some programming languages (such as Matlab), it is possible to develop a mean average calculation function which reads in a data array, calculates the average from the sum of the data divided by the number of data values and returns the calculated mean. Using the mbed, however, we cannot pass the array of data into the function, so instead we need to pass the pointer value which points to the data array.

Pointers are defined similarly to variables but by additionally using the * operator. For example, the following declaration defines a pointer called **ptr** which points to data of type **int**:

```
int *ptr;       // define a pointer which points to data of type int
```

The specific address of a data variable can also be assigned to a pointer by using the & operator, for example:

```
int datavariable=7;   // define a variable called datavariable with value 7
int *ptr;             // define a pointer which points to data of type int
ptr = &datavariable;  // assign the pointer to the address of datavariable
```

In program code the * operator can also be used to get the data from the given pointer address, for example:

```
int x = *ptr;     // get the contents of location pointed to by ptr and
                  // assign to x (in this case x will equal 7)
```

Pointers can also be used with arrays, because an array is really just a number of data values stored at consecutive memory locations. So if the following is defined:

```
int dataarray[]={3,4,6,2,8,9,1,4,6}; // define an array of arbitrary values
int *ptr;                            // define a pointer
ptr = &dataarray[0];                 // assign pointer to the address of
                                     // the first element of the data array
```

the following statements will therefore be true:

```
*ptr == 3;            // the first element of the array pointed to
*(ptr+1) == 4;        // the second element of the array pointed to
*(ptr+2) == 6;        // the third element of the array pointed to
```

So array searching can be done by moving the pointer value to the correct array offset. Program Example 10.5 implements a function for analyzing an array of data and returns the average of that data.

```
/* Program Example 10.5: Pointers example for an array average function
*/
#include "mbed.h"
Serial pc(USBTX, USBRX);                               // setup serial comms
char data[]={5,7,5,8,9,1,7,8,2,5,1,4,6,2,1,4,3,8,7,9}; //define some input data
char *dataptr;              // define a pointer for the input data
float average;              // floating point average variable

float CalculateAverage(char *ptr, char size);    // function prototype

int main() {
  dataptr=&data[0];     // point pointer to address of the first array element
  average = CalculateAverage(dataptr, sizeof(data));   // call function
  pc.printf("\n\rdata = ");
  for (char i=0; i<sizeof(data); i++) {       // loop for each data value
    pc.printf("%d ",data[i]);                 // display all the data values
  }
  pc.printf("\n\raverage = %.3f",average);     // display average value
}
// CalculateAverage function definition and code
float CalculateAverage(char *ptr, char size) {
```

```
  int sum=0;          // define variable for calculating the sum of the data
  float mean;         // define variable for floating point mean value
  for (char i=0; i<size; i++) {
    sum=sum + *(ptr+i);       // add all data elements together
  }
  mean=(float)sum/size;        // divide by size and cast to floating point
  return mean;
}
```

Program Example 10.5 Averaging function using pointers

C code feature

Looking at some key elements of Program Example 10.5, we can see that the pointer **dataptr** is assigned to the address of the first element of the **data** array. The **CalculateAverage( )** function takes in a pointer value that points to the first value of the data array and a second value that defines the size of the array. The function returns the floating point mean value. There is also an additional C/C++ keyword used here, **sizeof**, which deduces the size of a particular array. This gets the size (i.e. the number of data elements) in the array **data**. Also note that the calculation of the mean value is in the form of an integer divided by a char, yet we want the answer to be a floating point value. To implement this, we *cast* the equation as floating point by using **(float)**, which ensures that the resultant mean value is to floating point precision.

Pointers are used for a number of reasons, especially in C++ programs which rely heavily on functions, methods and passing data between these. Pointers can also be used to improve programming efficiency and speed, by directly accessing memory locations and data. In general, the use of pointers in the programming seen in this book is required because of a deficiency in the C/C++ capabilities, i.e. not being able to pass arrays to functions. Where possible, we try to avoid the use of pointers in order to keep code simple and readable, but sometimes they offer the best solution to a programming challenge.

10.6 Mini-Project: Accelerometer Data Logging on Exceeding Threshold

In this mini-project you are challenged to create a program that records the acceleration profile encountered in a simple vibrating cantilever. It is then possible to plot the acceleration data as shown in Figure 10.4. This project develops from the mini-project in Section 9.11. Design and implement a new project to the following specification:

1. Enable the mbed to access external SD card memory.
2. Attach a SPI accelerometer to a simple plastic cantilever with a flying lead to the mbed.
3. Program the accelerometer to cause an interrupt trigger when excessive acceleration is encountered.
4. Create an interrupt routine to log 100 data samples to a file on the SD card. You may wish to use the **fprintf( )** function to format accelerometer data in the text file.

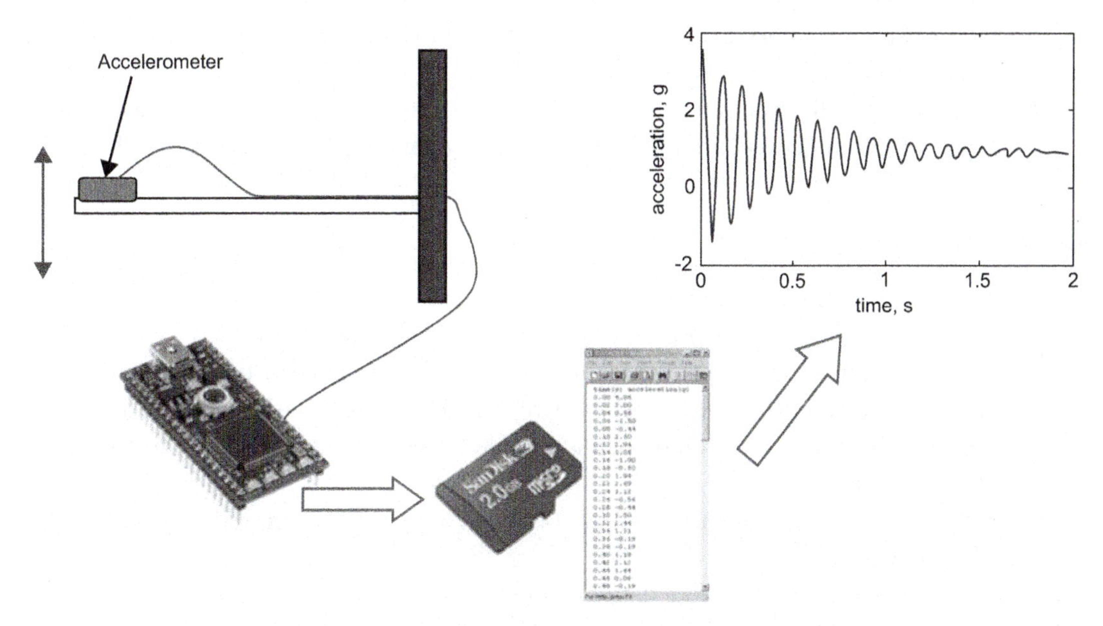

Figure 10.4:
Accelerometer data-logging mini-project

5. When a certain acceleration threshold is exceeded, the data should be logged. The text file can then be opened in a spreadsheet software program, such as Microsoft Excel, to plot the recorded acceleration waveform.

Note: to achieve a suitable sample period you may want to apply an mbed Ticker or Timer to ensure that the accelerometer logs data regularly. A sampling frequency of around 50 Hz should be sufficient to record a detailed acceleration waveform.

Chapter Review

- Microprocessors use memory for holding the program code (program memory) and the working data (data memory) in an embedded system.
- A coin or a logic flip-flop/bistable can be thought of as a single 1-bit memory device which retains its state until the state is actively changed.
- Volatile memory loses its data once power is removed, whereas non-volatile can retain memory with no power. Different technologies are used to realize these memory types, including SRAM and DRAM (volatile) and EEPROM and flash (non-volatile).
- The mbed has 512 kb of flash and 64 kb of SRAM in the LPC1768 IC, and a further 16-megabit USB memory area.
- Files can be created on the mbed for storing and retrieving data and formatted text.
- An external SD memory card can be interfaced with the mbed to allow larger memory.

- Pointers point to memory address locations to allow direct access to the data stored at the pointed location.
- Pointers are generally required owing to the fact that C/C++ does not allow arrays of data to be passed into functions, so a pointer to the array data must be passed instead.

Quiz

1. What does the term *bistable* mean?
2. How many bistables would you expect to find in the mbed's SRAM?
3. What are the fundamental differences between SRAM and DRAM type memory?
4. What are the fundamental differences between EEPROM and flash type memory?
5. Which C/C++ command would open a text file for adding additional text to the end of the current file.
6. Which C/C++ command should be used to open a text file called 'data.txt' and read the 12th character?
7. Give a practical example where data logging is required and explain the practical requirements with regard to timing, memory type and size.
8. Give one reason why pointers are used for direct manipulation of memory data.
9. Give the C/C++ code that defines an empty five-element array called *dataarray* and a pointer called *datapointer* which is assigned to the first memory address of the data array.
10. How is a pointer used to access and manipulate the different elements of a data array?

References

10.1. Grindling, G. and Weiss, B. (2007). Introduction to Microcontrollers. https://ti.tuwien.ac.at/ecs/teaching/courses/mclu/theory-material/Microcontroller.pdf

10.2. C++ stdio.h fopen reference. http://www.cplusplus.com/reference/clibrary/cstdio/fopen/

10.3. SD Group (Matsushita Electric Industrial Co. and SanDisk Corporation) (2006). SD Physical Layer Simplified Specification, Version 2.00. http://www.sdcard.org/developers/tech/sdcard/pls/Simplified_Physical_Layer_Spec.pdf

CHAPTER 11

An Introduction to Digital Signal Processing

Chapter Outline

11.1 What is a Digital Signal Processor?

Digital signal processing (DSP) refers to the computation of mathematically intensive algorithms applied to data signals, such as audio signal manipulation, video compression, data coding/decoding and digital communications. A digital signal processor, also informally called a DSP chip, is a special type of microprocessor used for DSP applications. A DSP chip provides rapid instruction sequences, such as *shift-and-add* and *multiply-and-add* (sometimes called *multiply-and-accumulate* or MAC), which are commonly used in signal processing algorithms. Digital filtering and frequency analysis with the Fourier transform requires many numbers to be multiplied and added together, so a DSP chip provides specific internal

Fast and Effective Embedded Systems Design. DOI: 10.1016/B978-0-08-097768-3.00011-8

hardware and associated instructions to make these operations rapid in operation and easier to code in software.

A DSP chip is therefore particularly suitable for number crunching and mathematical algorithm implementations. It is possible to perform DSP applications with any microprocessor or microcontroller, although specific DSP chips will outperform a standard microprocessor with respect to execution time and code size efficiency.

11.2 Digital Filtering Example

Filters are used to remove chosen frequencies from a signal, as shown in Figure 11.1, for example. Here, there is a signal with both low-frequency and high-frequency components. We may wish to remove either the low-frequency component, by implementing a *high-pass filter* (HPF), or the high-frequency component, by implementing a *low-pass filter* (LPF). A filter

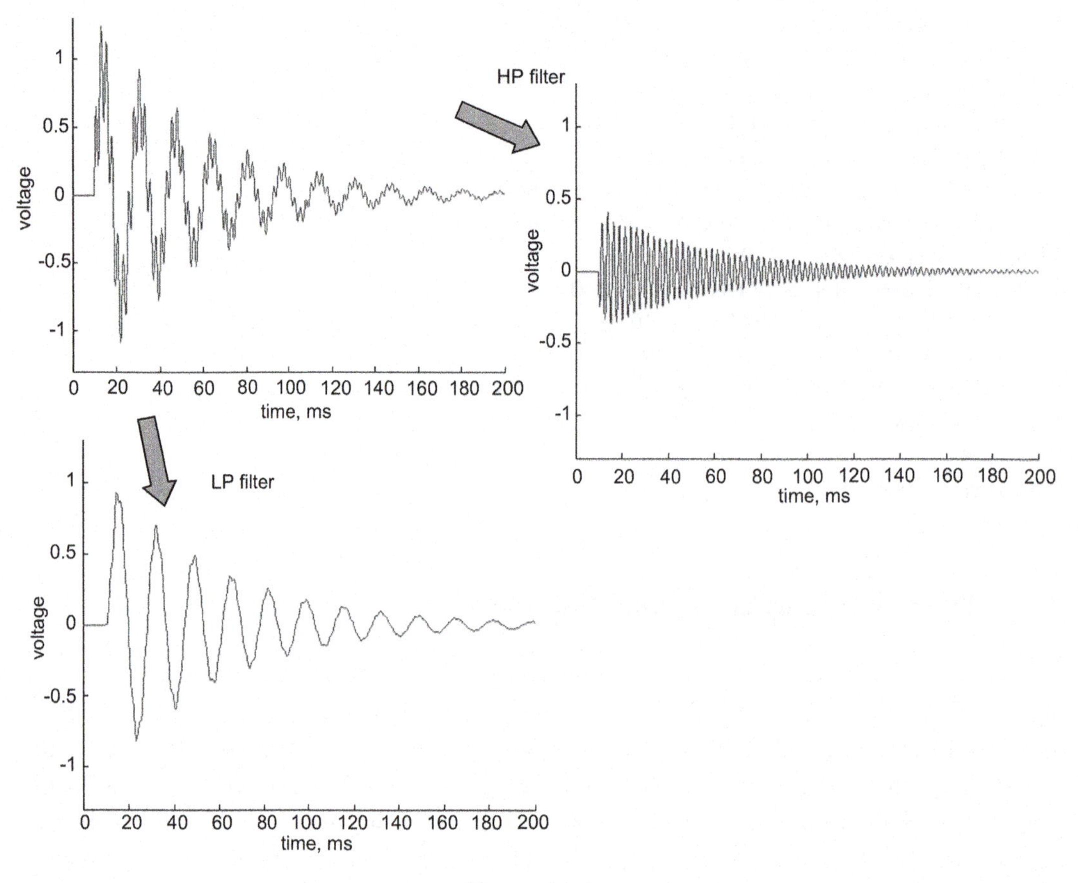

Figure 11.1:
High-pass and low-pass filtered signals

has a *cut-off frequency*, which determines which signal frequencies are in the *pass-band* (and are still evident after the filtering) and which are in the *stop-band* (which are removed by the filtering operation). The cut-off effect is not perfect, however, as frequencies in the stop-band are attenuated more the further away from the cut-off frequency they are. Filters can be designed to have different steepness of cut-off attenuation and so adjust the filter's *roll-off rate*. In general, the more complex the filter design, the steeper its roll-off can be. The filtering can be performed with an active or a passive analog filter, but the same process can also be performed in software with a DSP operation.

We will not go deeply into the mathematics of digital filtering, but the software process relies heavily on addition and multiplication. Figure 11.2 shows an example block diagram for a simple digital filtering operation.

Figure 11.2 shows that coefficient a_1 is multiplied by the most recent sample value. Coefficient a_2 is multiplied by the previous sample and a_3 is multiplied by the sample before that. The values of the *filter taps* determine whether the filter is high pass or low pass and what the cut-off frequency is. The *order* of the filter is given by the number of delays that are used, so the example shown in Figure 11.2 is a third order filter. The order of the filter determines the steepness of the filter roll-off curve; at one extreme a first order filter gives a very gentle roll-off, while at the other an eighth order is used for demanding anti-aliasing applications.

This filter is an example of a *finite impulse response* (FIR) filter, because it uses a finite number of input values to calculate the output value. Finding the required values for the filter taps is a complex process, but many design packages exist to simplify this process. As an

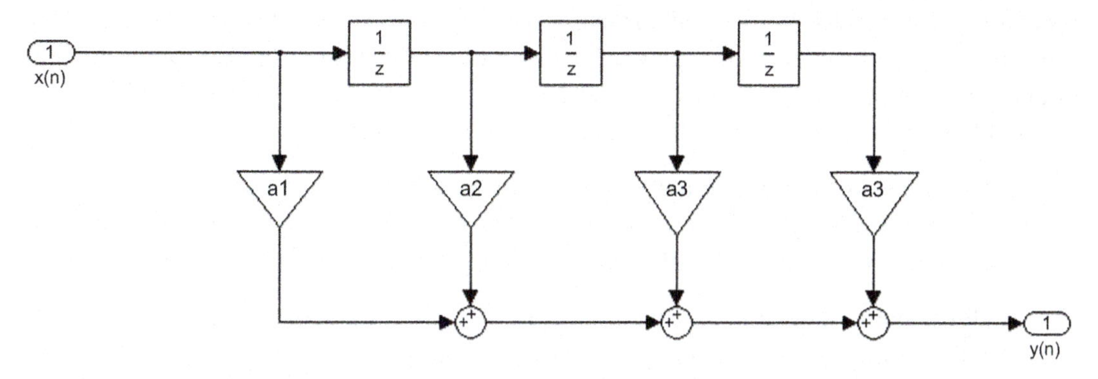

- x(n) is the input signal (the signal to be filtered)
- y(n) is the filtered signal
- a_1-a_4 are multiplication constants (filter coefficients or filter "taps")
- 1/z refers to a one sample delay

Figure 11.2:
Third order FIR filter

example, if the filter taps in Figure 11.2 all have a value of 0.25, then this implements a simple mean average filter (a crude LPF) which continuously averages the last four consecutive data values.

It can be seen that the filter has a number of multiply and addition processes, which can be grouped into a single MAC operation. A DSP chip has a special area of hardware for processing MAC operations. This is specifically designed so that MAC commands can be processed in a single clock cycle, i.e. much faster than on conventional processors. It should be noted that although the mbed is not a dedicated DSP processor, it is still powerful enough to perform many DSP operations.

11.3 An mbed DSP Example

11.3.1 Input and Output of Digital Data

We can develop an mbed program which reads signal data in via the analog-to-digital converter (ADC), processes the data digitally and then outputs the signal via the digital-to-analog converter (DAC). Since DSP algorithms need to sample and process data at regular fixed intervals, we will use the mbed Timer interrupts to enable a scheduled program. We will now write a program that reads audio data and implements low- and high-pass digital filters. In these examples, the data will be output from the mbed's DAC and through some simple electronic circuitry. The resulting audio output can then be routed to a loudspeaker amplifier (such as a set of portable PC speakers) so that you can listen to the processed signal.

First, we need a signal source to input to the mbed. A simple way to create a signal source is to use a host PC's audio output while playing an audio file of the desired signal data. A number of audio packages, such as Steinberg Wavelab, can be used to create wave audio files (with a .wav file extension); here, we will use three audio files as follows:

- 200hz.wav – an audio file of a 200 Hz sine wave
- 1000hz.wav – an audio file of a 1000 Hz sine wave
- 200hz1000hz.wav – an audio file with the 200 Hz and 1000 Hz audio mixed.

Each audio signal should be mono and around 60 seconds in duration. These files are available for download from the book website.

The audio files can be played directly from the host PC into a set of headphones, and the different sine wave signals can be heard. Now we can connect the audio signal to the mbed by connecting the host PC's audio cable to an mbed analog input pin. The signal can also be viewed on an oscilloscope. You will see that the signal oscillates positive and negative about 0 V. This is not much use for the mbed, as it can only read analog data between 0 and 3.3 V, so all negative data will be interpreted as 0 V.

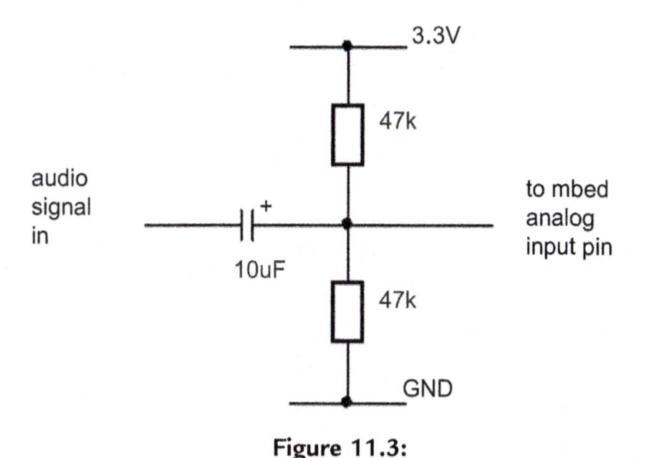

Figure 11.3:
Input circuit with coupling capacitor and bias resistors

Because the mbed can only accept positive voltage inputs, it is necessary to add a small coupling and biasing circuit to offset the signal to a midpoint of approximately 1.65 V. The offset here is often referred to as a direct current (DC) offset. The circuit shown in Figure 11.3 effectively couples the host PC's audio output to the mbed. Create a new project and enter the code of Program Example 11.1.

```
/* Program Example 11.1 DSP input and Output
*/
#include "mbed.h"
//mbed objects
AnalogIn Ain(p15);
AnalogOut Aout(p18);
Ticker s20khz_tick;

//function prototypes
void s20khz_task(void);
//variables and data
float data_in, data_out;

//main program start here
int main() {
  s20khz_tick.attach_us(&s20khz_task,50); // attach task to 50us tick
}

// function 20khz_task
void s20khz_task(void){
  data_in=Ain;
  data_out=data_in;
  Aout=data_out;
}
```

Program Example 11.1 DSP input and output

Program Example 11.1 first defines analog input and output objects (**data_in** and **data_out**) and a single Ticker object called **s20khz_tick**. There is also a function called **s20khz_task( )**.

The **main( )** function simply assigns the 20 kHz Ticker to the 20 kHz task, with a Ticker interval of 50 μs (which sets up the 20 kHz rate). The input is now sampled and processed at this regular rate, within **s20khz_task( )**.

Exercise 11.1

Compile Program Example 11.1 and use a two-channel oscilloscope to check that the analog input signal and the DAC output signal are similar. Use the oscilloscope to see how accurate the DAC output is with respect to the analog input signal for all three audio files. Consider amplitude, phase and the waveform profile.

You will also need to implement the input circuit shown in Figure 11.3.

11.3.2 Signal Reconstruction

If you look closely at the audio signals, particularly the 1000 Hz signal or the mixed signal, you will see that the DAC output has discrete steps. This is more obvious in the high-frequency signal as it is closer to the sampling frequency chosen, as shown in Figure 11.4a.

With many audio DSP systems, the analog output from the DAC is converted to a reconstructed signal by implementing an analog *reconstruction filter*, which removes all steps from the signal, leaving a smooth output. In audio applications, a reconstruction filter is usually designed to be an LPF with a cut-off frequency at around 20 kHz, because the human hearing range does not exceed 20 kHz. The reconstruction filter shown in Figure 11.5 can be implemented with the current project (which gives the complete DSP input/output circuit as shown in Figure 11.6). Note that after the LPF, a *decoupling capacitor* is also added to remove the 1.65 V DC offset from the signal. Once the DC offset has been removed, the signal can be routed to a loudspeaker amplifier to monitor the processed DAC output.

The mathematical theory associated with digital sampling and reconstruction is complex and beyond the scope of this book. As discussed in Chapter 4, in audio sampling systems it is often necessary to add an anti-aliasing filter prior to the analog-to-digital conversion. For simplicity this extra filter has not been implemented here. For those interested, the theory of sampling, aliasing and reconstruction is described well and in detail by a number of authors including Marven and Ewers (Reference 11.1) and Proakis and Manolakis (Reference 11.2).

11.3.3 Adding a Digital Low-Pass Filter

We will add a digital low-pass filter routine to filter out the 1000 Hz frequency component. This can be assigned to a switch input so that the filter is implemented in real time when a push-button is pressed.

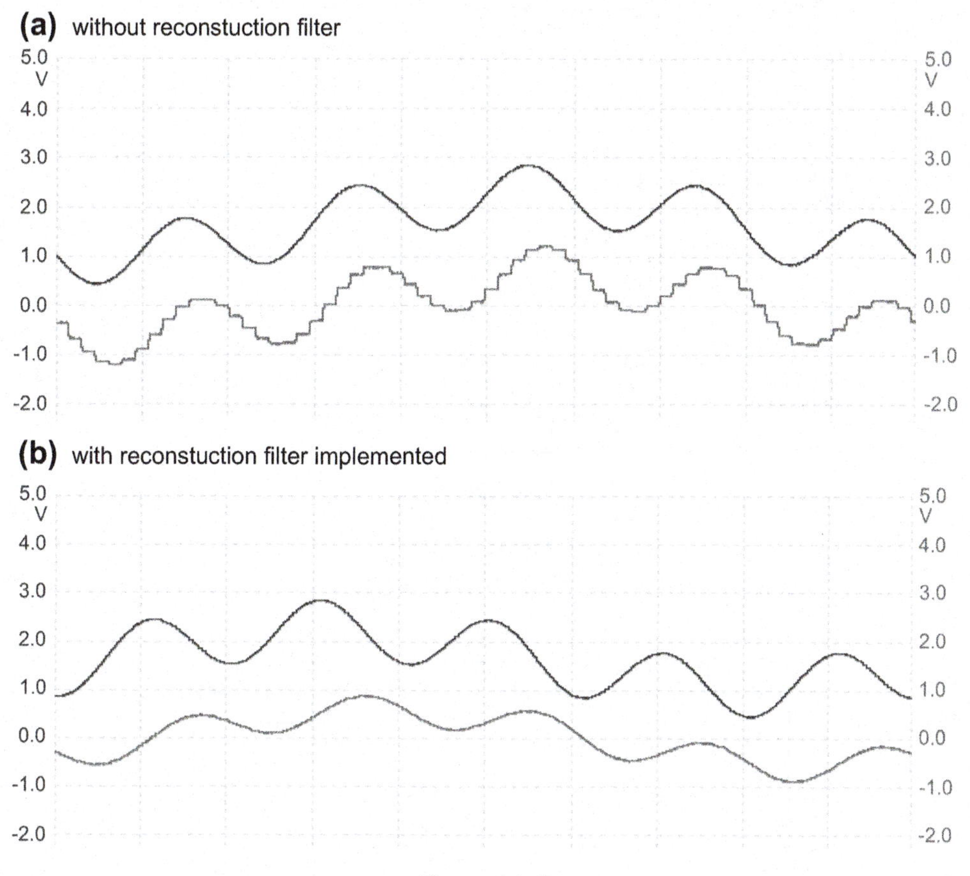

Figure 11.4:
Signal output

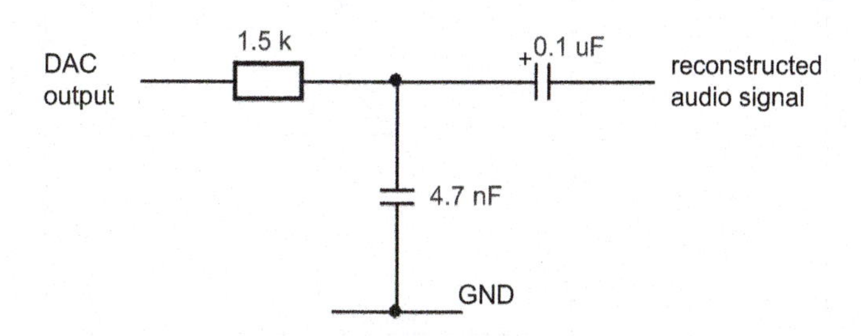

Figure 11.5:
Analog reconstruction filter and decoupling capacitor

This example will use a third order *infinite impulse response* (IIR) filter, as shown in Figure 11.7. The IIR filter uses recursive output data (i.e. data fed back from the output), as well as input data, to calculate the filtered output.

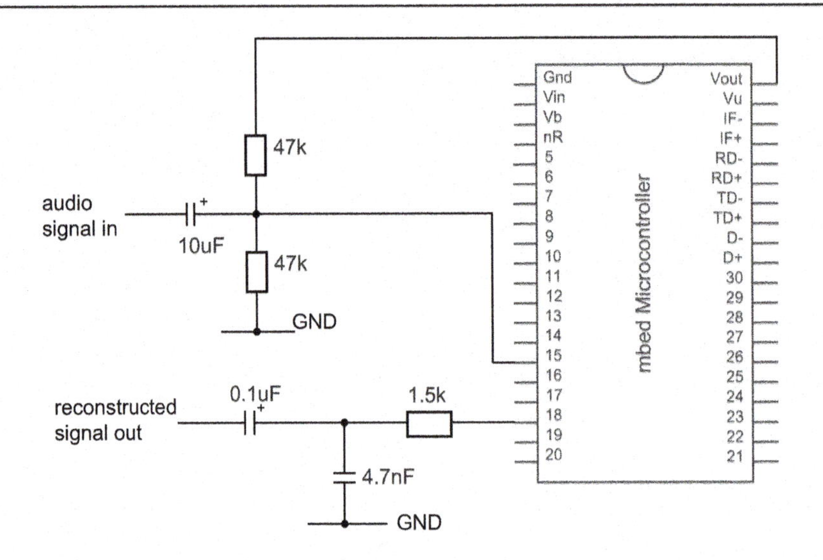

Figure 11.6:
DSP input/output circuit

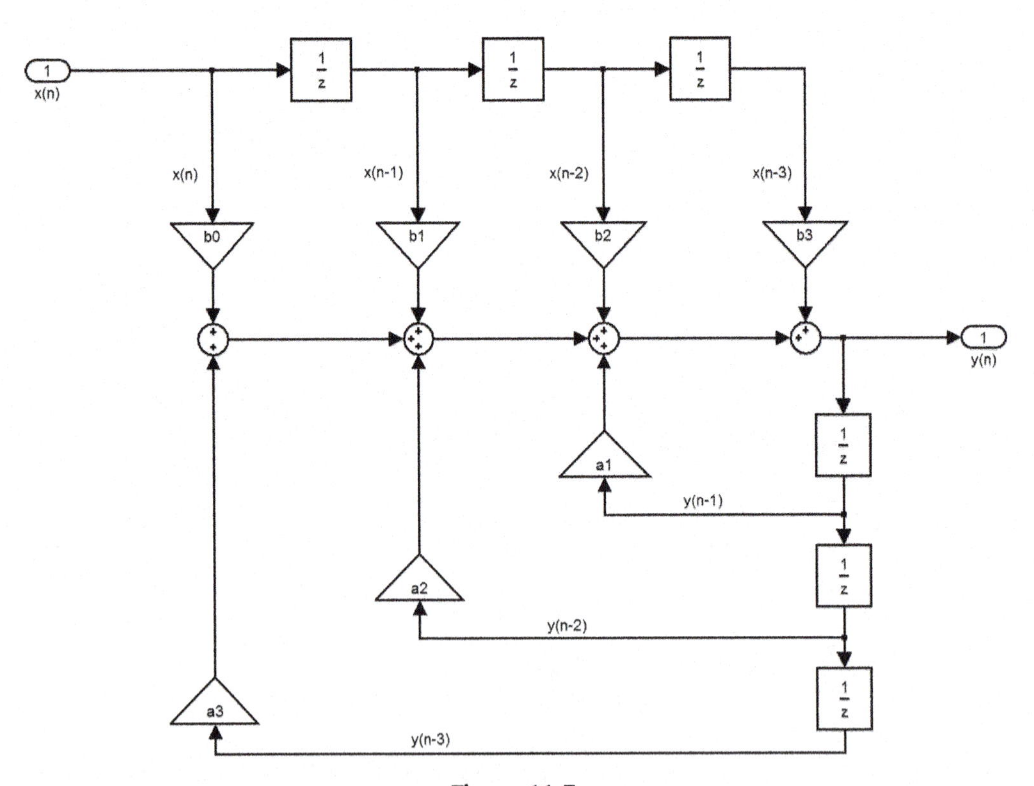

Figure 11.7:
Third order digital IIR filter

This filter results in the following equation for calculating the filtered value given the current input, the previous three input values and the previous three output values:

$$\begin{aligned} y(n) = {} & b_0x(n) + b_1x(n-1) + b_2x(n-2) + b_3x(n-3) \\ & + a_1y(n-1) + a_2y(n-2) + a_3y(n-3) \end{aligned} \quad 11.1$$

where $x(n)$ is the current data input value; $x(n-1)$ is the previous data input value, $x(n-2)$ is the data value before that, and so on; $y(n)$ is the calculated current output value; $y(n-1)$ is the previous data output value $y(n-2)$ is the data value before that, and so on; and a_{0-3} and b_{0-3} are coefficients (filter taps) that define the filter's performance.

We can implement this equation to achieve filtered data from the input data. The challenging task is determining the values required for the filter coefficients to give the desired filter performance. Filter coefficients can, however, be calculated by a number of software packages, such as the Matlab Filter Design and Analysis Tool (Reference 11.3), and those provided in Reference 11.4.

The LPF designed for a 600 Hz cut-off frequency (third order with 20 kHz sampling frequency) can be implemented as a C function as shown in Program Example 11.2. We chose a 600 Hz LPF because the cut-off is exactly half way between the two signal frequencies.

```
/*Program Example 11.2 Low-pass filter function
*/
float LPF(float LPF_in){

  float a[4]={1,2.6235518066,-2.3146825811,0.6855359773};
  float b[4]={0.0006993496,0.0020980489,0.0020980489,0.0006993496};
  static float LPF_out;
  static float x[4], y[4];

  x[3] = x[2]; x[2] = x[1]; x[1] = x[0];      // move x values by one sample
  y[3] = y[2]; y[2] = y[1]; y[1] = y[0];      // move y values by one sample

  x[0] = LPF_in;                              // new value for x[0]
  y[0] =  (b[0]*x[0]) + (b[1]*x[1]) + (b[2]*x[2]) + (b[3]*x[3])
          + (a[1]*y[1]) + (a[2]*y[2]) + (a[3]*y[3]);

  LPF_out = y[0];
  return LPF_out;                             // output filtered value
}
```

Program Example 11.2 Low-pass filter function

Here, we can see the calculated filter values for the **a** and **b** coefficients. These coefficients are deduced by the online calculator provided by Reference 11.4. Note also that we have used the **static** variable type for the internally calculated data values. The static definition ensures that data calculated are held even after the function is complete, so the recursive data values for previous samples are held within the function and not lost during program execution.

Exercise 11.2

Create a new mbed program based on Program Example 11.1, only this time add the function for the LPF seen in Program Example 11.2. Now in the 20 kHz task process the input data by feeding it through the LPF function, for example:

```
data_out=LPF(data_in);
```

Compile and run the code to check that high-frequency components are filtered from the mbed's analog output.

Your system should use the mbed with the input and output circuitry shown in Figure 11.6.

11.3.4 Adding a Push-Button Activation

We can now assign a conditional statement to a digital input, allowing the filter to be switched in and out in real-time. The following conditional statement implemented in the 20 kHz task will allow real-time activation of the digital filter:

```
data_in=Ain-0.5;
  if (LPFswitch==1){
     data_out=LPF(data_in);
  }
  else {
     data_out=data_in;
  }
  Aout=data_out+0.5;
```

You will notice that before performing the calculation, the mean value of the signal is subtracted, in the line:

```
data_in=Ain-0.5;
```

This is to *normalize* the signal to an average value of zero, so the signal oscillates positive and negative and allows the filter algorithm to perform DSP with no DC offset in the data. As the DAC anticipates floating point data in the range 0.0–1.0, we must also add the mean offset back to the data before we output, in the line:

```
Aout=data_out+0.5;
```

Exercise 11.3

Implement the real-time push-button activation in your LPF program. You can now use the oscilloscope or headphones to listen to the signal which includes both the 200 Hz and 1000 Hz signal played simultaneously. When the switch is pressed, the high-frequency component should be removed, leaving just the low-frequency component audible, or visible on the oscilloscope.

■

11.3.5 Digital High-Pass Filter

The implementation of an HPF function is identical to the low-pass function, but with different filter coefficient values. The filter coefficients for a 600 Hz HPF (third order with a 20 kHz sample frequency) are as follows:

```
float a[4]={1,2.6235518066,-2.3146825811,0.6855359773 };
float b[4]={0.8279712953,-2.4839138860,2.4839138860,-0.8279712953};
```

Exercise 11.4

Add a second switch to the circuit and a filter function to the program, to act as an HPF. This switch will enable filtering of the low-frequency component and leave the high-frequency signal. Implement the second function with a second conditional statement in the 20 kHz task, so that either the LPF or HPF can be activated in real time. You should now be able to listen to the simultaneous 200 Hz and 1000 Hz audio signal and filter either the low frequency or the high frequency, or both, dependent on which switch is pressed.

■

11.4 Delay/Echo Effect

Many audio effects use DSP systems for manipulating and enhancing audio. These can be useful for live audio processing (e.g. guitar effects) or for post-production. Audio production DSP effects include artificial reverb, pitch correction, dynamic range manipulation and many other techniques to enhance captured audio.

A feedback delay can be used to make an echo effect, which sees a single sound repeated a number of times. Each time the signal is repeated it is attenuated until it eventually decays away. Therefore, the speed of repetition and the amount of feedback attenuation can be manipulated. This effect is used commonly as a guitar effect, and for vocal processing and enhancement. Figure 11.8 shows a block diagram design for a simple delay effect unit.

To implement the system design shown in Figure 11.8, historical sample data must be stored so that it can be mixed back in with immediate data. Therefore, sampled digital data needs to be copied into a buffer (a large array) so that the feedback data is always available. The feedback gain determines how much buffer data is mixed with the sampled data.

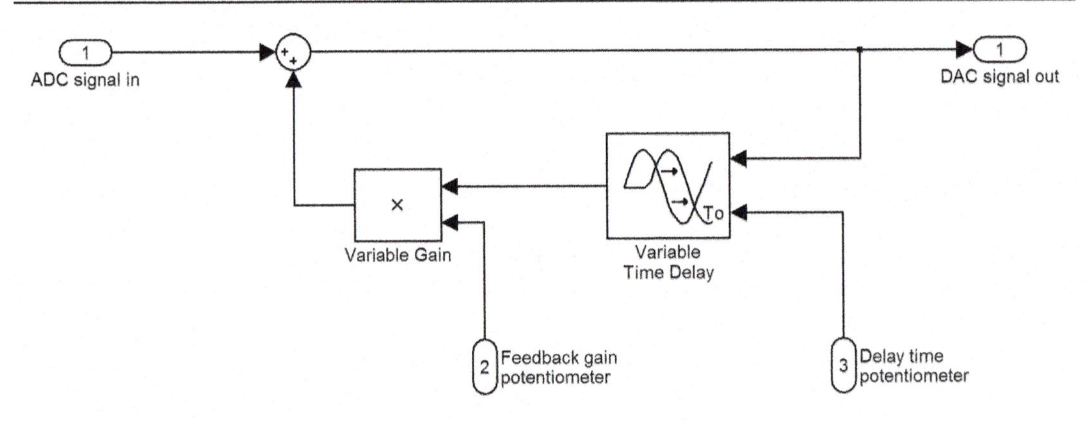

Figure 11.8:
Delay effect block diagram

For each sample coming in, an earlier value is mixed in also, but the length of time between the immediate and historical value can be varied by the delay time potentiometer. Effectively, this changes the size of the data buffer by resetting an array counter based on the value of the delay time input. If the delay time input is large, a large number of array data will be fed back before resetting, resulting in a long delay time. For smaller delay values the array counter will reset sooner, giving a more rapid delay effect.

```
/* Program Example 11.3 Delay / Echo Effect
*/
#include "mbed.h"
AnalogIn Ain(p15);                          //object definitions
AnalogOut Aout(p18);
AnalogIn delay_pot(p16);
AnalogIn feedback_pot(p17);
Ticker s20khz_tick;
void s20khz_task(void);                     // function prototypes
#define MAX_BUFFER 14000                    // max data samples
signed short data_in;                       // signed allows positive and negative
unsigned short data_out;                    // unsigned just allows positive values
float delay=0;
float feedback=0;
signed short buffer[MAX_BUFFER]={0};        // define buffer and set values to 0
int i=0;

//main program start here
int main() {
  s20khz_tick.attach_us(&s20khz_task,50);
}
// function 20khz_task
void s20khz_task(void){
  data_in=Ain.read_u16()-0x7FFF;             // read data and normalize
  buffer[i]=data_in+(buffer[i]*feedback);    // add data to buffer data
  data_out=buffer[i]+0x7FFF;                 // output buffer data value
```

```
    Aout.write_u16(data_out);          // write output
    if (i>(delay)){                    // if delay loop has completed
      i=0;                             // reset counter
      delay=delay_pot*MAX_BUFFER;      // calculate new delay buffer size
      feedback=(1-feedback_pot)*0.9;   // calculate feedback gain value
    }else{
      i=i+1;                           // otherwise increment delay counter
    }
}
```

Program Example 11.3 Delay/echo effect

Notice that the **data_in** and **data_out** values are defined as follows:

```
signed short data_in;        // signed allows positive and negative
unsigned short data_out;     // unsigned just allows positive values
```

The **short** data type (see Table B.4) ensures that the data is constrained to 16-bit values; this can be specified as **signed** (i.e. to use the range −32 768 to 32 767 decimal) or **unsigned** (to use the range 0 to 65535). We need the computation to take place with signed numbers. When outputting to the mbed's DAC, however, this needs an offset adding owing to the fact that DAC is configured to work with unsigned data.

The initial mbed hardware setup shown in Figure 11.6 can be used here. This demonstrates the advantage of a single hardware design that can operate multiple software features in a digital system. We will need to add potentiometers to mbed analog inputs in order to control the feedback speed and gain in real time, as shown in Figure 11.8. Program Example 11.3 defines the delay and feedback potentiometers being connected to mbed analog inputs on pins 16 and 17, respectively.

Exercise 11.5

Implement a new project with the code shown in Program Example 11.3. Initially, you will need to use a test signal which gives a short pulse followed by a period of inactivity to evaluate the echo effect performance. You should see a similar echo response to that shown in Figure 11.9. Verify that the delay and feedback gain potentiometers alter the output signal as expected. For testing purposes, a pulse signal called **pulse.wav** can be downloaded from the book website.

Figure 11.9 shows input and output waveforms for the delay/echo effect. It can be seen that for a single input pulse, the output is a combination of the pulse plus repeated echoes of that pulse with slowly diminishing amplitude. The rate of echo and the rate of attenuation are altered by adjusting the potentiometers as described. This project can be developed as a guitar

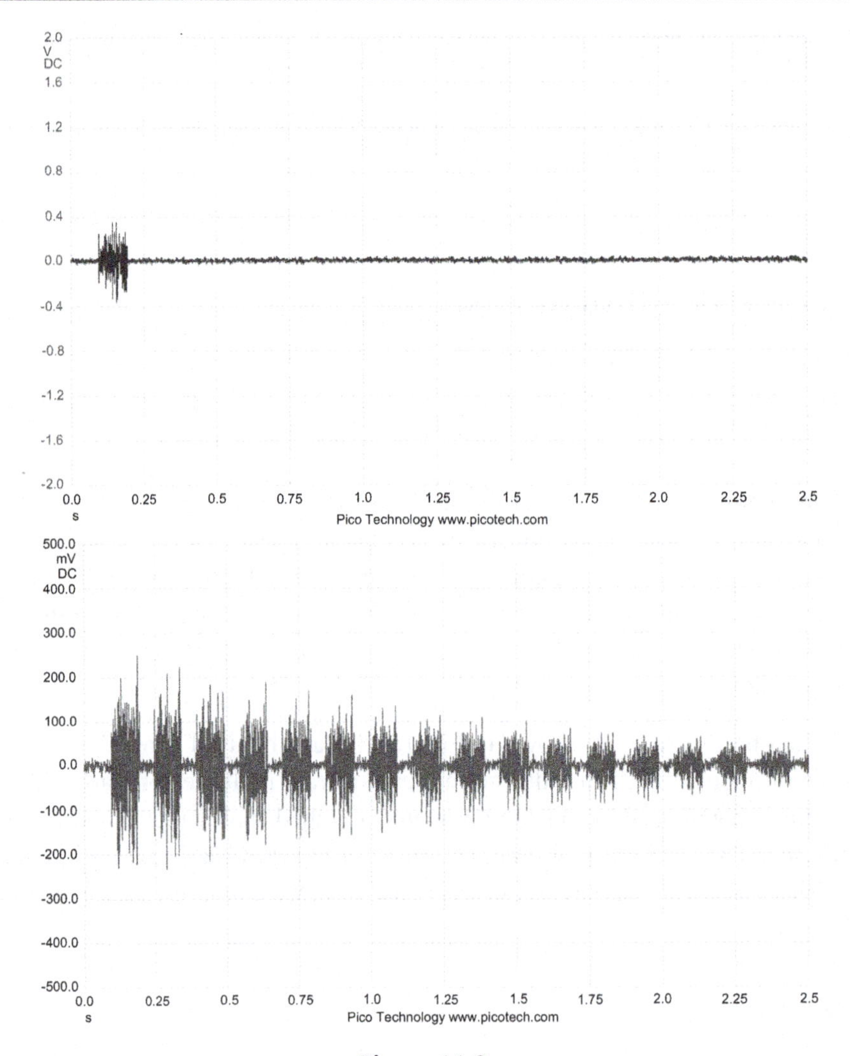

Figure 11.9:
Input signal (top) and output signal (bottom) for the digital echo effect system

effect unit by adding extra signal conditioning, variable amplification stages and enhanced output conditioning. Some good examples are given in Reference 11.5.

11.5 Working with Wave Audio Files

This chapter, so far, has analyzed a DSP application which captures data, manipulates the data and then outputs it. However, a number of signal processing applications rely on data that has previously been captured and stored in data memory. Continuing with examples relating to digital audio, this can be explored by analyzing wave audio files and evaluating the requirements in order to output a continuous and controlled stream of audio data.

11.5.1 The Wave Information Header

There are many types of audio data file, of which the wave (.wav) type is one of the most widely used. Wave files contain a number of details about the audio data followed by the audio data itself. Wave files contain uncompressed (i.e. raw) data, mostly in a format known as *linear pulse code modulation* (PCM). Since PCM specifically refers to the coding of amplitude signal data at a fixed sample rate, each sample value is given to a specified resolution (often 16-bit) on a linear scale. Time data for each sample is not recorded because the sample rate is known, so only amplitude data is stored. The wave header information details the actual resolution and sample frequency of the audio, so by reading this header it is possible to accurately decode and process the contained audio data. The full wave header file description is shown in Table 11.1.

It is possible to identify much of the wave header information simply by opening a .wav file with a text editor application, as shown in Figure 11.10. Here we see the ASCII characters for

Table 11.1: Wave file information header structure

Data name	Offset (bytes)	Size (bytes)	Details
ChunkID	0	4	The characters 'RIFF' in ASCII
ChunkSize	4	4	Details the size of the file from byte 8 onwards
Format	8	4	The characters 'WAVE' in ASCII
Subchunk1ID	12	4	The characters 'fmt' in ASCII
Subchunk1Size	16	4	16 for PCM
AudioFormat	20	2	PCM = 1. Any other value indicates data compressed format
NumChannels	22	2	Mono = 1; stereo = 2
SampleRate	24	4	Sample rate of the audio data in Hz
ByteRate	28	4	ByteRate = SampleRate * NumChannels * BitsPerSample/8
BlockAlign	32	2	BlockAlign = NumChannels * BitsPerSample/8. The number of bytes per sample block
BitsPerSample	34	2	Resolution of audio data
SubChunk2ID	36	4	The characters 'data' in ASCII
Subchunk2Size	40	4	Subchunk2Size = Number of samples * BlockAlign. The total size of the audio data in bytes
Data	44	–	The actual data of size given by Subchunk2Size

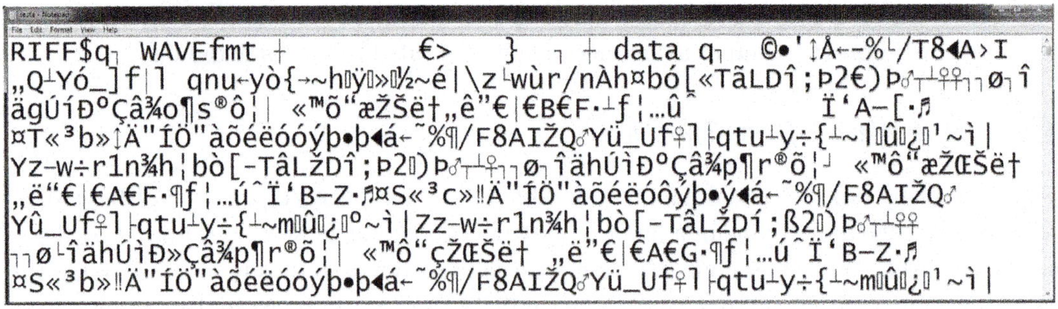

RIFF$q┐ WAVEfmt ┼ €> } ┐ ┼ data q┐ ©•'↕Å←-%└/T8◄A>I
„Q┴Yó_]f│1 qnu←yò{→~h▯ÿ▯»▯½~é|\z└wùr/nÀh¤bó[«TãLDî;Þ2€)Þ♂┬┴♀♀┐┐ø┐î
ägÚíÐ°Çâ¾o¶s®ô¦| «™õ"æŽŠë†„ê"€|€B€F·┴f¦…û^ Ï'A-[·♬
¤T«³b»↕Ä"ÍÖ"àõéëóóýþ•þ◄á←˜%¶/F8AIŽQ♂Yü_Uf♀1├qtu┴y÷{┴~1▯û▯¿▯¹~ì|
Yz-w÷r1n¾h¦bò[-TâLŽDî;Þ2▯)Þ♂┬┴♀┐┐ø┐îähÚìÐ°Çâ¾p¶r®õ¦┘ «™ô"æŽŒŠë†
„ë"€|€A€F·¶f¦…ú^Ï'B-Z·♬¤S«³c»‼Ä"ÍÖ"àõéëóôýþ•ý◄á←˜%¶/F8AIŽQ♂
Yû_Uf♀1├qtu┴y÷{┴~m▯û▯¿▯°~ì|Zz-w÷r1n¾h¦bò[-TâLŽDí;ß2▯)Þ♂┬┴♀♀
┐┐ø└îähÚìÐ»Çâ¾p¶r®õ¦| «™ô"çŽŒŠë† „ë"€|€A€G·¶f¦…ú^Ï'B-Z·♬
¤S«³b»‼Ä"ÍÖ"àõéëóóýþ•þ◄á←˜%¶/F8AIŽQ♂Yü_Uf♀1├qtu┴y÷{┴~m▯û▯¿▯¹~ì|

Figure 11.10:
Wave file opened in text editor

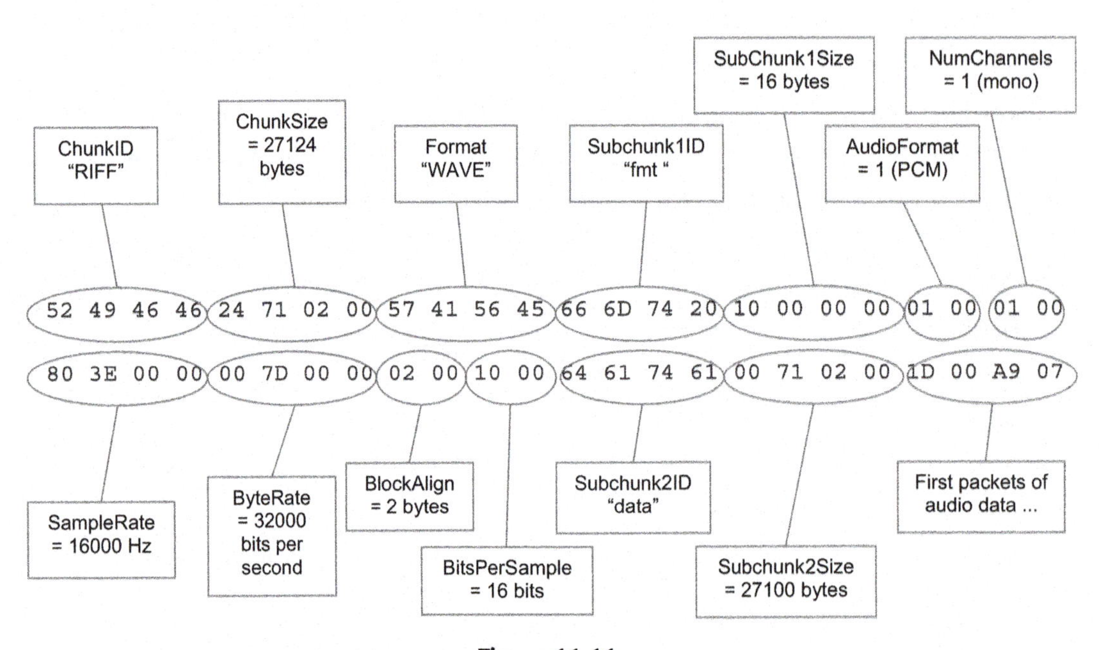

Figure 11.11:
Structure of the wave file header data

ChunkID ('RIFF'), Format ('WAVE'), Subchunk1ID ('fmt') and Subchunk2ID ('data'). These are followed by the ASCII data which makes up the raw audio.

Looking more closely at the header information for this file in hexadecimal format, the specific values of the header information can be identified, as shown in Figure 11.11. Note also that for each value made up of multiple bytes, the least significant byte is always given first and the most significant byte last. For example, the four bytes denoting the sample rate is given as 0x80, 0x3E, 0x00 and 0x00, which gives a 32-bit value of 0x00003E80 = 16 000 decimal.

11.5.2 Reading the Wave File Header with the mbed

In order to accurately read and reproduce wave data via a DAC, we need to interpret the information given in the header data, and configure the read and playback software features with the correct audio format (from **AudioFormat**), number of channels (**NumChannels**), sample rate (**SampleRate**) and the data resolution (**BitsPerSample**).

To implement wave file manipulation on the mbed, the wave data file should be stored on a Secure Digital (SD) memory card and the program should be configured with the correct SD reader libraries, as discussed in Chapter 10. The SD card is required predominantly because wave audio files are generally quite large, much too large to be stored in the mbed's internal memory, but also because the file access through the serial peripheral interface (SPI)/SD card interface is very fast.

Program Example 11.4 reads a wave audio file called **test.wav**, utilizes the **fseek( )** function used to move the file pointer to the relevant position, and then reads the header data and displays this to a host terminal. Note also that to read .wav files on the mbed, you should use the '8.3 filename' convention (sometimes referred to as *short file name* or SFN) for audio files. The SFN convention stipulates that filenames should not be longer than eight characters with a three-character extension.

```
/* Program Example 11.4 Wave file header reader
*/
#include "mbed.h"
#include "SDFileSystem.h"
SDFileSystem sd(p5, p6, p7, p8, "sd");
Serial pc(USBTX,USBRX);              // set up terminal link
char c1, c2, c3, c4;                 // chars for reading data in
int AudioFormat, NumChannels, SampleRate, BitsPerSample ;
int main() {
    pc.printf("\n\rWave file header reader\n\r");
      FILE *fp = fopen("/sd/sinewave.wav", "rb");
      fseek(fp, 20, SEEK_SET);         // set pointer to byte 20
      fread(&AudioFormat, 2, 1, fp);   // check file is PCM
      if (AudioFormat==0x01) {
      pc.printf("Wav file is PCM data\n\r");
      }
    else {
      pc.printf("Wav file is not PCM data\n\r");
    }
    fread(&NumChannels, 2, 1, fp);   // find number of channels
    pc.printf("Number of channels: %d\n\r",NumChannels);
      fread(&SampleRate, 4, 1, fp);    // find sample rate
      pc.printf("Sample rate: %d\n\r",SampleRate);
      fread(&BitsPerSample, 2, 1, fp); // find resolution
      pc.printf("Bits per sample: %d\n\r",BitsPerSample);
      fclose(fp);
}
```

Program Example 11.4 Wave header reader

Try reading the header of a number of different wave files and verify that the correct information is always read to a host PC. A wave audio file can be created in many simple audio packages, such as Steinberg Wavelab, or can be extracted from a standard music compact disc with music player software such as iTunes or Windows Media Player. When reading different wave files remember to update the **fopen( )** call to use the correct filename.

In Program Example 11.4 the **fread( )** function is used to read a number of data bytes in a single command. Examples of **fread( )** show that the memory address for the data destination is specified, along with the size of each data packet (in bytes) and the total number of data packets to be read. The file pointer is also given. For example, the command

```
fread(&SampleRate, 4, 1, fp);        // find sample rate
```

reads a single 4-byte data value from the data file pointed to by **fp** and places the read data at the internal memory address location of variable **SampleRate**.

Exercise 11.6

Extend Program Example 11.4 to display **ByteRate** and **Subchunk2Size**, which indicates the size of the raw audio data within the file.

11.5.3 Reading and Outputting Mono Wave Data

Having accessed the wave file and gathered important information about its characteristics, it is possible to read the raw audio data and output that from the mbed's DAC. To do this, we need to understand the format of the audio data. Initially, we will use an oscilloscope to verify that the data outputs correctly, but it is possible to output the audio data to a loudspeaker amplifier. For this example, we will use a 16-bit mono wave file of a pure sine wave of 200 Hz (Figure 11.12). The same sine wave as utilized in Section 11.3 can be used here.

The audio data in a 16-bit mono .wav file is arranged similarly to the other data seen in the header. The data starts at byte 44 and each 16-bit data value is read as two bytes with the least significant byte first. Each 16-bit sample can be outputted directly on pin 18 at the rate defined by **SampleRate**. It is very important to take good care of data input and output timing when working with audio, as any timing inaccuracies in playback can be heard as clicks or stutters. This can be a challenge because interrupts and timing overheads sometimes make it difficult for the audio to be streamed directly from the SD card at a constant rate. Therefore, a buffered system is used to enable accurate timing control.

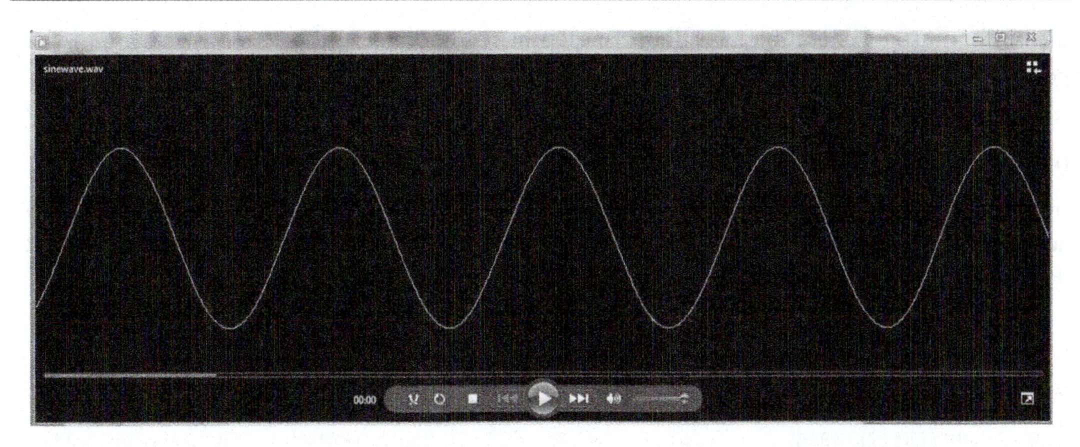

Figure 11.12:
A 200 Hz sine wave .wav file played in Windows Media Player

A good method to ensure accurate timing control when working with audio data is to use a *circular buffer*, as shown in Figure 11.13. The buffer allows data to be read in from the data file and read out from it to a DAC. If the buffer can hold a number of audio data samples, then, as long as the DAC output timer is accurate, it does not matter much if the time for data being read and processed from the SD card is somewhat variable. When the circular buffer write pointer reaches the last array element, it wraps around so that the next data is read into the first memory element of the buffer. There is a separate buffer read pointer for outputting data to the DAC, and this lags behind the write buffer.

It can be seen that two important conditions must be met for the circular buffer method to work; first, the data written to the buffer must be written at an equal or faster rate than data is read from the buffer (so that the write pointer stays ahead of the read pointer); and secondly,

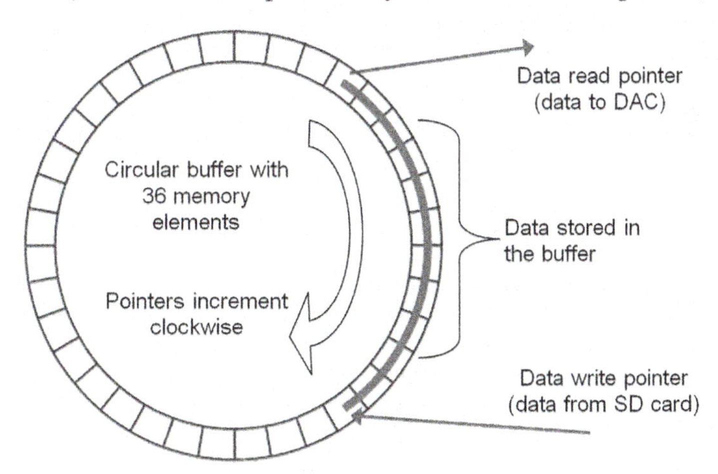

Figure 11.13:
Circular buffer example

the buffer must never completely fill up, or else the read pointer will be overtaken by the write pointer and data will be lost. It is therefore important to ensure that the buffer size is sufficiently large, or to implement control code to safeguard against data wrapping.

Program Example 11.5 reads a 16-bit mono audio file (called in this example **testa.wav**, which can be downloaded from the book website) and outputs at a fixed sample rate which is acquired from the wave file header. As discussed above, the circular buffer is used to iron out the timing inconsistencies found with reading from the wave data file.

```
/* Program Example 11.5: 16-bit mono wave player
*/
#include "mbed.h"
#include "SDFileSystem.h"
#define BUFFERSIZE 4096                          // number of data in circular buffer
SDFileSystem sd(p5, p6, p7, p8, "sd");
AnalogOut DACout(p18);
Ticker SampleTicker;
int SampleRate;
float SamplePeriod;                             // sample period in microseconds
int CircularBuffer[BUFFERSIZE];                 // circular buffer array
int ReadPointer=0;
int WritePointer=0;
bool EndOfFileFlag=0;

void DACFunction(void);                         // function prototype

int main() {
  FILE *fp = fopen("/sd/testa.wav", "rb");      // open wave file
  fseek(fp, 24, SEEK_SET);                      // move to byte 24
  fread(&SampleRate, 4, 1, fp);                 // get sample frequency
  SamplePeriod=(float)1/SampleRate;             // calculate sample period as float
  SampleTicker.attach(&DACFunction,SamplePeriod); // start output tick
  while (!feof(fp)) {         // loop until end of file is encountered
    fread(&CircularBuffer[WritePointer], 2, 1, fp);
    WritePointer=WritePointer+1;                // increment Write Pointer
    if (WritePointer>=BUFFERSIZE) {             // if end of circular buffer
       WritePointer=0;                          // go back to start of buffer
    }
  }
  EndOfFileFlag=1;
  fclose(fp);
}
// DAC function called at rate SamplePeriod
void DACFunction(void) {
  if ((EndOfFileFlag==0)|(ReadPointer>0)) {       // output while data available
    DACout.write_u16(CircularBuffer[ReadPointer]); // output to DAC
    ReadPointer=ReadPointer+1;                    // increment pointer
    if (ReadPointer>=BUFFERSIZE) {
      ReadPointer=0;                              // reset pointer if necessary
    }
  }
}
```

Program Example 11.5 Wave file player with utilizing a circular buffer

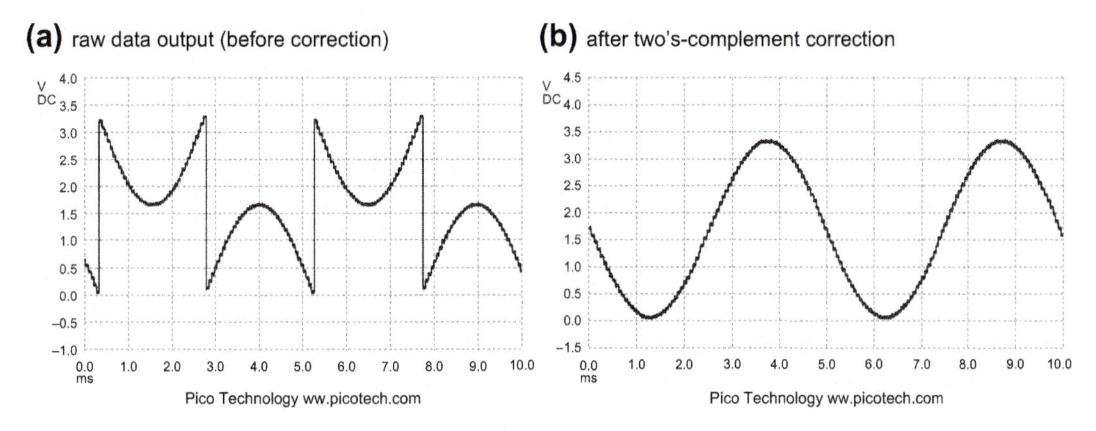

Figure 11.14:
Wave file sine wave

When Program Example 11.5 is implemented with a pure mono sine wave audio file, initially the results will not be correct. Figure 11.14a shows the actual oscilloscope trace of a sine wave read using this program example. Clearly, this is not an accurate reproduction of the sine wave data. The error occurs because the wave data is coded in 16-bit two's-complement form, which means that positive data occupies data values 0 to 0x7FFF and values 0x8000 to 0xFFFF represent the negative data. Two's complement arithmetic is introduced in Appendix A. A simple adjustment of the data is therefore required in order to output the correct waveform as shown in Figure 11.14b.

Exercise 11.7

Modify Program Example 11.5 to include two's complement correction and ensure that, when using an audio file of a mono sine wave, the output data observed on an oscilloscope is that of an accurate sine wave. Appendix A should help you to work out how to do this.

11.6 Summary on DSP

As shown in this chapter, DSP techniques require knowledge of a number of mathematical and data handling aspects. In particular, attention to detail of timing and data validity is required to ensure that no data overflow errors or timing inconsistencies occur. We have also looked at a new type of data file which holds signal data to be output at a specific and controlled rate. The ability to apply simple DSP techniques, whether on a dedicated DSP chip or on a general-purpose processor like the mbed, hugely expands our capabilities as embedded system designers.

11.7 Mini-Project: Stereo Wave Player

This project can be done in basic form, or with several extensions.

11.7.1 Basic Stereo Wave Player

A stereo wave player program can be developed by adding additional elements to Program Example 11.5. You will need to update the program to first analyze whether the wave data is stereo or not, and then implement a conditional feature for the case where stereo data is stored.

Stereo wave data is stored in consecutive samples for left and right stereo playback as shown in Figure 11.15. If using just the standard mbed analog output you will need to average the left and right channel data to mono before outputting to the DAC. Full stereo ADC and DAC chips, such as the Texas Instruments TLV320AIC23b, can be purchased and interfaced with the mbed through its serial ports.

11.7.2 Stereo Wave Player with PC Interface

Add a user interface that will display all the wave filenames stored on an SD card to a host PC terminal. The user can then select a file to be played by selecting a number that represents the chosen data file.

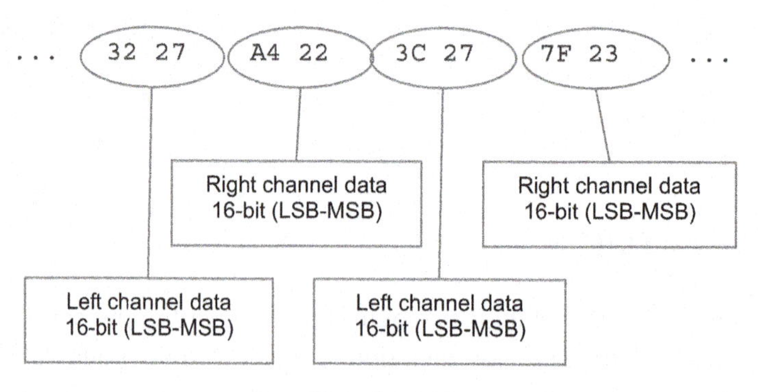

Figure 11.15:
Interleaved left and right channel stereo data

11.7.3 Portable Stereo Wave Player with Mobile Phone Display Interface

Experiment with using a liquid crystal display (LCD) mobile phone display and digital push-buttons to develop your own portable audio player.

Chapter Review

- DSP systems and algorithms are used for managing and manipulating streams of data and therefore require high precision and timing accuracy.
- A digital filtering algorithm can be used to remove unwanted frequencies from a data stream. Similar mathematical algorithms can be used for signal analysis, audio/video manipulation and data compression for communications.
- A DSP system communicates with the external world through analog-to-digital converters and digital-to-analog converters, so the analog elements of the system also require careful design.
- DSP systems usually rely on regularly timed data samples, so the mbed Timer and Ticker interfaces are useful for programming regular and real-time processing.
- Wave audio files hold high-resolution audio data which can be read from an SD card and output through the mbed DAC.
- Data management and effective buffering are required to ensure that timing and data overflow issues are avoided.

Quiz

1. What does the term MAC refer to and why is this important for DSP systems?
2. What differentiates a DSP microprocessor from a microcontroller?
3. What is the difference between an FIR and an IIR digital filter?
4. What are digital filter coefficients and how can they be acquired for a specific digital filter design?
5. Explain the role of analog biasing and anti-aliasing when performing an analog-to-digital conversion.
6. What is a reconstruction filter and where would this be found in a DSP system?
7. What is a circular buffer and why might it be used in DSP systems?
8. What are the potential effects of poor timing control in an audio DSP system?
9. A wave audio file has a 16-bit mono data value given by two consecutive bytes. What will be the correct corresponding voltage output, if this is output through the mbed's DAC, for the following data?
   (a) 0x35 0x04
   (b) 0xFF 0x5F
   (c) 0x00 0xE4

10. Draw a block diagram design of a DSP process for mixing two 16-bit data streams. If the data output is also to be 16 bit, what consequences will this process have on the output data resolution and accuracy?

References

11.1. Marven, C. and Ewers, G. (1996). A Simple Approach to Digital Signal Processing. Wiley Blackwell.

11.2. Proakis, J. G. and Manolakis, D. K. (1992). Digital Signal Processing: Principles, Algorithms and Applications. Prentice Hall.

11.3. The Mathworks (2010). FDATool – open filter design and analysis tool. http://www.mathworks.com/help/toolbox/signal/fdatool.html

11.4. Fisher, T. (2010). Interactive Digital Filter Design. Online Calculator. http://www-users.cs.york.ac.uk/~fisher/mkfilter/

11.5. Sergeev, I. (2010). Audio Echo Effect. http://dev.frozeneskimo.com/embedded_projects/audio_echo_effect

CHAPTER 12

Advanced Serial Communications

Chapter Outline

12.1 Introducing Advanced Serial Communication Protocols

We have already discussed a number of communication methods in earlier chapters, and in particular serial communications which allow fast messages to be sent and received with only a small number of physical connections, such as universal asynchronous receiver/transmitter (UART), serial peripheral interface (SPI) and inter-integrated circuit (I^2C). Each technique requires a protocol which defines the way in which communications are initiated and how the data messages are structured, effectively a code for untangling all the digital 0s and 1s, in order to understand a message as meaningful information.

Fast and Effective Embedded Systems Design. DOI: 10.1016/B978-0-08-097768-3.00012-X

There are a vast number of serial communications protocols and each is developed to a specific standard and usually with a very specific purpose or application in mind. For this reason, each serial communication method has its own advantages and disadvantages for certain applications. Universal serial bus (USB) communications, on the one hand, work very well for connecting high-level peripheral devices to personal computers (PCs) and have a simple 'plug-and-play' method for installation. Ethernet communications, on the other hand, are very high speed and have been developed to allow computer-to-computer communications, which led to the high-speed Internet networks that we nowadays often take for granted.

This chapter discusses a number of advanced serial communications technologies with respect to the mbed, to act as a doorway towards the development of more advanced projects relying on digital connectivity and data sharing.

12.2 Bluetooth Serial Communication

12.2.1 Introducing Bluetooth

Bluetooth is a novel form of digital radio communication, operating in the 2.402–2.480 GHz radio band. Bluetooth provides wireless data links between such devices as mobile phones, computers, wireless audio headsets and systems requiring the use of remote sensors. Bluetooth's main characteristics can be summarized as follows:

- The communications range is up to 100 m for Class 1 Bluetooth devices and up to 20 m for Class 2 Bluetooth devices.
- Bluetooth consumes relatively low power; Class 1 and 2 devices use around 100 mW and 2.5 mW, respectively.
- Data rates up to 3 Mbps can be achieved.
- Up to eight devices can be simultaneously linked.
- Spread-spectrum frequency hopping is applied, with the transmitter changing frequency in a pseudo-random manner 1600 times per second.

The Bluetooth standards, controlled by the Bluetooth Special Interest Group (Reference 12.1), dictate that when Bluetooth devices detect one another, they determine automatically whether they need to interact with each other. Each device has a Media Access Control (MAC) address which communicating devices can recognize and initialize interaction if required. Bluetooth systems in contact with each other form a *piconet*. Once communication is established, members of the piconet synchronize their frequency hopping, so they remain in contact. A single room can contain several piconets with a number of devices communicating.

12.2.2 Interfacing the RN-41 and RN-42 Bluetooth Modules

The Roving Networks RN-41 (shown in Figure 12.1) and RN-42 Bluetooth modules are serial devices that allow replacement of serial wires with a simple Bluetooth interface. For example,

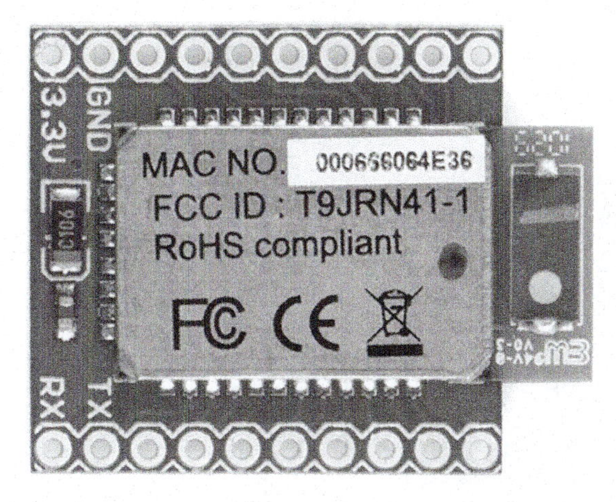

Figure 12.1:
Roving Networks RN-41 Bluetooth module. *(Image reproduced with permission of Sparkfun Electronics)*

most laptop computers have Bluetooth capability installed, so it is possible to replace the USB cable to an mbed with a Bluetooth device and allow wireless host terminal communication. The RN-41 and RN-42 modules have identical control and functionality; however, the RN-41 is a Class 1 Bluetooth device, whereas the RN-42 is Class 2. The RN-41 operates at baud rates between 1200 bps and 921 kbps and includes auto-detect and auto-connection features. In this section the RN-41 module is used; however, all software examples should work equally for an RN-42 device.

In order to communicate with a host PC from an RN-41 or RN-42 device, the host PC must first have Bluetooth enabled and the device added as a known Bluetooth module. When initializing the connection to the Bluetooth module, it will be necessary to specify a *passkey*, which is set by default to '1234' as specified in the Advanced User Manual for Roving Networks Bluetooth Devices (Reference 12.2).

12.2.3 Sending mbed Data over Bluetooth

The RN-41 has a number of configurable features. However, it is very simple to use out of the box as a standard serial interface with a simple TX/RX connection. The RN-41 therefore can be connected to one of the mbed's UART serial ports and configured with a simple mbed serial protocol. The wiring for an RN-41 connected to mbed serial port on pins 9 and 10 is shown in Figure 12.2.

It is possible to configure an mbed to send arbitrary data over a serial interface to verify that data is being received by a host terminal application such as Tera Term. Program Example 12.1 sets up serial communications with the RN-41 and sends data over the UART. The data is actually a continuous count through the ASCII values representing numerical characters 0–9. The onboard mbed light-emitting diodes (LEDs) are also configured to represent the count value. In the example we convert the ASCII byte to the relevant numerical value by bit

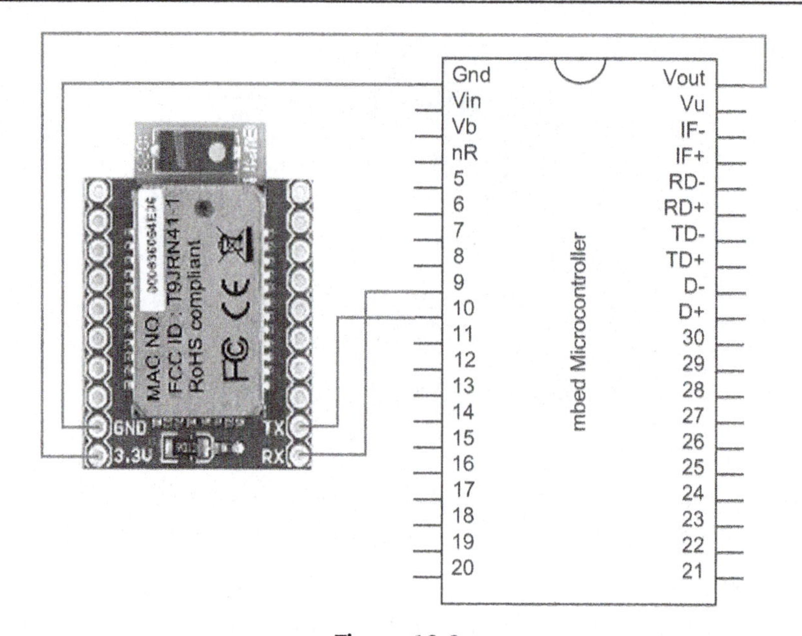

Figure 12.2:
Bluetooth RN-41 connected to the mbed *(Image reproduced with permission of Sparkfun Electronics)*

masking and clearing the higher 4 bits. This leaves just a single value between 0x00 and 0x09. In reality the bit masking of **x** is not essential because the **led BusOut** object can only take the lower four bits of the 8-bit anyway. We perform the bit mask here just for completeness, but you could remove it and check that functionality remains the same.

```
/* Program Example 12.1: Bluetooth serial test data program
*/
#include "mbed.h"
Serial rn41(p9,p10);
BusOut led(LED4,LED3,LED2,LED1);

int main() {
  rn41.baud(115200);             // set baud for RN-41
  while (1) {
    for (char x=0x30;x<=0x39;x++){ // ASCII numerical characters 0-9
      rn41.putc(x);              // send test char data on serial
      led = x & 0x0F;            // set LEDs to count in binary
      wait(0.5);
    }
  }
}
```

Program Example 12.1 Bluetooth serial spammer

If a Bluetooth connection is successfully set up on the host PC (as detailed in Section 12.2.2), the data sent over Bluetooth from Program Example 12.1 should be visible on a host terminal application.

Exercise 12.1

Power the mbed with a battery (acceptable voltage range 4.5–9 V) by connecting the battery outputs to pin 1 (GND) and pin 2 (VIN). This will allow you to remove the USB cable and demonstrate fully remote communication over Bluetooth.

■

12.2.4 Receiving Bluetooth Data from a Host Terminal Application

As well as sending data to a PC, it is possible to send data from the host PC to an mbed over Bluetooth. Program Example 12.2 will receive any data from a host PC terminal application; the remote mbed displays the lower 4 bits of the received byte on the four onboard LEDs. This program can be referred to as a 'sniffer' program as the mbed continuously monitors (or 'sniffs') the serial connection and performs an action when data becomes available.

```
/* Program Example 12.2: Bluetooth serial sniffer program
*/
#include "mbed.h"
Serial rn41(p9,p10);
BusOut led(LED4,LED3,LED2,LED1);

int main() {
  rn41.baud(115200);           // setup baud rate
  rn41.printf("Serial sniffer: outputs received data to LEDs\n\r");
  while (1) {
    if (rn41.readable()) {     // if data available
      char x=rn41.getc();      // get data
      led=x;                   // output LSByte to LEDs
    }
  }
}
```

Program Example 12.2 Bluetooth serial sniffer program

Program Example 12.2 should run and show wireless communications from the PC to the mbed. If the user presses numerical values between 0 and 9 on the PC keyboard, the corresponding binary representation should be shown on the remote mbed.

Exercise 12.2

Add a battery and a liquid crystal display (LCD) to the Bluetooth-enabled mbed. Update Program Example 12.2 to allow a user to type data that on the PC keyboard will appear on the wireless LCD.

■

12.2.5 Communicating between Two mbeds on Bluetooth

By implementing Bluetooth it is possible to have two or more mbeds communicating wirelessly with each other. The configuration for mbed-to-mbed communication is a little more involved, however, as it is necessary to perform bespoke RN-41 setup procedures in software during initialization. This therefore requires a much more thorough understanding of the previously mentioned user manual (Reference 12.2). In our example we will use two mbeds and send data from one to the other. We will therefore refer to the mbed sending the data as the master and the receiving mbed as the slave. The MAC address of the receiving system is required to successfully communicate data. The MAC address of the RN-41 module shown in Figure 12.1 can clearly be seen on the module and the passkey is given a default value specified in the RN-41 user manual (in this case '1234').

The master mbed communicates with the master RN-41 over UART. The master RN-41 needs to be configured to initialize a Bluetooth connection with the slave RN-41. The serial UART procedure for doing this is summarized as follows and discussed in more detail in the RN-41 user manual.

- Enter command mode by sending '$$$' to RN-41 module.
- Connect to a remote MAC address by sending 'C,<address>' where <address> is the MAC address of the remote slave module to be connected to.
- Exit connection mode by sending '—<cr>', where <cr> is the ASCII carriage return value, i.e. 0x0D.

Note that there are other useful commands which will generate status responses from the RN-41, such as:

- **'D'** – displays basic settings: Address, Name, UART Settings, Security, Pin code, Bonding, Remote Address
- **'GB'** – returns the Bluetooth Address of the device
- **'GK'** – returns the current connection status: 1 = connected, 0 = not connected
- **'SP,<text>'** – sets the security passkey for pairing (note, the default '1234' should already be set).

Program Example 12.3 shows an initialization routine for connecting a master RN-41 module to a remote slave RN-41 module. The slave RN-41 module in this example has a MAC address of 00066607ACC1.

```
/*Program Example 12.3: function to initialize paired Bluetooth connection with RN-41
modules
*/
void initialize_connection() {
  rn41.putc('$');  // Enter command mode
  rn41.putc('$');  //
```

```
  rn41.putc('$');  //
   wait(0.5);

  rn41.putc('C');  //
  rn41.putc(',');  // Send MAC address
  rn41.putc('0');  //
  rn41.putc('0');  //
  rn41.putc('0');  //
  rn41.putc('6');  //
  rn41.putc('6');  //
  rn41.putc('6');  //
  rn41.putc('0');  //
  rn41.putc('7');  //
  rn41.putc('A');  //
  rn41.putc('C');  //
  rn41.putc('C');  //
  rn41.putc('1');  //
  wait(0.5);

  rn41.putc('-');  // Exit command mode
  rn41.putc('-');  //
  rn41.putc('-');  //
  rn41.putc(0x0D); //
  wait(0.5);
}
```

Program Example 12.3 Function to initialize paired Bluetooth connection with RN-41 modules

Once a connection is initialized, it is possible to communicate data wirelessly from the master to the slave using simple **putc( )** commands. Program Example 12.4 shows the main code for sending a continuous count variable to the slave over Bluetooth. An override switch is also included to allow a user to override the count value and show remote control capabilities for the slave Bluetooth device.

```
/* Program Example 12.4: Paired Bluetooth master program
*/
#include "mbed.h"
Serial rn41(p9,p10);
BusOut led(LED4,LED3,LED2,LED1);
DigitalIn Din(p26);              // digital switch input on pin 14

char x;
void initialize_connection(void);

int main() {
  rn41.baud(115200);
  initialize_connection();
  while (1) {
    if (Din==1) {        // if digital input switched high
      x=0x0F;         // override with 0x0F
    } else {
```

```
        x++;                // else increment and
        if (x>0x0F) {       // output count value
          x=0;
        }
      }
      rn41.putc(x);         // send char data on serial
      led = x;              // set LEDs to count in binary
      wait(0.5);
    }
}

// add function initialize_connection code here...
```

Program Example 12.4 Paired Bluetooth master program

The slave Bluetooth system can use the same sniffer code as given in Program Example 12.2. The two mbed systems (similar to those shown in Figure 12.3) should now connect and share the same count data wirelessly.

Exercise 12.3

Add an LCD to each mbed device. Add also a unique sensor such as an accelerometer or ultrasonic range finder to each mbed. Implement a two-way communication program which allows the data from both sensors to be displayed on both displays simultaneously.

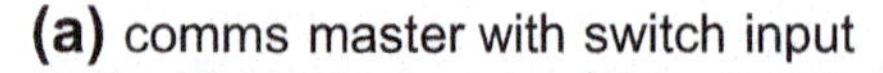

Figure 12.3:
Two battery-powered mbeds communicating via RN-41 Bluetooth modules

Bluetooth is an exciting technology that allows short-range wireless communication. This has many valuable applications where wires are intrusive, expensive or difficult to install. Recent enhancements to Bluetooth have enabled streaming of high-quality audio data and increased range, so the opportunities and applications for Bluetooth are continuously growing.

12.3 Introducing USB

The USB protocol was originally introduced to provide a more flexible interconnection system, whereby items could be added or removed without the need for reconfiguration of the whole system. USB is now ubiquitous and very familiar, widely used for its original purpose, but also to connect all manner of devices, for example digital cameras, MP3 players, webcams and memory sticks, to the PC.

A USB network has one host, and can have one or many *functions*, which in this case refers to USB-compatible devices that can interact with the host. It is also possible to include hubs, which can have a number of functions connected to them, and which in turn link back to the host. Three data rates are recognized: high speed at 480 Mbps, full speed at 12 Mbps and low speed at 1.5 Mbps. The last of these is for very limited capability devices, where only a small amount of data will be transferred. The specification for USB Version 2.0 is defined in Reference 12.3.

USB uses a four-wire interconnection. Two wires, labeled D+ and D−, carry the differential signal, and two are for power and earth. Within certain limits, USB functions can draw power from the bus, taking up to 100 mA at a nominal 5 V. This is supplied by the host. A higher power demand can also be requested; alternatively, the functions can be self-powered. When a device is first attached to the bus the host resets it, assigns it an address and interrogates it (a process known as enumeration). It thus identifies it and gathers basic operating information, for example device type, power consumption and data rate. All subsequent data transfers are initiated only by the host. It first sends a data packet which specifies the type and direction of data transfer, and the address of the target device. The addressed device responds as appropriate. There is usually then a handshake packet, to indicate success (or otherwise) of the transfer.

We have already used the mbed USB interface on a number of occasions: first to connect to the mbed and to download program binary files, but also to communicate serial data to and from a host PC, in order to interface with a terminal application. The mbed has two USB ports: one with a standard USB connector, which we have used already to connect to a PC, and a second on pins 31 and 32 (labeled D+ and D− on the mbed), as seen in Figure 2.1. Several USB library features are provided through the mbed website; these are particularly useful for emulating the mbed as a keyboard or mouse, or for working with musical instrument digital interface (MIDI) systems and advanced audio projects.

12.3.1 Using the mbed to Emulate a USB Mouse

Several mbed USB interfaces are defined within the mbed **USBDevice** library. In order to utilize the USB interfaces, the **USBDevice** library should be imported to a program from the following location:

http://mbed.org/users/samux/libraries/USBDevice/m24owv

Full documentation on all the USB interfaces can be found on the mbed website. One such interface is the **USBMouse** feature. With **USBMouse** it is possible to make the mbed behave like a standard USB mouse, sending position and button-press commands to the host. Program Example 12.5 implements a **USBMouse** interface and continuously sends relative position information to move the mouse pointer around four coordinates, which make up a square. The coordinates are defined by the two arrays **dx** and **dy**; the mouse is moved to these positions within a **for** loop. The mouse, when initialized to its default parameters, uses a relative coordinate system, so if dy = 40 and dx = 40, then the mouse pointer is instructed to move 40 pixels right and 40 pixels down. Negative coordinates for *x* and *y* planes move the pointer left and up, respectively.

```
/* Program Example 12.5: Emulating a USB mouse
*/
#include "mbed.h"                 // include mbed library
#include "USBMouse.h"             // include USB Mouse library
USBMouse mouse;                   // define USBMouse interface

int dx[]={40,0,-40,0};            // relative x position coordinates
int dy[]={0,40,0,-40};            // relative y position coordinates

int main() {
  while (1) {
    for (int i=0; i<4; i++) {     // scroll through position coordinates
      mouse.move(dx[i],dy[i]);    // move mouse to coordinate
      wait(0.2);
    }
  }
}
```

Program Example 12.5 Emulating a USB mouse

Exercise 12.4

Experiment with the other features of the **USBMouse** library. In particular, implement the following small programs:

1. Using the scroll library feature, automatically make pages scroll up and down. You will need to open a document or webpage to test the scroll feature.
2. Add a push-button to your mbed and implement a **mouseclick** feature.
3. Add an accelerometer to your mbed and use the read acceleration values to update the mouse pointer position.

12.3.2 Sending USB MIDI Data from an mbed

MIDI is a serial message protocol that allows digital musical instruments to communicate with digital signal processing systems which can turn those signals into sound. It was first developed in 1982 by the MIDI Manufacturers' Association. MIDI data contains a number of parameters, as described in detail in Reference 12.4. It is still a valuable method for enabling communications between electronic music systems, particularly given that in 1999 a new MIDI standard was developed to allow messaging through the more modern USB protocol.

An example instrument using a MIDI interface is a MIDI piano-style keyboard. The keyboard itself does not make any sound. Instead, the MIDI signals communicate with a sound module or computer with MIDI software installed, so that the signals can be turned into music. In a very simple MIDI system the most valuable information is:

- whether a musical note is to be switched on or off
- the pitch of the note.

The method for setting up and interfacing the mbed with MIDI sequencing software varies subtly depending on the software used. In most cases, however, the audio sequencer software (such as Ableton Live, Apple Logic or Steinberg Cubase) automatically recognizes the mbed MIDI interface and allows it to control a software instrument or synthesizer.

A MIDI interface can be created by first importing the **USBDevice** library (as mentioned in Section 12.3.1 above), importing the **USBMIDI.h** header file and initializing the interface (in this case named 'midi') as follows:

```
USBMIDI midi;                // initialize MIDI interface
```

A MIDI message to sound a note is activated by the following command:

```
midi.write(MIDIMessage::NoteOn(note));   // play musical note
```

This command uses the C++ *scope resolution operator* (::), which relates to a number of advanced C++ programming features. It is not our intention to explore advanced C++ programming concepts, so for the purpose of this example it is sufficient simply to accept the described syntax for sending a MIDI message from the mbed.

The value **note** represents notes on a piano keyboard; this is a 7-bit value so there are 128 possible notes that can be described by MIDI. The value zero represents a very low C note and, as there are 12 notes in the chromatic musical scale, octaves of C occur at multiples of 12. So the MIDI note value 60 represents middle C (also referred to as C4), which has a fundamental frequency of 261.63 Hz. An excerpt of the full MIDI note table is shown in Table 12.1.

Program Example 12.6 sets up an mbed MIDI interface to continuously step through the notes shown in Table 12.1. The Program Example also implements an analog input, which can be connected to a potentiometer, so that the speed of the steps can be manipulated.

Table 12.1: MIDI note values and associated musical notes and frequencies

MIDI note	48	49	50	51	52	53	54	55	56	57	58	59
Musical note	C3	C#3	D3	D#3	E3	F3	F#3	G3	G#3	A3	A#3	B3
Frequency (Hz)	130.1	138.6	146.8	155.6	164.8	174.6	185.0	196.0	207.7	220.0	233.1	246.9
MIDI note	**60**	**61**	**62**	**63**	**64**	**65**	**66**	**67**	**68**	**69**	**70**	**71**
Musical note	C4	C#4	D4	D#4	E4	F4	F#4	G4	G#4	A4	A#4	B4
Frequency (Hz)	261.6	277.2	293.7	311.1	329.6	349.2	370.0	392.0	415.3	440.0	466.2	493.9

```
/* Program Example 12.6: MIDI messaging with variable scroll speed
*/
#include "mbed.h"
#include "USBMIDI.h"
USBMIDI midi;                 // initialize MIDI interface
AnalogIn Ain(p19);            // create analog input

int main() {
  while (1) {
      for(int i=48; i<72; i++) {                // step through notes
         midi.write(MIDIMessage::NoteOn(i));    // note on
         wait(Ain);                             // pause
         midi.write(MIDIMessage::NoteOff(i));   // note off
         wait(2*Ain);                           // pause
      }
  }
}
```

Program Example 12.6 MIDI messaging with variable scroll speed

The output of Example 12.6 is shown in the MIDI control window of Apple Logic software, in Figure 12.4.

Exercise 12.5

Replace the analog input in Program Example 12.6 with a value measured by an ultrasonic range finder. It is therefore possible to change the speed of the note transitions by moving closer to or further away from the sensor.

Also experiment with modifying the note parameters, for example the distance measured by an ultrasonic range finder could be used to change the pitch (i.e. the note number) being played.

Other mbed USB library features, which are all supported by the **USBDevice** library, are shown in Table 12.2.

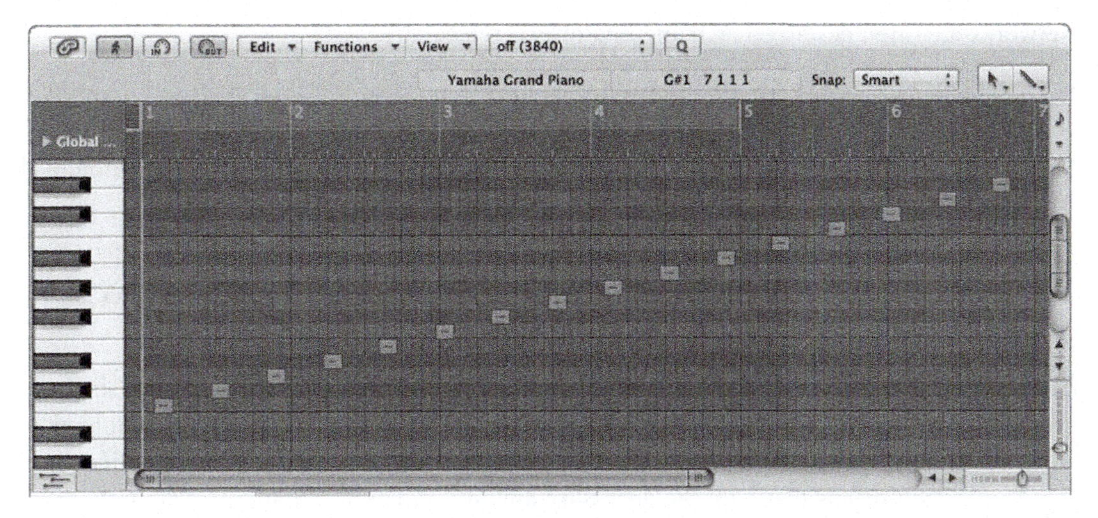

Figure 12.4:
MIDI notes generated by the mbed and recorded into Apple Logic software

Table 12.2: mbed USBDevice library

Mbed USB library	Description
USBMouse	Allows the mbed to emulate a USB mouse
USBKeyboard	Allows the mbed to emulate a USB keyboard
USBMouseKeyboard	A USB mouse and keyboard feature set combined in a single library
USBHID	Allows custom data to be sent and received from a human interface device (HID) allowing custom USB features to be developed without the need for host drivers to be installed
USBSerial	Emulates an additional standard serial port on the mbed, through the USB connections
USBMIDI	Allows send and receive of MIDI messages in communication with a host PC using MIDI sequencer software
USBAudio	Allows the mbed to be recognized as an audio interface allowing streaming audio to be read, output or analyzed and processed
USBMSD	Emulates a mass storage device over USB, allowing interaction with a USB storage device

12.4 Introducing Ethernet

12.4.1 Ethernet Overview

Ethernet is a serial protocol which is designed to facilitate network communications. Any device successfully connected to the Ethernet can potentially communicate with any other device connected to the network.

Networks are often described as being one of two types:

- local area network (LAN) – usually for devices connected together in close proximity, perhaps in the same building and often without Internet access
- wide area network (WAN) – describes a network of devices over a greater geographical area, usually connected by the Internet.

Ethernet communications are defined by the IEEE 802.3 standard (see Reference 12.5) and support data rates up to 100 Gigabits per second. Ethernet uses differential send (Tx) and receive (Rx) signals, resulting in four wires labeled RX+, RX–, TX+ and TX–.

Ethernet messages are communicated as serial data packets referred to as *frames*. Using frames allows a single message to hold a number of data values including a value defining the length of the data packet as well as the data itself. The Ethernet frame therefore defines its own size, which means that only the necessary amount of data is communicated, with no wasted or empty data bytes. Ethernet communications need to pass a large quantity of data at high rates, so data efficiency is a very important aspect. The use of frames allows an efficient method. Each frame also includes a unique source and destination MAC address. The frame is wrapped within a set of *preamble* and *start of frame* (SOF) bytes and a *frame check sequence* (FCS), which enables devices on the network to understand the function of each communicated data element. The standard 802.3 Ethernet frame is constructed as shown in Table 12.3.

The minimum Ethernet frame is 72 bytes; however, the preamble, SOF and FCS are often discarded once a frame has been successfully received. So a 72-byte message can be reported as having just 60 bytes by some Ethernet communications readers.

The Ethernet data takes the form of the *Manchester encoding* method, which relies on the direction of the edge transition within the timing window, as shown in Figure 12.5. If the edge transition within the timing frame is high to low, the coded bit is a 0; if the transition is low to high then the bit is a 1. The Manchester protocol is very simple to implement in integrated circuit hardware and, as there is always a switch from 0 to 1 or 1 to 0 for every data value, the clock signal is effectively embedded within the data. As shown in Figure 12.5, even when a stream of zeros (or ones for that matter) is being transmitted, the digital signal still shows transitions between high and low states.

Table 12.3: Ethernet frame structure

Preamble	Start of frame delimiter	Destination MAC address	Source MAC address	Length	Data	Frame check sequence	Interframe gap
7 – bytes of 10101010	1 byte of 10101011	6 bytes	6 bytes	2 bytes	46 – 1500 bytes	4 bytes	

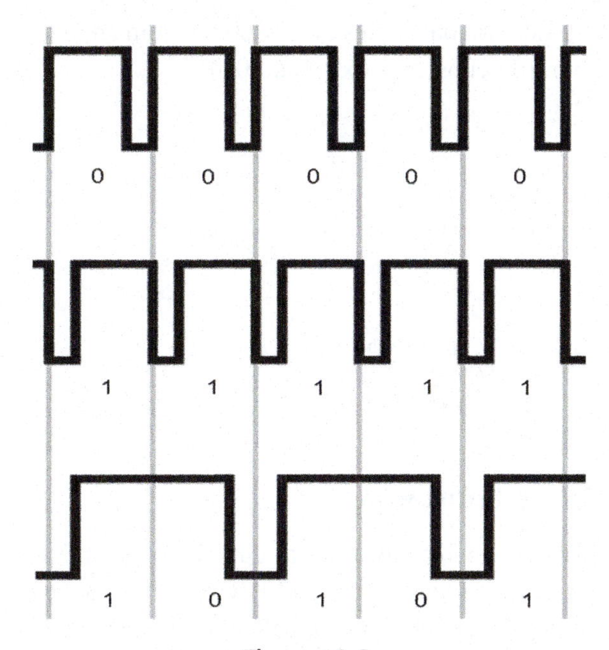

Figure 12.5:
Manchester encoding for Ethernet data

12.4.2 Implementing Simple mbed Ethernet Communications

The mbed Ethernet API is shown in Table 12.4. We can set up an mbed Ethernet system to send data and evaluate this on a logic oscilloscope to verify the Ethernet frame structure. Here, we are not initially concerned with the specific MAC addresses of sending or receiving devices, we simply want to send some known data and see it appearing on the logic scope.

Table 12.4: The mbed Ethernet API

Function	Usage
`Ethernet`	Create an Ethernet interface
`write`	Writes into an outgoing Ethernet packet
`send`	Send an outgoing Ethernet packet
`receive`	Receives an arrived Ethernet packet
`read`	Read from a received Ethernet packet
`address` `link`	Gives the Ethernet address of the mbed Returns the value 1 if an Ethernet link is present and 0 if no link is present
`set_link`	Sets the speed and duplex parameters of an Ethernet link

Program Example 12.7 sends two data bytes every 200 ms from an mbed's Ethernet port. The two byte values are arbitrarily chosen as 0xB9 and 0x46.

```
/* Program Example 12.7: Ethernet write
*/
#include "mbed.h"
#include "Ethernet.h"
Ethernet eth;                       // The Ethernet object
char data[]={0xB9,0x46};            // Define the data values
int main() {
  while (1) {
      eth.write(data,0x02);  // Write the package
      eth.send();            // Send the package
      wait(0.2);             // wait 200 ms
  }
}
```

Program Example 12.7 Ethernet write

Figures 12.6 and 12.7 show detailed analysis of the data packet on a fast logic analyzer (i.e. an oscilloscope that can measure at speeds of around 10 Gbps). The noise can easily be seen but the oscilloscope accurately converts the Ethernet signal to an idealized digital representation. In Figure 12.6, the 7-byte preamble and SOF data are identified.

The preamble and SOF delimiter are followed by destination and source MAC addresses (6 bytes each) and the 2 bytes denoting the data length, which is a minimum of 46 bytes (as described in Table 12.3). Note that even though in this example we have only sent two data bytes, the minimum data size of 46 bytes is sent. The remaining 44 data bytes are made up of empty (0x00) data, being followed by the four frame check sequence bytes. (This is not a particularly efficient use of Ethernet, which is usually required to send large data packets.) Looking at the end of the data packet, shown in Figure 12.7, we can also see the final zero padded data and the FCS data.

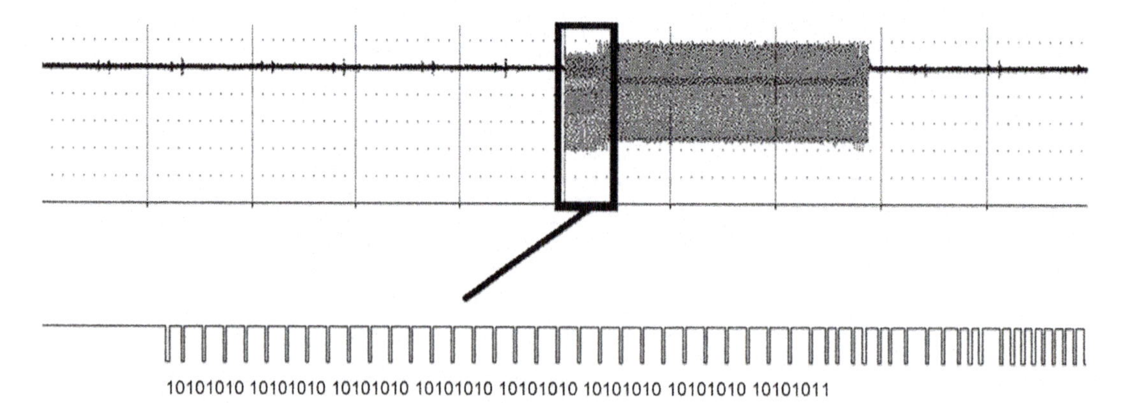

Figure 12.6:
Ethernet packet showing preamble and SOF data

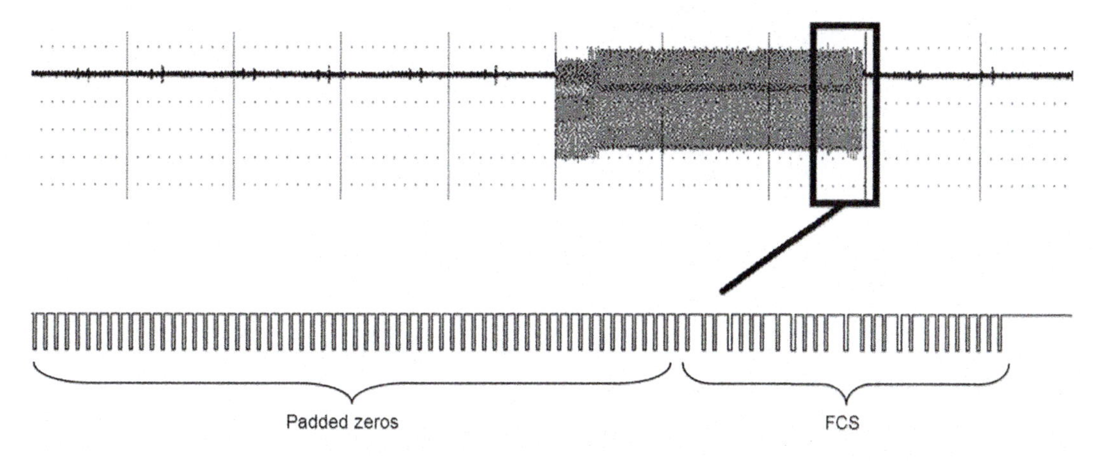

Figure 12.7:
Padded Ethernet data and frame check sequence

Exercise 12.6

Using a fast logic oscilloscope, identify the data making up the 0xB9 0x46 data values transmitted in Program Example 12.7. Experiment with different data values and array sizes, ensuring that each time the correct binary data can be observed on the oscilloscope.

12.4.3 Ethernet Communication between mbeds

An mbed Ethernet port can be used to read data. We can use the simple Ethernet write program of Program Example 12.7. We will also need a second mbed system to receive incoming data and display it to the host terminal screen, in order to verify that the correct data is being read.

The following program will allow an mbed to read Ethernet data traffic and display the captured data on the screen.

```
/* Program Example 12.8: Ethernet read
*/
#include "mbed.h"
Ethernet eth;               // Ethernet object
Serial pc(USBTX, USBRX);    // tx, rx for host terminal coms
char buf[0xFF];             // create a large buffer to store data
int main() {
  pc.printf("Ethernet data read and display\n\r");
  while (1) {
```

```
    int size = eth.receive();                    // get size of incoming data packet
    if (size > 0) {                              // if packet received
      eth.read(buf, size);                       // read packet to data buffer
      pc.printf("size = %d data = ",size);       // print to screen
      for (int i=0;i<size;i++) {                 // loop for each data byte
        pc.printf("%02X ",buf[i]);               // print data to screen
      }
      pc.printf("\n\r");
    }
  }
}
```

Program Example 12.8 Ethernet read

Note that Program Example 12.8 first defines a large data buffer to store incoming data. During the infinite loop the program uses the **eth.receive( )** function to evaluate the size of any Ethernet data traffic. If the data packet has a size greater than zero then it must be an Ethernet packet, so the display loop is entered. The size of the data package along with the read data is then displayed to the host terminal. To communicate successfully between the two mbeds, a crossed signal connection is required, as shown in Figure 12.8 and Table 12.5.

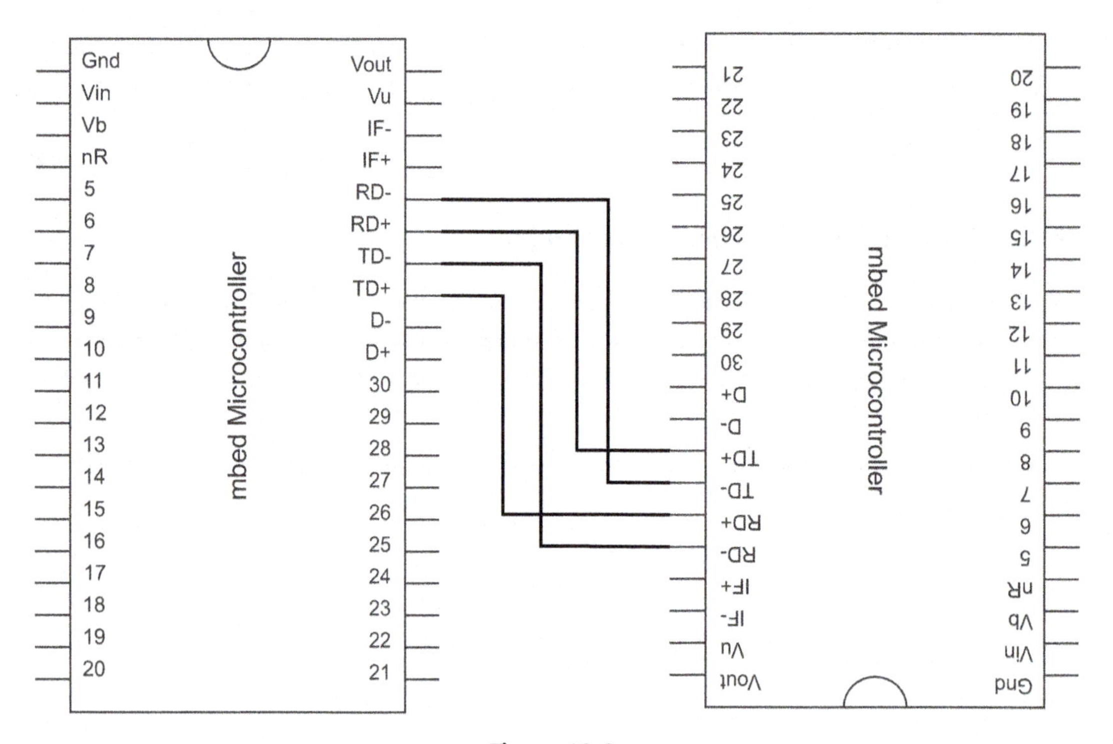

Figure 12.8:
mbed-to-mbed Ethernet wiring diagram

Table 12.5: mbed-to-mbed Ethernet wiring table

Data broadcasting mbed	Data receiving mbed
RD−	TD−
RD+	TD+
TD−	RD−
TD+	RD+

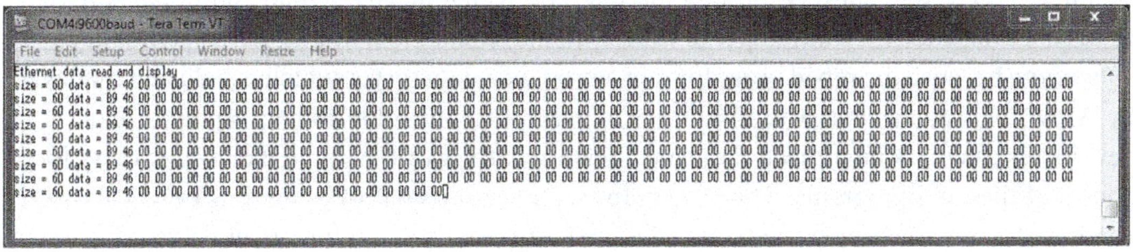

Figure 12.9:
Ethernet data successfully communicated between two mbeds and displayed on a host PC

The Ethernet-write mbed should run Program Example 12.7, and the data-receiving mbed runs 12.8. This is connected to Tera Term running on the host computer. When successfully completed, the host PC terminal application should display Ethernet data similar to that shown in Figure 12.9. Note that even though only two data bytes were transmitted the size of the data package received is 60 bytes, which is the minimum Ethernet data package size (72) minus the preamble, SOF and FCS bytes, which are of no real value once the message is received.

Exercise 12.7

Experiment with different data values and array sizes, ensuring that each time the correct binary data can be read by the host PC application.

■

12.5 Local Network and Internet Communications with the mbed

It is possible to use the mbed as a network-enabled system, i.e. for both LAN and WAN communications. The remaining sections of this chapter show some examples of the mbed being used to access online files, to host files itself, and to communicate data and control actions through local networks and the Internet. We must, however, advise you of the challenges associated with covering such a vast topic in this book. It is important to showcase the power and ability of the mbed, but in many cases here this relies on a very wide set of knowledge and skills related to Internet and network theory, as well as software engineering in other languages. So from here onwards in this chapter we will give some interesting examples but without the capability to evaluate them or their underpinning technologies in depth. Further support information is provided on the book website to assist with compiling and completing the Internet communications examples.

The Internet has transformed worldwide communications over the past 20 years, but rather than being just a simple network for sharing online information and communicating via email, the Internet nowadays facilitates advanced interactive applications. The mbed compiler, for example, is accessed entirely through the Internet, which means that no additional software needs to be installed for a developer to work with the mbed. This, furthermore, facilitates *cloud computing* because the mbed programs are stored on a data server which is held and managed by ARM. This means that it is no longer the user's responsibility to back up data, and developers can work on projects from any location in the world without the need to carry local copies of files and programs. This also helps with version control as there is only ever one version of a specific project, so there is less risk of someone working on an outdated version. Developers can download and archive their programs locally too.

With extended bandwidth and rapid data rates provided by the Internet, it is possible to keep communications open continuously and to stream data while still allowing multi-tasking of other online activities. It is possible nowadays to link small embedded systems to the Internet and hence to network devices that are under embedded control. Once the link is there, it can be used for many things: to monitor status or exert control, or even to cause program or data downloads. Examples include the washing machine that can alert the engineer to an impending fault, the vending machine that can tell the Head Office it is empty, the manufacturer who can download a new version of firmware to an installed burglar alarm, or the home owner who can switch on the oven from the office or check that the garage door is closed.

12.5.1 Using the mbed as an HTTP Client

Several mbed libraries exist to allow network communications. It is possible to use the mbed as an HTTP (hypertext transfer protocol) client in order to access data from the Internet. To do this we rely on the network interface libraries as shown in Table 12.6.

A standard Ethernet socket (RJ45) is required to connect the mbed Ethernet port to a network hub or router. The Sparkfun Ethernet breakout board (Reference 12.6) is used in this example and can be connected to the mbed as shown in Figure 12.10.

Program Example 12.9 enables the mbed to connect to a remote (online) text file and access a text string from within that file. The file is stored at http://www.embeddedacademic.com/mbed/mbedclienttest.txt. You can verify that it exists by accessing it through a standard Internet browser.

Table 12.6: Network interface libraries for HTTP client operation

Mbed library	Library import path
EthernetIF	http://mbed.org/users/donatien/programs/EthernetNetIf/5z422
HTTPClient	http://mbed.org/users/donatien/programs/HTTPClient/5yo73

Figure 12.10:
RJ45 Ethernet connection for mbed

```
/* Program Example 12.9: mbed HTTP client test
*/
#include "mbed.h"
#include "EthernetNetIf.h"
#include "HTTPClient.h"
EthernetNetIf eth(
   IpAddr(192,168,0,101),   //IP Address
   IpAddr(255,255,255,0),  //Network Mask
   IpAddr(192,168,0,1),     //Gateway
   IpAddr(192,168,0,1)      //DNS
);
HTTPClient http;
HTTPText txt;
Serial pc (USBTX,USBRX);
int main() {
  pc.printf("\r\nSetting up network connection...\n\r");
  eth.setup();
  pc.printf("\r\nSetup OK. Querying data...\r\n");
  // attempt to access file 'mbedclienttest.txt' through the Internet...
  HTTPResult r=http.get("http://www.embeddedacademic.com/mbed/mbedclienttest.txt", &txt);
  pc.printf("Result :\n\r\"%s\"\n\r", txt.gets());
}
```

Program Example 12.9 mbed HTTP client test

Figure 12.11:
Host terminal results for client test program

Table 12.7: Network interface libraries for HTTP server operation

Mbed library	Library import path
EthernetIF	http://mbed.org/users/donatien/programs/EthernetNetIf/5z422
HTTPServer	http://mbed.org/users/donatien/programs/HTTPServer/5yhmt

Note that the unique Internet protocol (IP) address for the mbed is set in the **EthernetIF** declaration; in this example it is chosen as 192.168.0.101. Predominantly, LANs use an IP with a default address of 192.168.0.0 and the router on that network will usually take the default address or the next available value above the default address (i.e. 192.168.0.1). Each system on the network needs its own unique IP address based on this default, so the chosen value of 192.168.0.101 is unlikely to clash with any other network systems (unless there are more than 100 other systems).

When the mbed is connected to an active Internet connection the messages shown in Figure 12.11 should be displayed on a host PC terminal application.

12.5.2 Using the mbed as an HTTP File Server

As well as accessing files on an external server, we can use the mbed to host files to be accessed from a remote PC. For this we need to import and utilize the mbed libraries detailed in Table 12.7.

Program Example 12.10 sets up the mbed as a file server.

```
/* Program Example: 12.10 mbed file server setup
*/
#include "mbed.h"
#include "EthernetNetIf.h"
#include "HTTPServer.h"
LocalFileSystem fs("webfs");
EthernetNetIf eth(
 IpAddr(192,168,0,101),    //IP Address
```

```
 IpAddr(255,255,255,0),    //Network Mask
 IpAddr(192,168,0,1),      //Gateway
 IpAddr(192,168,0,1)       //DNS
);
HTTPServer svr;
int main() {
   eth.setup();
   FSHandler::mount("/webfs", "/");  //Mount webfs path on root path
   svr.addHandler<FSHandler>("/");   //Default handler
   svr.bind(80);
 while(1)  {

    Net::poll();                    // poll for Internet data exchange requests
 }
}
```

Program Example 12.10   mbed file server setup

Notice the important lines, which initialize the C++ server handler routines:

```
FSHandler::mount("/webfs", "/");     //Mount webfs path on root path
  svr.addHandler<FSHandler>("/");    //Default handler
  svr.bind(80);
```

Also notice the `Net::poll();` line, which continuously polls the network in order to respond whenever an external request to access the mbed server arrives.

We need to create a text file to save onto the mbed, so that we can access it from the remote Internet browser. Create a .htm file called, for example, **HOME.HTM** and enter some example text, so that you will know when the Internet browser has correctly accessed the file. An example is shown in Figure 12.12. Create this file in a standard text editor and save it to the mbed via the standard USB cable connection.

Note that the mbed server libraries only support MS-DOS 8.3 type filenames. You will be able to access the mbed server from a remote browser by navigating to the specific IP address and webpage, in this example by entering the following address:

http://192.168.0.101/HOME.HTM

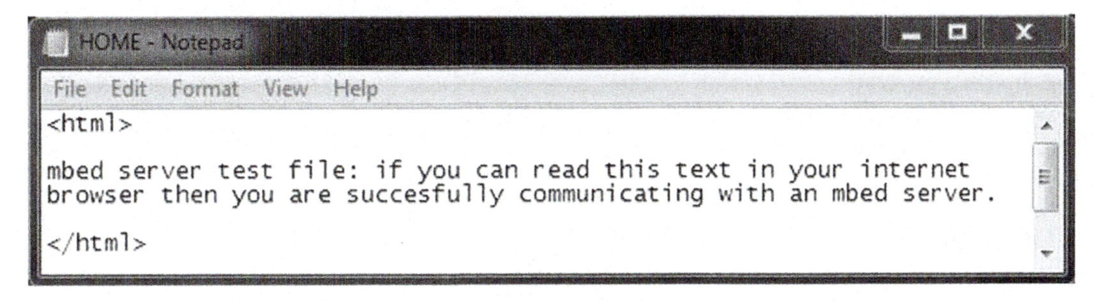

Figure 12.12:
HOME.HTM file to be stored on the mbed server and remotely accessed

We can now access the **HOME.HTM** file (stored on the mbed) from any other computer connected to the LAN. To do this we just need to open a web browser application (such as Internet Explorer) on a connected PC, and type in the navigation address shown above. Successful navigation to this network address should bring up the .htm text as contained in the **HOME.HTM** file.

12.5.3 Using Remote Procedure Calls to Modify mbed Outputs

Remote procedure calls (RPCs) define the process where an action on a device or computer is controlled from a remote location. Essentially, this means that, for example, RPCs can be made from a network PC to switch the mbed's LEDs on and off. Indeed, many of the mbed's outputs can be controlled and accessed by RPC calls from a remote network PC (see Reference 12.7).

The mbed HTTPServer library supports this operation. The key aspects for RPC control on the mbed are:

- Set up an mbed file server as described in Sections 12.5.1 and 12.5.2 above.
- Define mbed interfaces in the extended format with the 'name' defined within, e.g.

  ```
  DigitalOut led1(LED1, "led1");
  ```

- Make the required mbed interfaces available over RPC by adding the RPC base command, e.g.

  ```
  Base::add_rpc_class<DigitalOut>();
  ```

- Define the RPC Handler, e.g.

  ```
  svr.addHandler<RPCHandler>("/rpc");
  ```

- Manipulate mbed interfaces remotely by using the following browser address format:

  ```
  http://<mbed-ip-address>/rpc/<Object name>/<Method name> <Value>
  ```

If we take Program Example 12.10 and apply the required RPC features described above, we arrive at Program Example 12.11. This enables remote control of a **DigitalOut** object assigned to the onboard LED1:

```
/* Program Example 12.11 Remote Procedure Calls example
*/
#include "mbed.h"
#include "EthernetNetIf.h"
#include "HTTPServer.h"
LocalFileSystem fs("webfs");
EthernetNetIf eth(
  IpAddr(192,168,0,101),        //IP Address
  IpAddr(255,255,255,0),        //Network Mask
  IpAddr(192,168,0,1),          //Gateway
  IpAddr(192,168,0,1)           //DNS
);
```

```
HTTPServer svr;
DigitalOut led1(LED1, "led1");          // define mbed object
int main() {
  Base::add_rpc_class<DigitalOut>();    // RPC base command
  eth.setup();                          // Ethernet setup
  FSHandler::mount("/webfs", "/");      // Mount /webfs path on root
  svr.addHandler<FSHandler>("/");       //Default handler
  svr.addHandler<RPCHandler>("/rpc");   // Define RPC handler
  svr.bind(80);
  while(1) {
    Net::poll();
  }
}
```

Program Example 12.11 Remote procedure calls example

The value of the mbed **led1** object can now be changed via a web browser on a network PC, i.e. remotely. This is done by entering into a web browser the RPCs as shown in Table 12.8.

Exercise 12.8

Implement a program to enable manipulation of a pulse width modulation (PWM) duty cycle by an RPC command. You will need to define a PWM object with the extended format, for example:

```
PwmOut pulse(p21, "pulse");
```

You will also need to include the PWM RPC base command as follows:

```
Base::add_rpc_class<PwmOut>();
```

Set the PWM period at the start of the program and check that the duty cycle, as observed on an oscilloscope, can be manipulated remotely.

Connect the PWM output to a servo motor and show that the servo motor position can be controlled by RPC commands.

12.5.4 Controlling the mbed using a Remote JavaScript Interface

Manipulating mbed objects by entering address commands in the Internet browser is not an ideal method for communicating with the mbed on a network. It is slow to type the command,

Table 12.8: Commands for RPC control of mbed LED1

Action	Remote procedure call
Remotely switch led1 ON	http://192.186.0.101/rpc/led1/write 1
Remotely switch led1 OFF	http://192.186.0.101/rpc/led1/write 0

Table 12.9: RPC library and header file details

Mbed library	Library import path	Header files to include
RPCInterface	http://mbed.org/users/MichaelW/libraries/RPCInterface/likpmz	"RPCVariable.h", "SerialRPCInterface.h"

easy to enter the wrong address and not user intuitive. A better solution is to use a *graphical user interface* (GUI) with buttons and displays to allow manipulation and control of mbed objects. It is also useful to be able to display the status of mbed objects on a webpage, to allow remote monitoring.

The mbed JavaScript interface (described in more detail at Reference 12.8) allows communication between the mbed and a Java-enabled browser, using the RPC protocol. This type of program requires the library shown in Table 12.9 to be imported and the described header files also to be included in the project.

Program Example 12.12 sets up two RPC variables, one to hold LED status data and the other to receive remote button-press data.

```
/* Program Example 12.12 Using RPC variables for remote mbed control
*/
#include "mbed.h"
#include "EthernetNetIf.h"
#include "HTTPServer.h"
#include "RPCVariable.h"
#include "SerialRPCInterface.h"
LocalFileSystem fs("webfs");
EthernetNetIf eth(
  IpAddr(192,168,0,101),//IP Address
  IpAddr(255,255,255,0),//Network Mask
  IpAddr(192,168,0,1),  //Gateway
  IpAddr(192,168,0,1)   //DNS
);
HTTPServer svr;
DigitalOut Led1(LED1);        // define mbed object
DigitalIn Button1(p21);       // button
int RemoteLEDStatus=0;
RPCVariable<int> RPC_RemoteLEDStatus(&RemoteLEDStatus,"RemoteLEDStatus");
int RemoteLED1Button=0;
RPCVariable<int> RPC_RemoteLED1Button(&RemoteLED1Button,"RemoteLED1Button");

int main() {
  Base::add_rpc_class<DigitalOut>();      // RPC base command
  eth.setup();                            // Ethernet setup
  FSHandler::mount("/webfs", "/");        // Mount /webfs path root path
  svr.addHandler<FSHandler>("/");         // Default handler
  svr.addHandler<RPCHandler>("/rpc");     // Define RPC handler
```

```
    svr.bind(80);
    printf("Listening...\n");
    while (true) {
      Net::poll();
    if ((Button1==1)|(RemoteLED1Button==1)) {
      Led1=1;
      RemoteLEDStatus=1;
    } else {
      Led1=0;
      RemoteLEDStatus=0;
   }
  }
}
```

Program Example 12.12 Using RPC variables for remote mbed control

The RPC variables are created by the following lines of code:

```
int RemoteLEDStatus=0;
RPCVariable<int> RPC_RemoteLEDStatus(&RemoteLEDStatus,"RemoteLEDStatus");
int RemoteLED1Button=0;
RPCVariable<int> RPC_RemoteLED1Button(&RemoteLED1Button,"RemoteLED1Button");
```

It can be seen in Program Example 12.12 that if the push-button connected to mbed pin 21 is pressed, LED1 is lit. The RPC variable **RemoteLED1Button** can also activate LED1, however, if it becomes set. Furthermore, the RPC variable **RemoteLEDStatus** is set to maintain the same value as LED1.

A Java application (a *Java Applet*) can be built and linked to a .htm file, which is served by the mbed. The .htm file generates the webpage shown in Figure 12.13 when accessed on the mbed by a remote browser. Creating the .htm file and Java application code is outside the scope of this book; however, these specific examples can be downloaded from the book website to allow you to implement the example. In this example, the small LED graphic will always be in the same state as LED1 on the mbed.

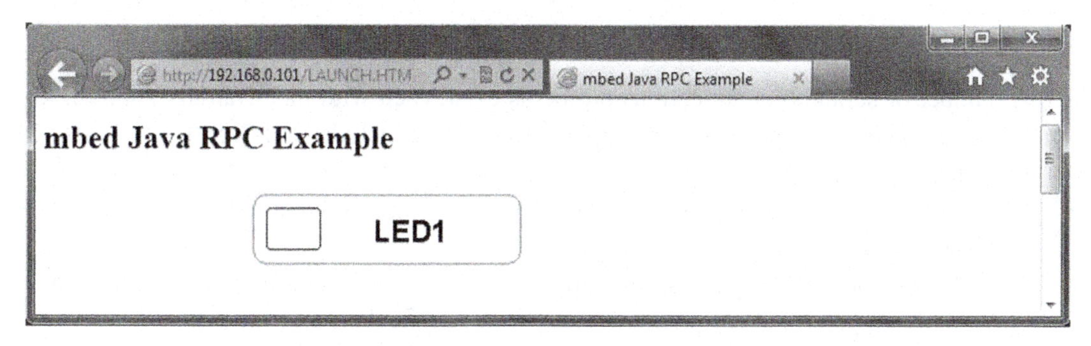

Figure 12.13:
Remote interface Java Applet: LED1 = off

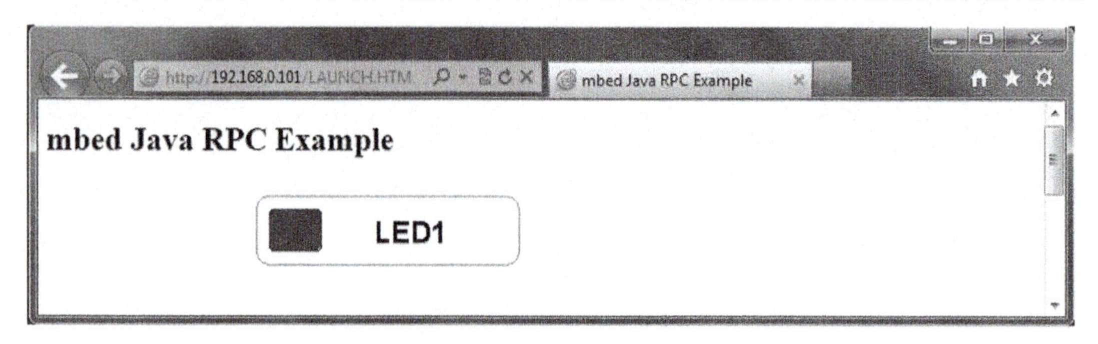

Figure 12.14:
Remote interface Java Applet: LED1 = on

If the user performs a mouse click within the button area on the webpage, the mbed's LED1 will also toggle. This is because within the Java Applet it is possible to connect the user mouse click action to the RPC variable named **RemoteLED1Button**. Clicking the button area on the webpage sends **RemoteLED1Button** high and hence causes the **if** statement in Program Example 12.12 to be entered, lighting the LED on the mbed.

Whenever LED1 on the mbed is lit, the LED indicator displayed on the Java webpage becomes filled in, as shown in Figure 12.14. This is because the Java Applet is capable of using a conditional drawing command to fill the LED indicator whenever **RemoteLEDStatus** is set high.

Here, we are now controlling the mbed LED1 output by a remote web application, and the Java Application also maintains a correct status indication of the mbed's LED. This example is quite complex and programming skills with C/C++, HTML and Java, beyond the scope of this book, are required to build on this example. This does, however, serve as an example of the power of the mbed as a device which can be uniquely accessed and controlled through networks and the Internet.

Chapter Review

- A number of serial communications protocols exist, each with its advantages and disadvantages and each aimed at specific applications. These include UART, SPI, I^2C, USB and Ethernet.
- Bluetooth communications allow standard serial messages to be communicated in a wireless manner.
- The USB protocol is designed specifically for allowing plug-and-play communications between a computer and peripheral devices such as a keyboard or mouse.
- There are several mbed USB libraries allowing the mbed to operate as a mouse or a keyboard, or as an audio or MIDI interface, for example.

- Ethernet is a high-speed serial protocol which facilitates networked systems and communications between computers, either within a local network or through the World Wide Web.
- The mbed can communicate through Ethernet to access data from files stored on a data server computer.
- The mbed can be configured to act as an Ethernet file server itself, allowing data stored on the mbed to be accessed through a network.
- The mbed RPC interface and libraries allow variables and outputs to be manipulated from an external remote interface, written in Java, for example.

Quiz

1. What is the communication range of a Class 1 Bluetooth device?
2. What does the term MAC address refer to?
3. What does MIDI stand for, and how is this relevant to USB communications?
4. What is the MIDI note value for C2 (fundamental frequency of 65.3 Hz)?
5. Describe the Manchester digital communication format for Ethernet signals.
6. Sketch the following Ethernet data streams, as they would appear on an analog oscilloscope, labeling all points of interest:
   (a) 0000
   (b) 0101
   (c) 1110
7. What are the minimum and maximum Ethernet data packet sizes, in bytes?
8. Draw a table listing the advantages and disadvantages of using USB or Ethernet for communicating between two mbeds.
9. What does the term RPC refer to?

References

12.1. The official Bluetooth website. http://www.bluetooth.com
12.2. Roving Networks Advanced User Manual. Version 4.77, 21 November 2009. rn-bluetooth-um
12.3. Universal Serial Bus Specification, Revision 2.0. April 2000. Compaq, Hewlett-Packard, Intel, Lucent, Microsoft, NEC, Philips.
12.4. Summary of MIDI Messages. http://www.midi.org/techspecs/midimessages.php
12.5. Ethernet Working Group. IEEE 802.3. http://www.ieee802.org/3/
12.6. Breakout Board for RJ45. http://www.sparkfun.com/products/716
12.7. Interfacing Using RPC. http://mbed.org/cookbook/Interfacing-Using-RPC
12.8. Interfacing with JavaScript. http://mbed.org/cookbook/Interfacing-with-JavaScript

CHAPTER 13

An Introduction to Control Systems

Chapter Outline

13.1 Control Systems

We have already implemented a number of *electromechanical control systems*, which can be broadly defined as electronic and/or software systems that ar e designed to control a designated physical variable. For example, in Chapter 4 we discussed a system that moved a servo to a desired position by applying the correct pulse width modulation (PWM) waveform. Electromechanical control systems are very common, with applications ranging from home and office equipment, to industrial machinery, automotive vehicles, robotics and aerospace. Think of how much control there must be in a humble inkjet printer: drawing in and guiding the paper, shooting the print cartridges backwards and forwards with extraordinary precision, and squirting just the right amount of ink, of the right color, in just the right place. The accuracy of a control system is a very important aspect, as errors and inaccuracy can have significant relevance to performance, wear and safety.

The whole topic of control systems is a very detailed and broad field, which draws on advanced mathematical theory as well as electronics, digital communications, applied

Fast and Effective Embedded Systems Design. DOI: 10.1016/B978-0-08-097768-3.00013-1

mechanics, and sensor and actuator design. It is not possible to give a thorough insight into control systems in this book, but the mbed is a valuable device which can be used to implement some simple control algorithms and communication methods. This chapter therefore acts as a starting point for the reader who may wish to go on to study or implement control systems at a later date.

13.1.1 Closed and Open Loop Control Systems

Closed loop control systems employ internal error analysis to achieve accurate control of actuators, given continuous sensor input readings. For example, this means that the system measures the position of an actuator, compares it with a signal representing the position it is meant to be at (called the *setpoint*), and from this information tries to move the actuator to the correct position. The fundamental components of a closed loop system are therefore an actuator, which causes some form of action; a sensor, which measures the effect of the action; and a controller, to compute the difference between setpoint and actual position. Although we say position, any physical variable can be controlled. For example, a heater element might be used to heat an oven to a specified temperature and a temperature sensor used to measure the actual heat of the oven. If the closed loop controller tells the oven to be at 200°C then the heater will provide heat until the sensor signals the controller that 200°C has been achieved. If the sensor measurement shows that the actuator has achieved its desired action (i.e. 200°C is achieved), then the error between the desired setpoint and the actual reading is zero and no further action is required. If, however, the sensor reads that the desired action has not been achieved (e.g. the temperature is only 190°C), then the control system knows that more action is required to meet the desired setpoint.

An *open loop control system* is one that employs no analysis sensor and relies on a calibrated actuator setting to achieve its desired action. For example, a 5 V electric motor might rotate at different speeds over the voltage input range, say 0 V gives 0 revolutions per second (rps) and 5 V gives 100 rps. If we want to rotate the motor at 50 rps we might simply assume that the response characteristic of the motor is linear (or have a calibrated look-up table for its response profile) and provide 2.5 V. However, we will not actually know if the motor is rotating at the desired speed; for different operating temperatures, with different friction loading and with component drift over time, the speed achieved given 2.5 V input cannot be accurately predicted. Here, an improved closed loop system with a speed sensor could be used to modify the input voltage and ensure that an accurate 50 rps is achieved.

Advantages of closed loop systems are improved accuracy of a controlled variable and the ability to automatically make continuous adjustment. In fact, many actuators are quite erratic and non-linear in terms of performance, so in many cases closed loop control is a necessity rather than a desire. Closed loop systems built around fast microcontrollers can also allow advanced control of systems that were previously thought uncontrollable. For example, the

Figure 13.1:
The Segway personal transporter (www.segway.com). *(Image reproduced with permission of Segway Inc.)*

Segway personal transportation systems (Figure 13.1, Reference 13.1) use sensitive gyroscopes to ensure that the standing platform always remains horizontal. If weight shifts forward (and the gyroscope shows an imbalance) then the motor in the wheels moves forwards marginally to compensate and stops moving when a horizontal position has been achieved again. The microcontroller, sensors and actuators inside read and compute so rapidly that this process can be performed faster than a human's motion can displace their weight, so the controller can always ensure that the platform stays stable.

13.1.2 Closed Loop Cruise Control Example

Another example of closed loop control is *cruise control* in many automotive vehicles. Here, the driver chooses a desired speed and the vehicle automatically maintains that speed until the driver disables the mode, either by switching it off or by applying the brake.

A key point to note is that the actuator performs a different operation from the specific sensor reading that is being used for comparison with the setpoint. For example, the setpoint might be a vehicle speed of 80 kph, but the actuator is actually the electronic throttle valve, which determines the amount of air and hence fuel that is drawn into the engine on each piston cycle. So a complex chain of events occurs, from automatically moving the throttle, to sucking air and fuel into the engine, engine combustion, torque being applied to the wheels, the vehicle moving, and the vehicle motion finally being measured and converted to a horizontal speed value in kph. Many factors affect the accuracy of this process, for example changes in air pressure, fuel type, engine wear, tire size and tire pressure, as well as road friction and environmental factors such as drag and gravity. An open loop system could never be possible for cruise control.

A closed loop control system can be quite simple. All we need to do, given a microprocessor-based system, is to write a control algorithm which takes a setpoint (desired speed chosen by the driver) and calculates the error between the setpoint and the actual speed. The algorithm then uses this error to calculate the required action. For example, if the error is negative (actual speed is less than the desired setpoint) then the action must be positive to increase the amount of fuel to the engine, increase the torque to the wheels and increase the car speed. If the error is positive, then the action must be negative, to reduce the car speed to the setpoint. The use of *negative feedback*, where the measured position is subtracted from the setpoint (in order to calculate the error), is therefore found in most closed loop control algorithms.

The mechanical system, between the actuator and the sensor, involves a chain of events with an inherent response time. For a cruise control system, the response time is the time it takes for an increase in fuel injected to the engine to cause an increase in speed of the vehicle – in some cases this might be of the order of seconds, or at least hundreds of milliseconds. The slow response time can cause problems for an automatic control system, because once the desired setpoint is achieved, it is likely to overshoot. With cruise control, once the desired speed has been achieved it will take a small amount of time for the engine to reduce its torque output (as fuel will already have been injected into the engine), so the controller could overshoot and cause an error in the opposite direction. The controller will automatically then attempt to reverse the action and reduce the error back towards zero again, but may overshoot in the opposite direction too. In a poorly designed cruise control system the controller may never even achieve a satisfactory and steady 80 kph as desired by the driver. Indeed, the system may oscillate continuously between, say, 78 and 82 kph, which would be an unpleasant driving experience.

It follows that the performance of the control system depends on the effectiveness of the control algorithm employed. Design and tuning of the control algorithm is a delicate process, requiring knowledge of the system characteristics and usually experimental testing to understand response times. When final calibration is performed effectively, however, very stable and responsive control systems can be developed.

13.1.3 Proportional Control

The simplest form of closed loop control uses a proportional algorithm to form a linear relationship between the measured error and the actuator control. Figure 13.2 describes the electromechanical process of a closed loop cruise control system. The driver tells the microcontroller the speed at which they would like the vehicle to cruise. At the same time, the actual vehicle speed is calculated from a wheel speed sensor reading. The speed

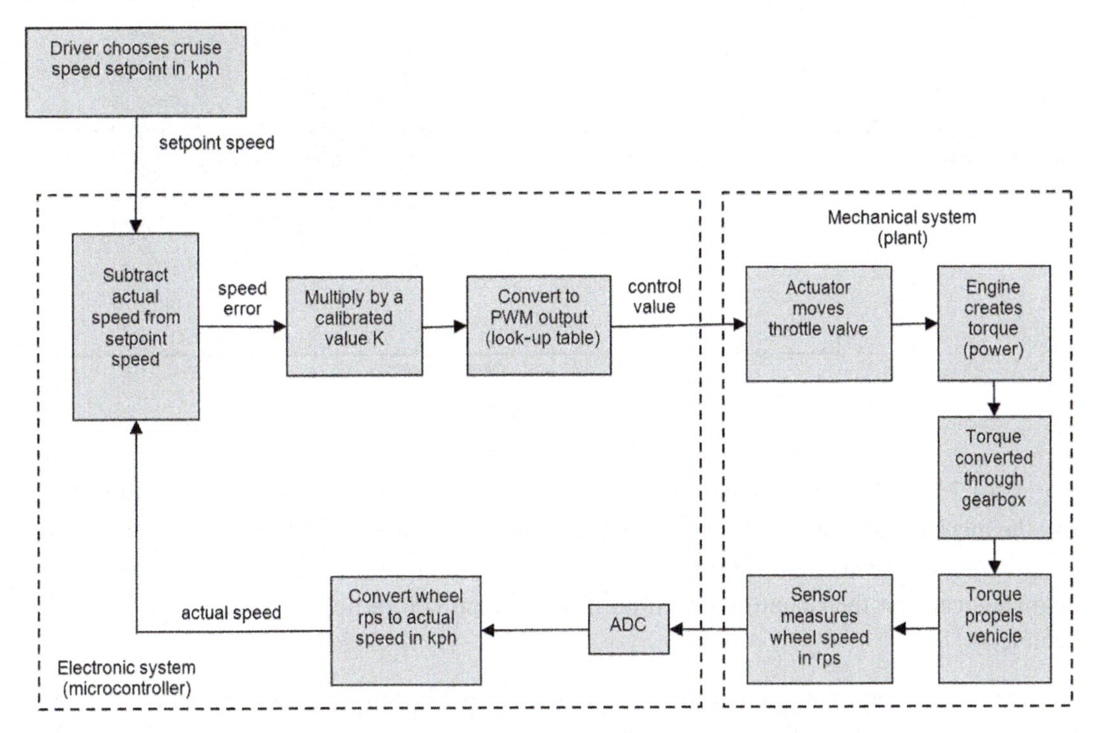

Figure 13.2:
Closed loop cruise control example. PWM: pulse width modulation;
ADC: analog-to-digital converter

setpoint is compared to the actual vehicle speed to calculate the error, which is then multiplied by a calibrated value *K*. The calculated value is then used to set a PWM output of the microcontroller, and there may be a conversion look-up table here to ensure that a suitable PWM output is set. The PWM output controls a solenoid, which can move the engine's throttle position. The position of the throttle valve determines how much air and fuel are drawn into the engine, and hence the amount of torque or power generated. The engine torque, via the vehicle gearbox, determines how much power is delivered to the wheels, which subsequently dictates the speed of the vehicle. If a setpoint is chosen then, with a continuously sampling microcontroller, the control loop will automatically self-adjust so that a speed error of zero is achieved. In reality, a zero error is not quite achieved, but the control loop will get as close as possible. If suddenly the setpoint is changed, for example from 80 kph to 90 kph, there is as an immediate *step change* in the setpoint. It will take some time for the vehicle to speed up to the new setpoint, and the time taken for the *step response* is dependent on a number of factors. It will depend predominantly on the value of *K*, the physical response time of the mechanical system and the speed of the microprocessor software loop.

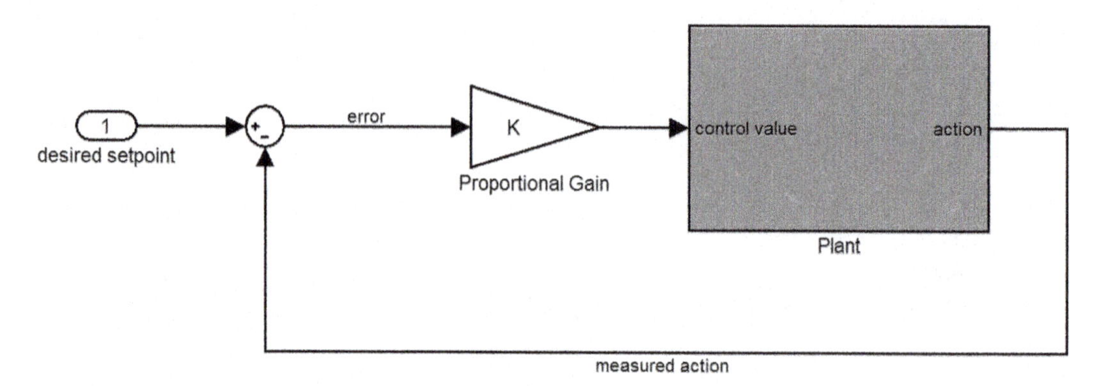

Figure 13.3:
Proportional control system

Figure 13.2 also shows the elements of the electronic control system (i.e. the microcontroller) and the mechanical system, which is sometimes referred to as a *plant*. In many cases the plant itself is modeled mathematically and implemented in another microcontroller, so that engineers can test their control algorithm in simulation before implementing it on a real vehicle.

A more general representation of the closed loop control system is shown in Figure 13.3. We can see that when the error is positive (i.e. the measured value is above the setpoint), the control value is reduced (owing to negative feedback) and vice versa for negative errors. Furthermore, when the error is large, the control value is changed by a large amount; when the error is small, the control value is adjusted by only a small amount. The value of the gain determines how quickly the system responds to errors.

The fact that the control value is altered proportionally is a good thing when the error is large; however, this causes problems when the error is small. As the error gets smaller, the control value adjustment get smaller too, which results in the desired setpoint never actually being achieved. With proportional only control there is always therefore a *steady-state error*. The steady-state error can be reduced by making the gain larger, but unfortunately this could easily result in overshoot and a longer settling period, and even continuous instability if the gain is too high. Figure 13.4 shows an example step response of a proportional control system with high and low gain values.

To overcome this error with controlled overshoot and stability it is necessary to add integral and derivative terms to the control algorithm.

13.1.4 Proportional Integral Derivative Control

Proportional integral derivative (PID) control is similar to proportional control, but with the addition of algorithm components relating to the integral and derivative values of the error

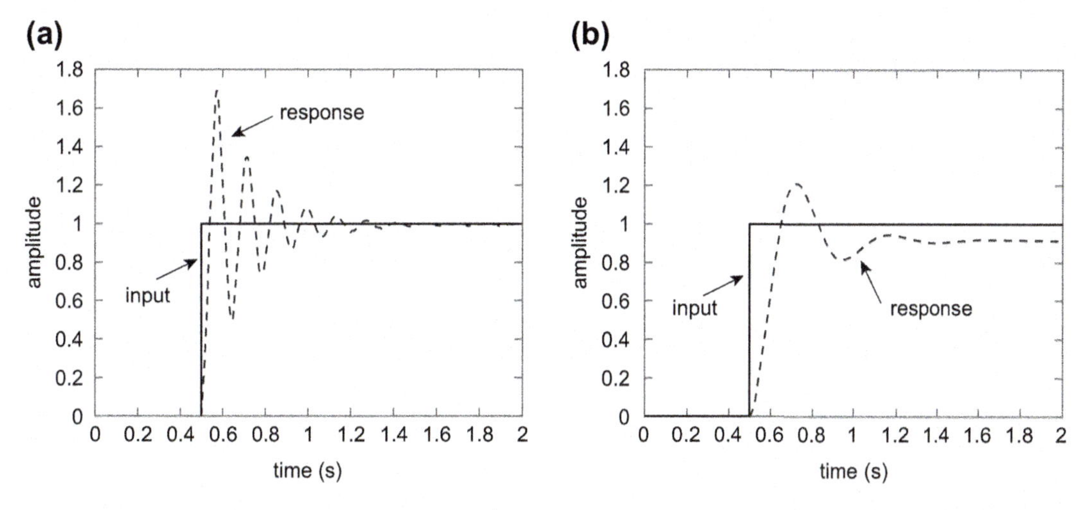

Figure 13.4:
Step response with (a) high proportional gain; (b) low proportional gain

data. This adds an element of history to the algorithm, rather than it being responsive to the current error value alone. A PID control system is shown in Figure 13.5.

Figure 13.6 shows that the error is integrated and multiplied by an integral gain value as well as differentiated and multiplied by a differential gain. These are then summed together with the proportional gain term to give a control value which takes into account not only

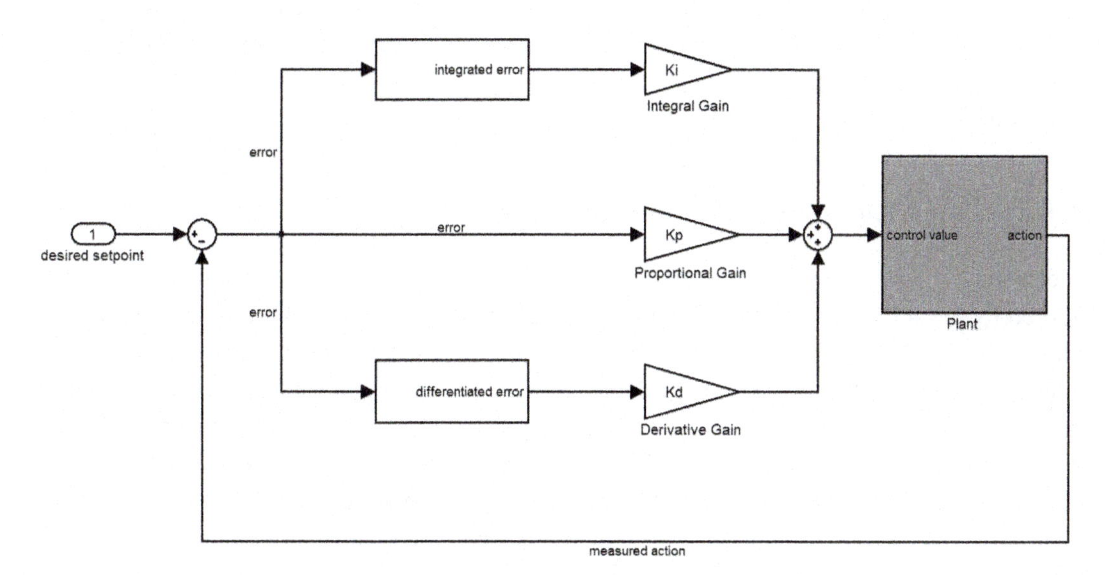

Figure 13.5:
PID control system

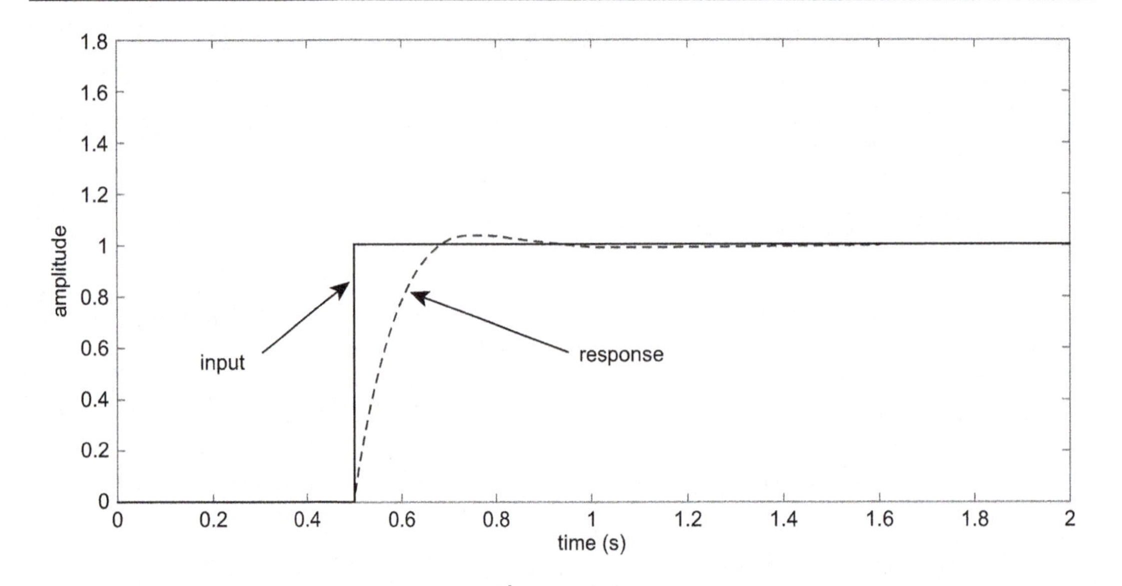

Figure 13.6:
PID controller tuned to give step response with low overshoot and low steady-state error

current position, but also the rate of change of the position and the longer term history of change. Now the tuning of the system becomes even more complex, but the results can be an accurate control system with low steady error and low overshoot. The gain terms can be calculated by experimentation and by employing software tools. Complex mathematical methods can help to determine the best PID gain values (see Reference 13.2, for example), although some fine manual tuning is often required also. Figure 13.6 shows the step response of a PID control system which has its gain values tuned to give a fast response time with low overshoot and low steady-state error.

13.2 Closed Loop Digital Compass Example

A closed loop system requires a sensor and an actuator, whereby the sensor's position or state is altered by the actuator's action. A simple practical example is that of a digital compass integrated circuit chip, positioned on a servo. As the servo rotates, the compass readings will alter accordingly. We can design a closed loop system to automatically adjust the servo position to a desired compass reading and use a simple algorithm to ensure accurate positioning and smooth movement. Implementing a full PID closed loop controller is rather complex and outside the scope of this book. However, it is possible to implement a simple proportional control algorithm for the purpose of introducing closed loop control on the mbed.

In this example, we will use the mbed with an HMC6352 digital compass module with a 360 degree rotation servo. You will also need a 6 V battery supply for the servo. It is important to

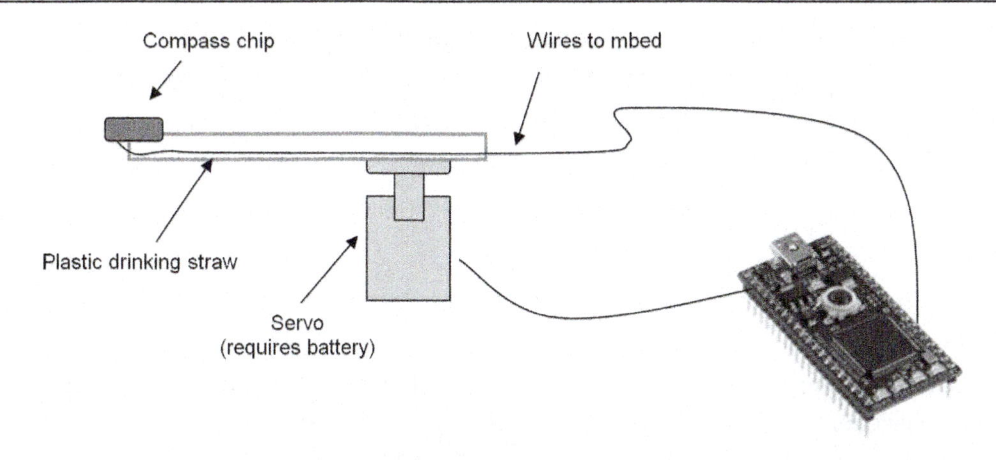

Figure 13.7:
Closed loop compass system

think carefully about connections and wiring to ensure that the wires do not become tangled as the servo rotates. The compass sensor also needs to be kept as far away as possible from any unwanted magnetic fields. An example test setup is shown in Figure 13.7.

13.2.1 Using the HMC6352 Digital Compass

Honeywell's HMC6352 compass is another highly integrated sensor, similar in concept to the ADXL345 accelerometer and TMP102 temperature sensor introduced in Chapter 7. Fully integrated onto a single integrated circuit (IC), it contains sensors, signal conditioning, microprocessor and inter-integrated circuit (I^2C) serial interface. The two sensors, mounted at right angles to sense the components of the surrounding magnetic field, are based on the magnetoresistive principle. This relies on the fact that the apparent resistance of a conductor can be seen to change in the presence of a magnetic field. We use it on a tiny breakout board, as shown in Figure 13.8. Full information is given in Reference 13.3.

Figure 13.8:
HMC6352 mounted on a breakout board. *(Image reproduced with permission of Sparkfun Electronics)*

Table 13.1: Pin connections for HMC6352 digital compass

HMC6352 pin	mbed pin
SCL (I^2C)	27
SDA (I^2C)	28
VCC (3.3 V)	40
GND (0 V)	1

The HMC6352 compass can be connected to the mbed via an I^2C interface, as described by the pin connections in Table 13.1. Before building a complete system, we will first write a program to ensure that the compass initializes and reads data correctly.

As detailed in the HMC6352 datasheet (Reference 13.3), the base address for the compass is 0x42. The datasheet also describes that in order to initialize the IC we write an ASCII 'G' (0x47) command to inform the device that we wish to write to a particular RAM address. This is followed by the operation mode register address (0x74), followed by the operation mode control byte. We will initialize the compass to update in continuous mode at a rate of 20 Hz, and we also want to enable the Set/Reset function. This gives a data byte value of 0x72. The compass setup can therefore be executed by an I^2C write message containing the three command bytes as a single array, i.e:

```
cmd[0] = 0x47;          // 'G' write to RAM address
cmd[1] = 0x74;          // Operation mode register address
cmd[2] = 0x72;          // Op mode = 20H, no S/R, continuous
```

A test program to set up and read the HMC6352 is shown in Program Example 13.1.

```
/* Program Example 13.1: HMC6352 Compass Setup and Read
*/
#include "mbed.h"
// mbed objects
I2C compass(p28, p27);       // sda, scl
Serial pc(USBTX, USBRX);     // tx, rx
// variables
const int addr = 0x42;       // define the I2C write Address
char cmd[3];                 // command array for read and write
float pos;                   // measured position

// main code
int main() {
  cmd[0] = 0x47;         // 'G' write to RAM address
  cmd[1] = 0x74;         // Operation mode register address
  cmd[2] = 0x72;         // Op mode = 20H, no S/R, continuous
  compass.write(addr,cmd, 3);  // Send operation
  while (1) {
    compass.read(addr, cmd, 2);  // read the two-byte echo result
```

```
        pos = 0.1 * ((cmd[0] << 8) + cmd[1]);  //convert to degrees
        if (pos>180){
            pos=pos-360;
        }
        pc.printf("deg = %.1f\n", pos);
        wait(0.3);
    }
}
```

Program Example 13.1 HMC6352 compass setup and read

To read data from the compass we use an infinite loop and output data values to a host PC terminal application. Inside the infinite loop we simply perform a 2-byte read which returns the 16-bit data value (in tenths of a degree). To form a single number from these 2 bytes we shift the first byte left by 8 (as it is the more significant byte) and add the second byte; this forms a 16-bit number. In the same line of code this is then multiplied by 0.1, to give a result in degrees. It is actually more useful to have degrees in the range −180 to +180 (rather than 0 to 360), so values greater than 180 have the 360 subtracted.

With the compass IC connected to mbed pins 27 and 28, Program 13.1 should initialize and run, allowing you to verify that correct compass readings are shown on the PC terminal. Zero degrees indicates north and 180 degrees (or −180 degrees) indicates south, and hence 90 degrees represents due east and −90 degrees represents due west. Be careful to ensure that large magnets or power sources do not interfere with the compass's sensitivity. If you are in a steel-framed building or a well-equipped laboratory, you may find that the earth's field is totally overwhelmed by what is around you!

13.2.2 Implementing a 360 Degree Rotation Servo

We can implement an open loop position controller with a 360 degree servo and a potentiometer to control. Servos with 360 degree motion can be sourced (as detailed in Appendix D), or a standard 180 degree motion servo can be modified to allow complete rotation, as described by Reference 13.4. The 360 degree servo requires the same mbed connections as the standard 180 degree servo (as shown in Figure 4.10). However, the 360 degree version is a more erratic device and more challenging to control. When connecting a PWM signal to the device a low duty cycle signal will cause the servo to continuously turn anti-clockwise, whereas a higher duty cycle will cause the servo to rotate continuously clockwise. There will be a single PWM duty value which is a 'zero' point at which the servo holds stationary.

We will now implement a scheduled program to allow different rate functions. In particular, we will use a fast rate function to control the servo positioning and to read the sensor data (100 Hz), but a slower rate function to send data to the host terminal application (5 Hz). We will therefore use two Ticker objects and two task functions to control and implement the program timing. In this program we will use a potentiometer to allow open loop control of the servo; we will also read and display compass data to the screen to quantify the direction in which the compass is pointing. We therefore need to define an analog input (to read the potentiometer) and a PWM output (to actuate the servo). We will also need to implement the same HMC6352 initialization and read commands as used in Program Example 13.1.

Start a new project and enter the code shown in Program Example 13.2.

```
/* Program Example: 13.2 Open Loop Compass
*/
#include "mbed.h"
//***** mbed objects *****
I2C compass(p28, p27);       // sda, scl
PwmOut PWM(p25);
AnalogIn Ain(p20);
Serial pc(USBTX, USBRX);     // tx, rx
Ticker s100hz_tick;          // 100 Hz (10ms) ticker
Ticker s5hz_tick;            // 5 Hz (200ms) ticker
//***** variables *****
const int addr = 0x42;       // define the I2C write Address
char cmd[3];
float pos;                   // measured position
float ctrlval;               // PWM control value
//***** function prototypes *****
void s100hz_task(void);      // 100 Hz task
void s5hz_task(void);        // 5 Hz task
//***** main code *****
int main() {
 // initialize and setup data
  PWM.period(0.02);
  cmd[0] = 0x47;              // 'G' write to RAM address
  cmd[1] = 0x74;              // Operation mode register address
  cmd[2] = 0x72;              // Op mode = 20H, S/R, continuous
  compass.write(addr,cmd, 3);// Send operation
  s100hz_tick.attach(&s100hz_task,0.01);    // attach 100 Hz task
  s5hz_tick.attach(&s5hz_task,0.2);         // attach 5 Hz task
  while(1){
  // loop forever
    }
}

//***** function 100hz_task *****
void s100hz_task(void) {
  compass.read(addr, cmd, 2);   // read the two-byte compass data
  pos = 0.1 * ((cmd[0] << 8) + cmd[1]);     //convert to degrees
  if (pos>180)
  pos=pos-360;               convert to ±180deg
  ctrlval=Ain;      // set control value (also try ctrlval=Ain/4)
  PWM=ctrlval;      // output control value to PWM
}
```

```
//***** function 5hz_task *****
void s5hz_task(void) {
   pc.printf("deg = %.1f  PWM = %.4f\n", pos, ctrlval);
}
```

Program Example 13.2 Open loop compass

If a potentiometer input is now connected to pin 20 and the PWM output on pin 25 is connected to a powered servo, your compiled code should allow the potentiometer to move the servo clockwise and anti-clockwise. Position data will also be sent to a host PC running Tera Term.

You will notice that the motion of the servo is quite sensitive and erratic; it is actually very difficult to achieve an exact stationary position. Indeed, the analog input gives values between 0.0 and 1.0, but we know from our experience of servos in Chapter 4 that the servo only uses the PWM duty cycle range of about 5 to 20%. It is therefore possible to decrease the open loop sensitivity by implementing either a scaling factor or some hard limits to the possible PWM control values.

Exercise 13.1

1. Modify Program Example 13.2 so that the control value **ctrlval** is only a fraction of **Ain** by dividing by a fixed number. Try different values and see whether the control and stability can be improved.
2. Looking at the Tera Term output, find the PWM value (**ctrlval**) that causes the servo to hold stationary. This is a constant for the servo you are using. Make a note of this 'zero' value, as we will need it later on, at which time we will call it **PWM_zero.**

13.2.3 Implementing a Closed Loop Control Algorithm

We can use a simple proportional control algorithm to implement a closed loop system. To demonstrate this we will set our desired compass position to be 0 degrees, i.e. the compass will always point north (much like a real compass). The servo will move the position of the compass automatically to ensure that it does this. First, we will need to define a few variables for calculating error, the desired setpoint (= 0) and associated PWM control value, as follows:

```
float setpos=0; // desired setpoint position = zero (North)
float error;    // calculated error
float ctrlval;  // PWM control value
float kp=0.0002;        // proportional gain (to be tuned)
float PWM_zero=0.075;   // zero value (as deduced in Exercise 13.1)
```

Program Example 13.3 implements the closed loop algorithm, which retains much of the code from Program Example 13.2.

```
/* Program Example 13.3: Closed loop compass program
*/
#include "mbed.h"
// mbed objects
  I2C compass(p28, p27);      // sda, scl
  PwmOut PWM(p25);
  AnalogIn Ain(p20);
  Serial pc(USBTX, USBRX);      // tx, rx
  Ticker s100hz_tick;           // 100 Hz (10ms) ticker
  Ticker s5hz_tick;             // 5 Hz (200ms) ticker
// variables
  const int addr = 0x42; // define the I2C write Address
  char cmd[3];
  float pos;              // measured position
  float setpos=0;         // setpoint position = zero (North)
  float error;            // calculated error
  float ctrlval;          // PWM control value
  float kp=0.0002;          // proportional gain
  float PWM_zero=0.075;     // zero value
// function prototypes
  void s100hz_task(void);  // 100 Hz task
  void s5hz_task(void);    // 5 Hz task
// main code
  int main() {
 // initialize and setup data
   PWM.period(0.02);
   cmd[0] = 0x47;                 // 'G' write to RAM address
   cmd[1] = 0x74;                 // Operation mode register address
   cmd[2] = 0x72;                 // Op mode = 20H, S/R, continuous
   compass.write(addr,cmd, 3);  // Send operation
 // assign timers
   s100hz_tick.attach(&s100hz_task,0.01); //attach 100 Hz task to 10ms tick
   s5hz_tick.attach(&s5hz_task,0.2);      //attach 5Hz task to 200ms tick
   while(1){
 // loop forever
    }
}

// function 100hz_task
  void s100hz_task(void) {
   compass.read(addr, cmd, 2); // read the two-byte echo result
 //convert data to degrees
   pos = 0.1 * ((cmd[0] << 8) + cmd[1]);
   if (pos>180)
    pos=pos-360;
   error = setpos - pos;          // get error
   ctrlval = (kp * error);        // calculate ctrlval (proportional)
   ctrlval = ctrlval + PWM_zero;  // add control value to zero position
   PWM = ctrlval;                 // output to PWM
}
```

```
// function 5hz_task
  void s5hz_task(void) {
   pc.printf("deg = %.1f error=%.1f ctrlval=%.4f\n",pos,error,ctrlval);
  }
```

Program Example 13.3 Closed loop compass program

The 100 Hz task includes the closed loop control algorithm. Immediately after the **pos** variable has been calculated, the error between the actual and desired setpoint positions, i.e. the proportional control value **ctrlval**, is calculated as follows:

```
error = setpos - pos;         // get error
ctrlval = (kp * error);       // calculate ctrlval (proportional)
```

The control value is a calculated position that automatically adjusts to make up for the difference between the desired and actual in a smooth and controlled manner. However, we first need to add the zero position to the control value in the following line:

```
ctrlval = ctrlval + PWM_zero;  // add control value to zero position
```

The zero position, **PWM_zero**, is a simple calibratable value which we deduced in the second part of Exercise 13.1. It defines the PWM duty cycle that holds the servo in a stationary position. It now acts as an offset to the control value, so a negative error causes the servo to rotate to one side of the zero position, and a positive value causes it to rotate to the other side. When there is zero error in the control algorithm the servo stays still. In this example the zero offset value is calibrated as **PWM_zero=0.075**, although your own servo may require a different value.

Exercise 13.2

1. Modify the value of **Kp** and see how this affects the system stability. In general, if **Kp** is too large then overshoot and instability are observed, whereas if **Kp** is too small, then the servo response is slow and inaccurate.
2. Investigate the required accuracy of the **PWM_zero** value. Are three decimal places sufficient or does refining this value to four or five decimal places enhance the controller's performance?

■

Exercise 13.3

1. **LCD output** – Add a liquid crystal display (LCD) to your system to display position, error and PWM data. You can also use the 6 V servo battery pack as a portable supply voltage for the mbed. This will allow you to disconnect the mbed from the host PC and have a fully mobile digital compass.
2. **Position control** – Develop your system to allow a user to input a desired position (in degrees) in a terminal application, and the servo will immediately move to that position.

■

13.3 Communicating Control Data over the Controller Area Network

13.3.1 The Controller Area Network

The Controller Area Network (CAN) is another type of serial communications protocol that was developed within the automotive industry to allow a number of electronic units on a single vehicle to share essential control data. A vehicle nowadays uses many microcontrollers for autonomous control systems. Each microcontroller system is referred to as an electronic control unit (ECU) and these include the engine management ECU, an anti-braking system (ABS) ECU, a dashboard ECU, active suspension and the radio/CD player, for example. Each of these ECUs manages its own control strategies, but they all need to access information relevant to their own operation, for example drawn from engine speed, throttle position, brake pedal position and engine temperature. However, an automotive vehicle generates a high level of electromagnetic interference and a wide temperature and humidity range; this is a hostile environment for any signal and indeed for any electronic device. Moreover, very high reliability is essential as vehicles are safety-critical systems. The serial standards developed for the benign environment of home or office, such as UART, SPI and I^2C, are completely inappropriate for this hostile environment, and a new standard was therefore needed. Initially, CAN was developed by the German company Bosch. They published Version 2.0 of the standard in 1991, and in 1993 it was adopted by the International Organization for Standardization (ISO), as ISO 11898. At the time of writing, CAN specification Version 2.0B can be downloaded from the Bosch website (Reference 13.5). The CAN standard has a high level of data security, it is inevitably complex and just the briefest overview is given here. The main features are listed below.

- Communication is asynchronous, half duplex, with (for a given system) a fixed bit rate. The maximum for this is 1 Mbit/s.
- The configuration is 'peer to peer', i.e. all nodes are viewed as equals. There is, however, a mechanism for prioritization. Master and slave designation is not used.
- Logic values on the bus are defined as 'dominant' or 'recessive', where dominant overrides recessive. Physical interconnect is not otherwise defined.
- The bus access is flexible. With all nodes being peers, any can start a message. An ingenious arbitration process is applied in the case of simultaneous access, which does not lead to loss of time or data. The arbitration process recognizes prioritization.
- There is an unlimited number of nodes.
- Bus nodes do not have addresses, but apply 'message filtering' to determine whether data on the bus is relevant to them.
- Data is transferred in frames, which have a complex format. This starts with identifier bits, during which arbitration can take place. Eight data bytes are allowed per frame.
- There is an exceptionally high level of data security, with exhaustive error checking. A node that recognizes that it is faulty can disconnect itself from the bus.

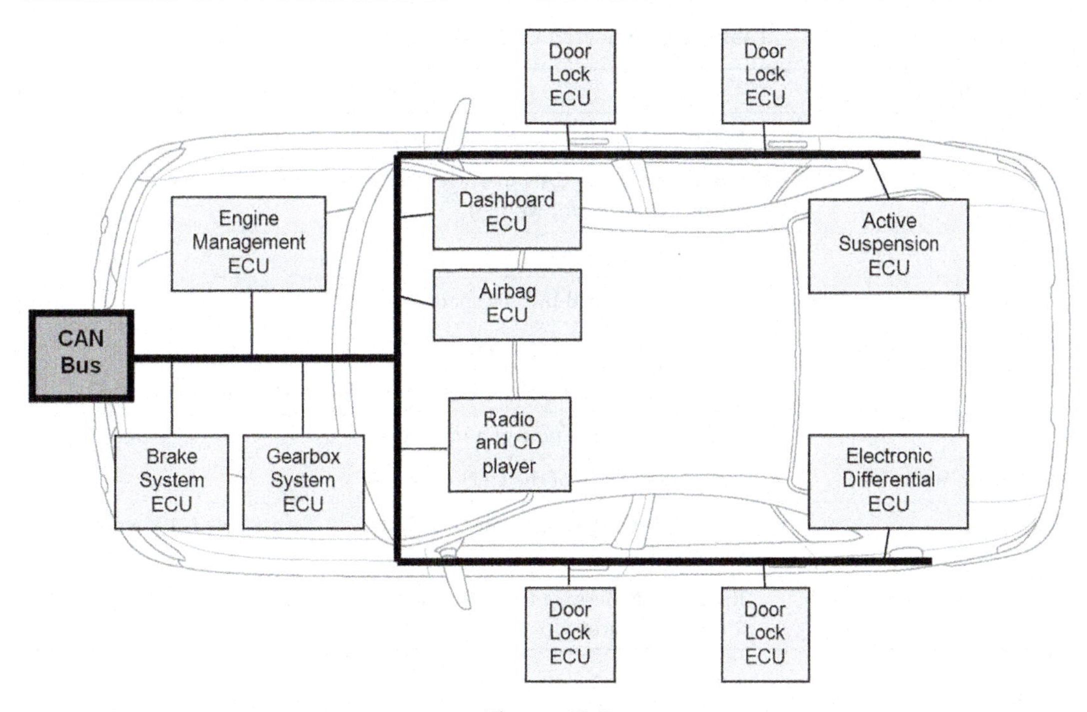

Figure 13.9:
An example vehicle CAN bus network

Figure 13.9 shows an example vehicle CAN bus network. The types of control data that need sharing along the bus can vary. For example, the dashboard ECU will need to access data including engine speed, engine temperature, vehicle speed and diagnostic information, which will be predominantly provided by sensors attached to the engine management and gearbox ECUs, as well as door open/closed data from the door ECUs. Furthermore, if this vehicle is to implement an automatic door lock strategy once a threshold vehicle speed has been achieved, then the door lock ECUs will need to access vehicle speed data from the CAN bus.

So it can be seen that a very reliable and thorough communications protocol for sharing control data around an automotive vehicle is required, and CAN provides this. CAN is also used in industrial and factory applications where noise issues and reliability requirements are similar to those of the automotive industry.

13.3.2 CAN on the mbed

The mbed has two CAN controller interfaces on pins 9–10 and 30–29; note that only the latter two pins appear as a CAN interface in Figure 2.1. The pins refer to receive (RD or RXD) and transmit (TD or TXD) signals, respectively. The CAN interface can be used to write data words out of a CAN port and will return the data received from another CAN device. The

Table 13.2: Selected CAN API functions

Function	Usage
`CAN` `CANMessage`	Creates a CAN interface connected to specific pins
	Creates an empty CAN message (for reading) or a CAN message with specific content (for writing):
	`id` the message ID
	`data` space for 8-byte payload
	`len` length of data in bytes
	`format` defines if the message has standard or extended format
	`type` defines the type of a message
`frequency` `write` `read`	Set the frequency of the CAN interface
	Write a CANMessage to the bus. Returns 0 if write failed, 1 if write was successful
	Read a CANMessage from the bus. Returns 0 if no message arrived, 1 if message arrived

CAN clock frequency can also be configured. A selection of the CAN application programming interface (API) functions is shown in Table 13.2.

When using the mbed CAN API, we first have to define a **CAN** object and then define a message (**CANMessage**) to be written to or read from the CAN bus. The mbed CAN controllers alone, however, cannot be used to communicate directly with a CAN network. To do this we need to use the mbed CAN interface to control a specific *CAN transceiver* IC such as the Microchip MCP2551. This is shown in Figure 13.10 and described in Reference 13.6.

The CAN bus implementation used is differential, and connects to the **CANH** and **CANL** pins. An external resistor connected at **Rs** controls the slew rate of the data signal, with a slower rate minimizing electromagnetic interference, although for simple testing this can be connected directly to ground. The mbed connects from its CAN controllers to the transceiver's **RXD** and **TXD** pins.

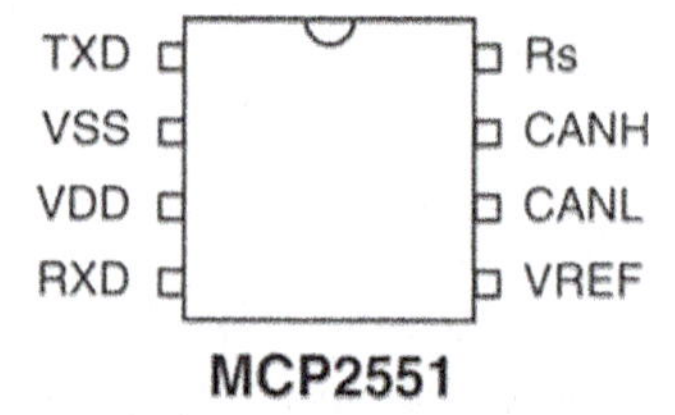

Figure 13.10:
Microchip MCP2551 CAN transceiver IC

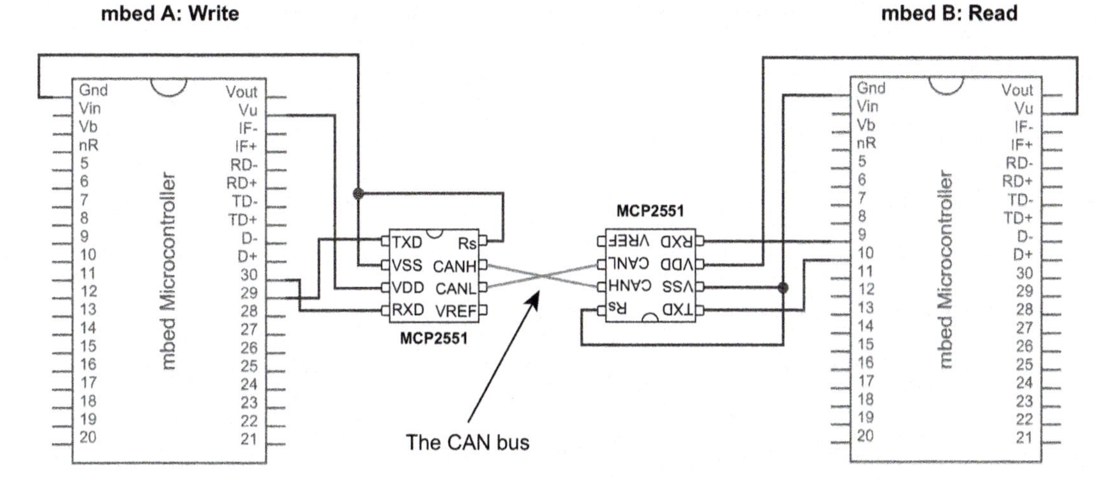

Figure 13.11:
Hardware build for mbed-to-mbed CAN communication

To demonstrate CAN with the mbed we will communicate some data between two mbeds over a CAN network. One mbed will send data to the CAN bus while the other will receive data from the CAN bus. The hardware connections required to create a simple CAN network test are shown in Figure 13.11 and Table 13.3. Notice that the 'CAN bus' is defined as the wires CANH and CANL, which sometimes require a termination resistance of 100–200 Ω between them over long wires. In this simple, noise-free example the short connections mean that a termination resistor is not necessary. Note also that we use the mbed CAN controller on pins 30 and 29 for the send system and the CAN controller uses pins 9 and 10 – these are arbitrarily chosen, predominantly to make the hardware connection diagram neat!

Having built the CAN test system described in Figure 13.11 and Table 13.3, Program Example 13.4 can be implemented to regularly send data values to the CAN bus.

```
/* Program Example 13.4: CAN data write – sends an incrementing count value to the CAN
bus every second
*/
#include "mbed.h"
Serial pc(USBTX, USBRX);        // tx, rx for Tera Term output

DigitalOut led1(LED1);          // status LED
CAN can1(p30, p29);             // CAN interface
char counter = 0;
int main() {
  printf("send... ");
  while (1) {
    // send value to CAN bus and monitor return value to check if CAN
    // message was sent successfully. If so display, increment and toggle
```

```
    if (can1.write(CANMessage(1, &counter, 1))) {
       pc.printf("Message sent: %d\n", counter);    // display
    counter++;                                     // increment
       led1 = !led1;                               // toggle status LED
    }else{
    can1.reset();
     }
     wait(1);
  }
}
```

Program Example 13.4 CAN data write

Here, one very important line of code is the **if** statement, which performs a number of actions:

```
if (can1.write(CANMessage(1, &counter, 1))) {
```

First, this line creates a **CANMessage** with **id=1**, holding the data from the address of variable **counter** and of length 1 byte. The message is then written to the CAN bus and the

Table 13.3: Pin connections for mbed-to-mbed CAN communications

MCP2551 pin	Description	Pin connection
	Write system	
1 TXD	Transmit Data Input	mbed p29 CAN TD
2 VSS	Ground	mbed p1 GND
3 VDD	Supply Voltage	mbed p39 Vu (5 V)
4 RXD	Receive Data Output	mbed p30 CAN RD
5 Rs	Reference Output Voltage	mbed p1 GND
6 CANL	CAN Low-Level Voltage I/O	CAN bus low
7 CANH	CAN High-Level Voltage I/O	CAN bus high
8 VREF	Slope-Control Input	–
	Read system	
1 TXD	Transmit Data Input	mbed p10 CAN TD
2 VSS	Ground	mbed p1 GND
3 VDD	Supply Voltage	mbed p39 Vu (5 V)
4 RXD	Receive Data Output	mbed p9 CAN RD
5 Rs	Reference Output Voltage	mbed p1 GND
6 CANL	CAN Low-Level Voltage I/O	CAN bus low
7 CANH	CAN High-Level Voltage I/O	CAN bus high
8 VREF	Slope-Control Input	–

return value of the **write** operation is monitored. Only if the **write** operation returns a successful response (by returning a value 1) does the counter increment and light-emitting diode (LED) toggle.

Program Example 13.5 creates an empty **CANMessage**. It continuously monitors the CAN bus and displays any read CAN messages to Tera Term on a host PC.

```
/* Program Example 13.5: CAN data read – reads CAN messages from the CAN bus
*/
#include "mbed.h"
Serial pc(USBTX, USBRX);      // tx, rx for Tera Term output
DigitalOut led2(LED2);        // status LED
CAN can1(p9, p10);            // CAN interface
int main() {
  CANMessage msg;              // create empty CAN message
  printf("read...\n");
  while(1) {
    if(can1.read(msg)) {      // if message is available, read into msg
      printf("Message received: %d\n", msg.data[0]);  // display message data
      led2 = !led2;                                   // toggle status LED
    }
  }
}
```

Program Example 13.5 CAN data read

You can now implement the hardware connections shown and Program Examples 13.4 and 13.5 to show two mbeds communicating over CAN.

Exercise 13.4

1. Enhance the CAN write program to send a second data value to the CAN bus. Give the second data value a different ID value and make it two bytes in size and at a different message rate (say every 5 seconds). Evaluate how the CAN reader program interprets the CAN messages it receives.
2. Implement a procedure in the CAN reader program so that it can interpret the two different CAN messages based on the message ID, and uniquely display a text message in Tera Term based on the ID of the message received.

For a practical implementation of a CAN messaging system, such as for an automotive system, a *CAN specification* will define the messages that can be communicated between ECUs. Each message, for example vehicle speed, engine speed, engine temperature and door open/closed status, will have a unique ID as well as a definition of the size of the data and any conversion equation required in order to turn the raw data into a sensible value. Each ECU will be pre-programmed with the CAN specification so that when a message is received it can immediately know what the message is and how to interpret the data.

Several CAN interface devices have been developed and are available commercially for example to read diagnostic information from vehicles. It is also possible to use an mbed to read and write data messages to and from a vehicle CAN bus with the correct interface, and hence even control a vehicle this way. Of course, to do this you will need to know the CAN specification for the vehicle that is being interfaced with. A number of interesting examples and an mbed CAN interface board can be found in Reference 13.7.

Chapter Review

- Closed loop control systems use negative feedback of sensor data to ensure that a specific actuator position is accurate.
- When responding to a step change in desired setpoint, a closed loop control system may respond too slowly or overshoot, or have a steady-state error. Tuning the control algorithm can ensure a best-possible step response.
- The time taken for the step response is dependent on the control algorithm design, the physical response time of the mechanical system and the speed of the microprocessor software loop.
- Proportional, integral and derivative control can be tuned to give smooth control of many electromechanical systems.
- Control data can be shared on a network of electronic control units (ECUs) over the Controller Area Network (CAN) protocol.
- An automotive vehicle may have a number of ECUs for controlling, for example, engine management systems, brake systems, the dashboard, airbag control and door locks. These can all be connected by CAN.
- CAN is an extremely reliable communications protocol with high tolerance to noise and interference.
- To implement CAN on the mbed, additional CAN transceiver ICs must be used to create the CAN bus.

Quiz

1. What are the main differences between an open loop and a closed loop control system?
2. What do the terms setpoint, plant and negative feedback refer to in closed loop control?
3. Where and why might closed loop control be required in the design of a helicopter?
4. What does 'the step response' of a control system refer to?
5. What benefits does CAN communication have over other serial protocols such as I^2C and USB?
6. Give five example ECUs that may be found on a modern car.

References

13.1. The Segway website. http://www.segway.com

13.2. Copeland, B. R. (2008). The Design of PID Controllers using Ziegler–Nichols Tuning. http://www.eng.uwi.tt/depts/elec/staff/copeland/ee27B/Ziegler_Nichols.pdf

13.3. 2-Axis Compass with Algorithms. HMC6352. Document #900307. Rev. D. January 2006. http://www.magneticsensors.com/

13.4 Servo modification for 360 degree rotation. http://www.embeddedtronics.com/servo.html

13.5. The CAN section of the Bosch website. www.can.bosch.com

13.6. Microchip Technology (2003). MCP2551 High-Speed CAN Transceiver Data Sheet. Document DS21730E.

13.7. mbed CAN-Bus Demo Board. http://mbed.org/forum/news-announce

CHAPTER 14

Letting Go of the mbed Libraries

Chapter Outline

14.1 Introduction

The mbed library contains many useful functions, which allow us to write simple and effective code. This seems a good thing, but it is also sometimes limiting. What if we want to use a peripheral in a way not allowed by any of the functions? Therefore it is useful to understand how peripherals can be configured by direct access to the microcontroller's registers. In turn, this leads to a deeper insight into some aspects of how a microcontroller works. As a by-product, and because we will be working at the bit and byte level, this study develops further skills in C programming.

Fast and Effective Embedded Systems Design. DOI: 10.1016/B978-0-08-097768-3.00014-3

It is worth issuing a very clear health warning at this early stage: this chapter is in some ways more complex than any of the others that have gone before. It introduces some of the complexity of the LPC1768 microcontroller, which lies at the heart of the mbed, a complexity which the mbed designers rightly wish to keep from you. Your own curiosity, ambition or your professional needs may, however, lead you to want to work at this deeper level.

Working with the complexity of this chapter may result in two opposing feelings. One is a sense of gratitude to the writers of the mbed libraries, that they have saved you the complexity of controlling the peripherals directly. At the opposite extreme, getting to grips with the chapter should also be a very liberating experience, like throwing away the water wings after you have learnt to swim. At the moment you probably think that you cannot write any program unless you have the library functions at the ready. When you are through with this chapter you will realize that you are no longer dependent on the library; you use it when you want, and write your own routines when you want – the choice becomes yours!

This chapter may be read in sequence with all other chapters. Alternatively, it can be read in different sections as extensions of earlier chapters. We will refer to Reference 2.3, the LPC1768 datasheet, and even more Reference 2.4, its user manual. Because we are now working at microcontroller level, rather than mbed level, we will have to take more care about how the microcontroller pins connect with the mbed pins. Therefore you will also need to have the mbed schematics, Reference 2.2, ready.

14.2 Control Register Concepts

It is useful at this stage to understand a little more about how the microcontroller central processing unit (CPU) interacts with its peripherals. Each of these has one or more *system control registers* which act as the doorway between the CPU and the peripheral. To the CPU, these registers look just like memory locations. They can usually be written to and read from, and each has its own address within the memory map. The clever part is that each of the bits in the register is wired across to the peripheral. They might carry control information, sent by the CPU to the peripheral, or they might return status information from the peripheral. They might also provide an essential path for data. The general idea of this is illustrated in Figure 14.1. The microcontroller peripherals also usually generate interrupts, for example to flag when an analog-to-digital conversion is complete or a new word has been received by a serial port.

Early in any program, the programmer must write to the control registers, to set up the peripherals to the configuration needed. This is called *initialization*, and turns the control registers from a general-purpose and non-functioning piece of hardware into something that is useful for the project in hand. In the mbed, this task is undertaken in the mbed utilities; in this chapter, we move to doing this work ourselves. In all of this, an important question arises: what happens in that short period *after* power has been applied, but *before* the peripherals

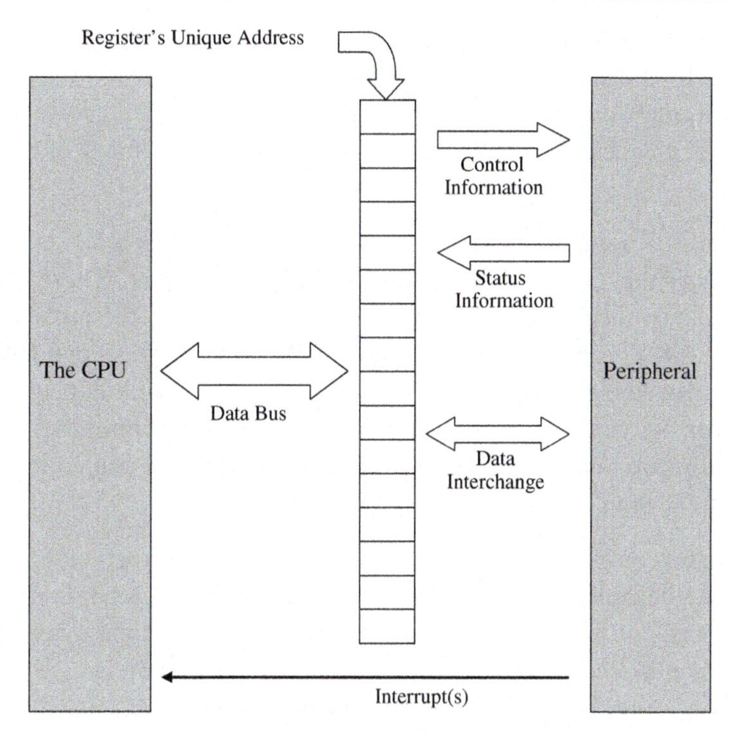

Figure 14.1:
The principle of a control register

have been set up by the program? At this time an embedded system is powered, but not under complete control. Fortunately, all makers of microcontrollers design in a *reset condition* for each control register. This generally puts a peripheral into a predictable, and often inactive, state. We can look this up, and write the initialization code accordingly. On some occasions, the reset state may be the state that we need.

A central theme of this chapter is the exploration and use of some of the LPC1768 control registers. Although we draw information from the LPC1768 data, the chapter is self-contained, with all necessary data included. However, we show which table in the manual the data is taken from, so it is easy to cross-refer. We hope this gives you the confidence and ability to move on to explore and apply the other registers.

14.3 Digital Input/Output

14.3.1 mbed Digital Input/Output Control Registers

The digital input/output (I/O) is a useful place to start our study of control registers, as this is simpler than any other peripherals on the LPC1768. This has its digital I/O arranged nominally as five 32-bit ports; yes, that implies a stunning 160 bits. Only parts of these are used, however, for example Port 0 has 28 bits implemented, Port 2 has 14, and Ports 3 and 4 have only two each. In the end, among its 100 pins, the LPC1768 has around 70 general purpose I/O pins available. However, the mbed has only 40 accessible pins, so only a subset of the microcontroller pins actually appears on the mbed interconnect, many of which are shared with other features.

As we have seen, it is possible to set each port pin as an input or as an output. Each port has a 32-bit register that controls the direction of each of its pins. These are called the **FIODIR** registers. To specify which port the register relates to, the port number is embedded within the register name, as shown in Table 14.1. For example, **FIO0DIR** is the direction register for Port 0. Each bit in this register then controls the corresponding bit in the I/O port, for example bit 0 in the direction register controls bit 0 in the port. If the bit in the direction register is set to 1, then that port pin is configured as an output; if the bit is set to 0, the pin is configured as an input.

It is sometimes more convenient not to work with the full 32-bit direction register, especially when we might just be thinking of one or two bits within the register. For this reason it is also possible to access any of the bytes within the larger register, as single-byte registers. These registers have a number code at their end. For example, **FIO2DIR0** is byte 0 of the Port 2 direction register, also seen in Table 14.1. From the table you can see that the address of the whole word is shared by the address of the lowest byte.

Table 14.1: Example digital I/O control registers

Register name	Register function	Register address
FIO*n*DIR	Sets the data direction of each pin in Port *n*, where *n* takes value 0 to 4. A port pin is set to output when its bit is set to 1, and as input when it is set to 0. Accessible as word. Reset value = 0, i.e. all bits are set to input on reset	—
FIO0DIR	Example of above for Port 0	0x2009C000
FIO2DIR	Example of above for Port 2	0x2009C040
FIO*n*DIR*p*	Sets the data direction of each pin in byte *p* of Port *n*, where *p* takes value 0 to 3. A port pin is set to output when its bit is set to 1. Accessible as byte	—
FIO0DIR0	Example of above, Port 0 byte 0	0x2009C000
FIO0DIR1	Example of above, Port 0 byte 1	0x2009C001
FIO2DIR0	Example of above, Port 2 byte 0	0x2009C040
FIO0PIN0 FIO2PIN0	Sets the data value of each bit in least significant byte of Port 0 or 2. Accessible as byte. Reset value = 0	0x2009C014 0x2009C054

A second set of registers, called **FIOPIN**, holds the data value of the microcontroller's pins, whether they have been set as input or output. Again, two are seen in Table 14.1, with the same naming pattern as the **FIODIR** registers. If a port bit has been set as an output, then writing to its corresponding bit in its **FIOPIN** register will control the logic value placed on that pin. If the pin has been set as input, then reading from that bit will tell you the logic value asserted at the pin. The **FIODIR** and **FIOPIN** registers are the only two register sets we need to worry about for our first simple I/O programs.

14.3.2 A Digital Output Application

A digital output can be configured and then tested simply by flashing a light-emitting diode (LED). We did this near the beginning of the book, in Program Example 2.1. We will look at the complete method for making this happen, working directly with the microcontroller registers. This first program, Program Example 14.1, will use Port 2 pin 0 as the digital output. The mbed schematic (Reference 2.2) shows that this pin is routed to the mbed pin 26.

Follow through Program Example 14.1 with care. Notice that for once we are *not* writing `#include "mbed.h"`! At the program's start the names of two registers from Table 14.1 are defined to equal the contents of their addresses. This is done by defining the addresses as pointers (see Section B.8.2), using the * operator. The format of this is not entirely simple, and for the present purposes we can use it as shown. Further information can, however, be found in Reference 14.1. Having defined these pointers, we can now refer to the register names in the code rather than worrying about the addresses. The C keyword **volatile** is also applied. This is used to define a data type whose value may change outside program control. This is a common occurrence in embedded systems, and is particularly applicable to memory-mapped registers, such as we are using here, which can be changed by the external hardware.

Within the **main( )** function the data direction register of Port 2, byte 0 is set to output, by setting all bits to Logic 1. A **while** loop is then established, just as in Program Example 2.1. Within this loop, pin 0 of Port 2 is set high and low in turn. Check the method of doing this, if you are not familiar with it, as this shows one way of manipulating a single bit within a larger word. To set the bit high, it is ORed with Logic 1. All other bits are ORed with Logic 0, so will not change. To set it low, it is ANDed with binary 1111 1110. This has the effect of returning the least significant bit (LSB) to 0, while leaving all other bits unchanged. The 1111 1110 value is derived by taking the logic inversion of 0x01, using the C ~ operator.

The delay function should be self-explanatory, and is explored more in Exercise 14.1. Its function prototype appears early in the program.

```
/*Program Example 14.1: Sets up a digital output pin using control registers, and
flashes an led.

*/
// function prototypes
void delay(void);
//Define addresses of digital i/o control registers, as pointers to volatile data
#define FIO2DIR0    (*(volatile unsigned char *)(0x2009C040))
#define FIO2PIN0    (*(volatile unsigned char *)(0x2009C054))

int main() {
  FIO2DIR0=0xFF;  // set port 2, lowest byte to output
   while(1) {
    FIO2PIN0 |= 0x01;  // OR bit 0 with 1 to set pin high
    delay();
    FIO2PIN0 &= ~0x01; // AND bit 0 with 0 to set pin low
    delay();
   }
  }
//delay function
  void delay(void){
   int j;             //loop variable j
   for (j=0;j<1000000;j++) {
    j++;
    j--;              //waste time
   }
  }
```

Program Example 14.1 Manipulating control registers to flash an LED

Connect an LED between an mbed pin 26 and 0 V, and compile, download and run the code. Press reset and the LED should flash.

Exercise 14.1

With an oscilloscope, carefully measure the duration of the delay function in Program Example 14.1. Estimate the execution time for the j++ and j−− instructions. Adjust the delay function experimentally so that it is precisely 100 ms. Then create a new 1 s delay function, which works by calling the 100 ms one 10 times. Now write a library delay routine, which gives a delay of *n* ms, where *n* is the parameter sent.

■

14.3.3 Adding a Second Digital Output

Following the principles outlined above, we can add further digital outputs. Here, we add a second LED and make a flashing pattern. Port 2 pin 1 is used, which connects to mbed pin 25. Add a second LED to this pin, preferably of a different color. We have already defined the direction control and pin I/O registers for Port 2, so we do not need to add any new registers. We do, however, add a variable to allow us to generate an interesting flashing pattern.

Copy Program Example 14.1 to a new program, and add into it the following variable declaration, prior to the main program function:

```
char i;
```

Also add the **for** loop of Program Example 14.2 at the end of the while loop. Notice that now we are toggling bit 1 of the **FIO2PIN0** register. We do this as before by ORing with 0x02, and then ANDing with the inverse of 0x02. This is where our second LED is connected.

```
for (i=1;i<=3;i++){
  FIO0PIN0 |= 0x02;       // set port 2 pin 1 high (mbed pin 25)
  delay();
  FIO0PIN0 &= ~0x02;     // set port 2 pin 1 low
  delay();
}
```

Program Example 14.2 (code fragment) Controlling a second LED output

Run the program. It should flash the first LED once, followed by three flashes of the second. If you have LEDs of different colors this will give you a pattern something like:

green – red – red – red – green – red – red – red – green – ...

Exercise 14.2

In the mbed circuit diagrams (Reference 2.2) identify the LPC1768 pins that drive the onboard LEDs. Rewrite Program Example 14.2 so that the onboard LEDs are activated, instead of using external ones.

14.3.4 Digital Inputs

We can create digital inputs simply by setting a port bit to input, using the correct bit in an **FIODIR** register. Program Example 14.3 now develops the previous example, by including a digital input from a switch. The state of the switch changes the loop pattern, determining which LED flashes three times and which flashes just once in a cycle. This then gives a control system which has outputs that are dependent on particular input characteristics. It uses the outputs already used, and adds bit 0 of Port 0 as a digital input. Looking at the schematic, we can see that this is pin 9 of the mbed.

It should not be difficult to follow Program Example 14.3 through. As before, we see the necessary register addresses defined. Directly inside the main function Port 0 byte 0 is set as a digital input, noting that a Logic 0 sets the corresponding pin to input. We have set all of byte 0 to input by sending 0x00; we could of course set each pin within the byte individually if needed. Moreover, this setting is not fixed; we can, if needed, change a pin from input to output as a program executes. After all this, a reading of Table 14.1 reminds us that the reset value of all ports is as input; therefore, this little bit of code is not actually necessary – try removing it when you run the program. However, it is good practice to reassert values that are said to be in place due to the reset; it gives you the confidence that the value is in place, and it is a definite statement in the code of a setting that you want.

The **while** loop then starts, and at the beginning of this we see the **if** statement testing the digital input value. The **if** condition uses a bit mask to discard the value of all the other pins on Port 0 byte 0 and simply returns a high or low result dependent on pin 8 alone. The variables **a** and **b** will hold values which will change depending on the switch position. A little later, we see the **a** and **b** values used to define the port values for the green and red LEDs. If **b** has been set to 0x01 before the **for** loop, then the red LED will flash three times; if it has been set to 0x02, then the green LED will flash three times.

```
/* Program Example 14.3: Uses digital input and output using control registers, and
flashes an LED. LEDS connect to mbed pins 25 and 26. Switch input to pin 9.
                                                                                  */

// function prototypes
void delay(void);
//Define Digital I/O registers
#define FIO0DIR0 (*( volatile unsigned char *)(0x2009C000))
#define FIO0PIN0 (*( volatile unsigned char *)(0x2009C014))
#define FIO2DIR0 (*(volatile unsigned char *)(0x2009C040))
#define FIO2PIN0 (*(volatile unsigned char *)(0x2009C054))
//some variables
char a;
char b;
char i;

int main() {
  FIO0DIR0=0x00;              // set all bits of port 0 byte 0 to input
  FIO2DIR0=0xFF;              // set port 2 byte 0 to output
    while(1) {
     if (FIO0PIN0&0x01==1){   // bit test port 0 pin 0 (mbed pin 9)
      a=0x01;                 // this reverses the order of LED flashing
      b=0x02;                 // based on the switch position
     }
     else {
      a=0x02;
      b=0x01;
     }
     FIO2PIN0 |= a;
     delay();
     FIO2PIN0 &= ~a;
     delay();
```

```
      for (i=1;i<=3;i++){
       FIO2PIN0 |= b;
       delay();
       FIO2PIN0 &= ~b;
       delay();
      }
    }                         //end while loop
  }
void delay(void){             //delay function.
//program continues
```

Program Example 14.3 Combined digital input and output

To run this program, set up the circuit of Figure 3.6, except that the green and red LEDs should connect to mbed pins 25 and 26, respectively, and the switch input is on pin 9. Compile and run. You should see that the position of the switch toggles the flashing LEDs between the patterns

green – red – red – red – green – red – red – red – green – ...

and

green – green – green – red – green – green – green – red – green – ...

Exercise 14.3

Rewrite Program Example 14.3 so that it runs on the exact circuit of Figure 3.6.

14.4 Getting Deeper into the Control Registers

To continue with this chapter we need to get further into the use of control registers. This section therefore looks at some of the registers that control features across the microcontroller – informally called 'global' registers – that will be used in later sections. These relate to the allocation of pins, setting clock frequency and controlling power. This is by no means a complete survey, and we will not even see or make use of many of the features of this microcontroller.

14.4.1 Pin Select and Pin Mode Registers

One of the reasons that modern microcontrollers are so versatile is that most pins are multifunctional. They can be allocated to different peripherals and used in different ways. With the mbed library, this flexibility is (quite reasonably) more or less hidden from the user;

Table 14.2: PINSEL1 register

PINSEL1	Pin name	Function when 00	Function when 01	Function when 10	Function when 11	Reset value
1:0	P0.16	GPIO Port 0.16	RXD1	SSEL0	SSEL	00
3:2	P0.17	GPIO Port 0.17	CTS1	MISO0	MISO	00
5:4	P0.18	GPIO Port 0.18	DCD1	MOSI0	MOSI	00
7:6	P0.19[1]	GPIO Port 0.19	DSR1	Reserved	SDA1	00
9:8	P0.20[1]	GPIO Port 0.20	DTR1	Reserved	SCL1	00
11:10	P0.21[1]	GPIO Port 0.21	RI1	Reserved	RD1	00
13:12	P0.22	GPIO Port 0.22	RTS1	Reserved	TD1	00
15:14	P0.23[1]	GPIO Port 0.23	AD0.0	I2SRX_CLK	CAP3.0	00
17:16	P0.24[1]	GPIO Port 0.24	AD0.1	I2SRX_WS	CAP3.1	00
19:18	P0.25	GPIO Port 0.25	AD0.2	I2SRX_SDA	TXD3	00
21:20	P0.26	GPIO Port 0.26	AD0.3	AOUT	RXD3	00
23:22	P0.27[1,2]	GPIO Port 0.27	SDA0	USB_SDA	Reserved	00
25:24	P0.28[1,2]	GPIO Port 0.28	SCL0	USB_SCL	Reserved	00
27:26	P0.29	GPIO Port 0.29	USB_D+	Reserved	Reserved	00
29:28	P0.30	GPIO Port 0.30	USB_D_	Reserved	Reserved	00
31:30	–	Reserved	Reserved	Reserved	Reserved	00

[1]Not available on 80-pin package.
[2]Pins P0.27 and P0.28 are open drain for I^2C bus compliance.
Redrawn from NXP Semiconductors UM10360 LPC17xx User Manual, Table 79: Pin function select register 1 (PINSEL1 – address 0x4002 C004) bit description, with permission of NXP.

the libraries tidily make the allocations for you, without you even knowing. If they did not, you would be faced with a bewildering choice of possibilities every time you tried to develop an application. As our expertise grows, however, it is good to know that some of these possibilities are available.

Two important sets of registers used in the LPC1768 are called **PINSEL** and **PINMODE**. The **PINSEL** register can allocate each pin to one of four possibilities. An example of part of one register that we will be using soon, **PINSEL1**, is shown in Table 14.2. This controls the upper half of Port 0. The first column shows the bit number within the register; each line details two bits. The second column shows the microcontroller pin that is being controlled. The two bits under consideration can have four possible combinations; each of these connects the pin in a different way. These are shown in the next four columns. Do not worry if some of the abbreviations shown have little meaning to you; we will pick out the ones we need, when we need them.

Table 14.3: PINMODE0 register

PINMODE0	Symbol	Value	Description	Reset value
1:0	P0.00MODE		Port 0 pin 0 on-chip pull-up/down resistor control	00
		00	P0.0 pin has a pull-up resistor enabled	
		01	P0.0 pin has repeater mode enabled	
		10	P0.0 pin has neither pull-up nor pull-down	
		11	P0.0 pin has a pull-down resistor enabled	
3:2	P0.01MODE		Port 0 pin 1 control, see P0.00MODE	00
	continued to P0.15MODE			

Redrawn from NXP Semiconductors UM10360 LPC17xx User Manual, Table 87: Pin Mode select register 0 (PINMODE0 – address 0x4002 C040) bit description, with permission of NXP.

Let us take as an example the line showing the effect of bits 21:20, i.e. line 11 of the table; these control bit 26 of Port 0. Column 3 shows that if the bits are 00, the pin is allocated to Port 0 bit 26, i.e. the pin is connected as general-purpose I/O. Importantly, the final column shows that this is also the value when the chip is reset. In other words, as long as we only want to use digital I/O, we do not need to worry about this register at all, as the reset value is the value we want. If the bits are set to 01, the pin is allocated to input 3 of the analog-to-digital converter (ADC). If set to 10, the pin is used for analog output, i.e. the digital-to-analog converter (DAC) output. If set to 11, the pin is allocated as Receiver input for universal asynchronous receiver/transmitter (UART) 3.

Turning to the **PINMODE** registers, partial details of one of these are shown in Table 14.3. This is **PINMODE0**, which controls the input characteristics of the lower half of Port 0. The pattern is the same for every pin, so there is no need for repetition. It is easy to see that pull-up and pull-down resistors (as seen in Figure 3.5) are available, and this is explored in Program Example 14.4. The repeater mode is a neat little facility which enables pull-up resistor when the input is a Logic 1, and pull-down if it is low. If the external circuit changes so that the input is no longer driven, then the input will hold its most recent value.

Exercise 14.4

Change the hardware for Program Example 14.3 so that you use an SPST (single-pole, single-throw) switch, for example a push-button, instead of the toggle (single-pole, double-throw, SPDT) switch. Connect it first between pin 9 and ground, and run the program with no change. This should run as before, because you are depending on the pull-up resistor being in place owing to the reset value of the **PINMODE** register. In diagrammatic terms, you have moved from the switch circuit of Figure 3.5a to that of Figure 3.5b. Now change the setting of **PINMODE0** so that the pull-down resistor is enabled, and connect the switch between pin 9 and 3.3 V, i.e. applying Figure 3.5c. The program should again work, but with the changed input mode selection.

■

14.4.2 Power Control and Clock Select Registers

Power control and clock frequency are very closely linked. Every clock transition causes the circuit of the microcontroller to take a tiny pulse of current; the more transitions, the more the current taken. Hence, a processor or peripheral running at a high clock speed will cause high power consumption; one running at a low clock frequency will consume less power. One with its clock switched off, even if it is powered up, will (if a purely digital circuit using CMOS technology) take negligible power. To conserve power, it is possible to turn off the clock source to many of the LPC1768 peripherals. This power management is controlled by the **PCONP** register, seen in part in Table 14.4. Where a bit is set to 1, the

Table 14.4: Power Control register PCONP (part of)

Bit	Symbol	Description	Reset value
0	–	Reserved	NA
1	PCTIM0	Timer/Counter 0 power/clock control bit	1
2	PCTIM1	Timer/Counter 1 power/clock control bit	1
3	PCUART0	UART0 power/clock control bit	1
4	PCUART1	UART1 power/clock control bit	1
5	–	Reserved	NA
6	PCPWM1	PWM1 power/clock control bit	1
7	PCI2C0	I^2C0 interface power/clock control bit	1
8	PCSPI	SPI interface power/clock control bit	1
9	PCRTC	RTC power/clock control bit	1
10	PCSSP1	SSP 1 interface power/clock control bit	1
11	–	Reserved	NA
12	PCADC	A/D converter (ADC) power/clock control bit Note: Clear the PDN bit in the AD0CR before clearing this bit, and set this bit before setting PDN	0
continued to bit 28			

Redrawn from NXP Semiconductors UM10360 LPC17xx User Manual, Table 46: Power Control for Peripherals register (PCONP – address 0x400F C0C4) bit description, with permission of NXP.

peripheral is enabled; when set to 0 it is disabled. It is interesting to note that some peripherals, such as the serial peripheral interface (SPI), are reset in the enabled mode, and others, such as the ADC, are reset disabled.

Aside from being able to switch the clock to a peripheral on or off, there is some control over the peripheral's clock frequency itself. This will control the peripheral's speed of operation as well as its power consumption. This clock frequency is controlled by the **PCLKSEL** registers. Peripheral clocks are derived from the clock that drives the CPU, which is called **CCLK**. For the mbed, **CCLK** normally runs at 96 MHz. Partial details of **PCLKSEL0** are shown in Table 14.5, which shows that two bits are used per peripheral to control the clock frequency to each. The four possible combinations are shown in Table 14.6. This shows that the **CCLK** frequency itself can be used to drive the peripheral. Alternatively, it can be divided by 2, 4 or 8. **CCLK** is derived from the main oscillator circuit, and can be manipulated in a number of interesting ways; however, we do not go into the details of that in this book.

Table 14.5: Peripheral Clock Selection register PCLKSEL0

Bit	Symbol	Description	Reset value
1:0	PCLK_WDT	Peripheral clock selection for WDT	00
3:2	PCLK_TIMER0	Peripheral clock selection for TIMER0	00
5:4	PCLK_TIMER1	Peripheral clock selection for TIMER1	00
7:6	PCLK_UART0	Peripheral clock selection for UART0	00
9:8	PCLK_UART1	Peripheral clock selection for UART1	00
11:10	–	Reserved	NA
13:12	PCLK_PWM1	Peripheral clock selection for PWM1	00
15:14	PCLK_I2C0	Peripheral clock selection for I^2C0	00
17:16	PCLK_SPI	Peripheral clock selection for SPI	00
19:18	–	Reserved	NA
21:20	PCLK_SSP1	Peripheral clock selection for SSP1	00
23:22	PCLK_DAC	Peripheral clock selection for DAC	00
25:24	PCLK_ADC	Peripheral clock selection for ADC	00
27:26	PCLK_CAN1	Peripheral clock selection for CAN1[1]	00
29:28	PCLK_CAN2	Peripheral clock selection for CAN2[1]	00
31:30	PCLK_ACF	Peripheral clock selection for CAN acceptance filtering[1]	00

[1]PCLK_CAN1 and PCLK_CAN2 must have the same PCLK divide value when the CAN function is used.

Redrawn from NXP Semiconductors UM10360 LPC17xx User Manual, Table 40: Peripheral Clock Selection register (PCLKSEL0 – address 0x400F C1A8) bit description, with permission of NXP.

Table 14.6: Peripheral Clock Selection register bit values

PCLKSEL0 and PCLKSEL1 individual peripheral's clock select options	Function	Reset value
00	PCLK_peripheral = CCLK/4	00
01	PCLK_peripheral = CCLK	
10	PCLK_peripheral = CCLK/2	
11	PCLK_peripheral = CCLK/8, except for CAN1, CAN2 and CAN filtering when "11" selects = CCLK/6	

Redrawn from NXP Semiconductors UM10360 LPC17xx User Manual, Table 42: Peripheral Clock Selection register bit values, with permission of NXP.

14.5 Using the DAC

We now turn to controlling the DAC through its registers, trying to replicate and develop the work done in Chapter 4. Remind yourself of the general block diagram of the DAC, as seen in Figure 4.1. It is worth mentioning here that the positive reference voltage input to the LPC1768, shared by both ADC and DAC, is called V_{REFP}, and appears on pin 12 of the microcontroller. A careful look at the mbed circuit diagrams (Reference 2.2) shows that this is connected to the power supply 3.3 V, filtered to limit noise input. This is a sensible connection, and we are unable to change it. The negative reference voltage input, called V_{REFN}, appears on pin 15 of the microcontroller. On the mbed it is connected directly to ground.

14.5.1 mbed DAC Control Registers

As with all peripherals, the DAC has a set of registers that control its activity. In terms of the 'global' registers that we have just seen, the DAC power is always enabled, so there is no need to consider the **PCONP** register. The *only* pin that the DAC output is available on is Port 0 pin 26, so we must allocate this pin appropriately through the **PINSEL1** register, as seen in Table 14.2. The DAC output is labeled AOUT here. It is no surprise to see in the mbed schematics that this pin is connected to mbed pin 18, the only mbed analog output.

The only register specific to the DAC that we will use is the **DACR** register. This is comparatively simple to grasp, and is shown in Table 14.7. We can see that the digital input to the DAC must be placed in here, in bits 6–15. Most of the rest of the bits are unused, apart from the bias bit, explained in the table.

Applying Equation 4.1 to this 10-bit DAC, its output is given by:

$$V_O = (V_{REFP} \times D)/1024 = (3.3 \times D)/1024 \qquad 14.1$$

where D is the value of the 10-bit number placed in bits 15–6 of the **DACR** register.

Table 14.7: The DACR register

Bit	Symbol	Value	Description	Reset value
5:0	–		Reserved, user software should not write ones to reserved bits. The value read from a reserved bit is not defined	NA
15:6	VALUE		After the selected settling time after this field is written with a new VALUE, the voltage on the AOUT pin (with respect to V_{SSA}) is VALUE $\times$ $((V_{REFP} - V_{REFN})/1024) + V_{REFN}$	0
16	BIAS[1]	0	The settling time of the DAC is 1 μs max and the maximum current is 700 μA. This allows a maximum update rate of 1 MHz	0
		1	The settling time of the DAC is 2.5 μs max, and the maximum current is 350 μA. This allows a maximum update rate of 400 kHz	0
31:17	–		Reserved, user software should not write ones to reserved bits. The value read from a reserved bit is not defined	NA

[1]The settling times noted in the description of the BIAS bit are valid for a capacitance load on the AOUT pin not exceeding 100 pF. A load impedance value greater than that value will cause settling time longer than the specified time.
Redrawn from NXP Semiconductors UM10360 LPC17xx User Manual, Table 539: D/A Converter register (DACR – address 0x4008 C000) bit description, with permission of NXP.

14.5.2 A DAC Application

Let's try now a simple program to drive the DAC, accessing it through the microcontroller control registers. Program Example 14.4 replicates the simple sawtooth output, which we first achieved in Program Example 4.2. The program follows the familiar pattern of defining register addresses, and then setting these appropriately early in the main function. The **PINSEL1** register is set to select DAC output on port bit 0.26. An integer variable, called **dac_value**, is then repeatedly incremented and transferred to the DAC input, in register **DACR**. It has to be shifted left six times, to place it in the correct bits of the **DACR** register. Between each new value of DAC input a delay is introduced. The effect of this is explored in Exercise 14.5.

```
/* Program Example 14.4: Sawtooth waveform on DAC output. View on oscilloscope. Port
0.26 is used for DAC output, i.e. mbed Pin 18
                                                                                */
// function prototype
void delay(void);
// variable declarations
int dac_value;                //the value to be output
//define addresses of control registers, as pointers to volatile data
#define DACR (*(volatile unsigned long *)(0x4008C000))
```

```
#define PINSEL1 (*(volatile unsigned long *)(0x4002C004))

int main(){
 PINSEL1=0x00200000; //set bits 21-20 to 10 to enable analog out on P0.26
   while(1){
    for (dac_value=0;dac_value<1023;dac_value=dac_value+1){
   DACR=(dac_value<<6);
    delay();
    }
   }
  }

  void delay(void)  //delay function.
//program continues
```

Program Example 14.4 Sawtooth output on the DAC

Compile the program and run it on an mbed; no external connections are needed. View the output from the mbed pin 18 on an oscilloscope.

Exercise 14.5

Measure the period of the sawtooth waveform. How does it relate to the delay value you measured in Exercise 14.1? Try varying the period by varying the delay value, or removing it altogether. Can you estimate how long a single digital-to-analog conversion takes?

14.6 Using the ADC

We now turn to controlling the ADC through its registers, trying to replicate and develop the work done in Chapter 5. It is worth glancing back at Figure 5.1, as this represents many of the features that we will need to control. This will include selecting (or at least knowing the value of) the voltage reference, clock speed and input channel, starting a conversion, detecting a completion and reading the output data. The ADC has a set of registers that control all this activity. The LPC1768 has eight inputs to its ADC, which appear – in order from input 0 to input 7 – on pins 9–6, 21, 20, 99 and 98 of the microcontroller. A study of the mbed circuit shows that the lower six of these are used, connected to pins 15–20 inclusive of the mbed.

14.6.1 mbed ADC Control Registers

The LPC1768 has a number of registers that control its ADC, particularly in its more sophisticated operation. However, we will only apply two of these, the ADC control register (**ADCR**) and the Global Data Register (**ADGDR**). These are detailed in Tables 14.8 and 14.9. As we have seen, the ADC can also be powered down; indeed, on microcontroller reset it is

Table 14.8: The AD0CR register

Bit	Symbol	Value	Description	Reset value
7:0	SEL		Selects which of the AD0.7:0 pins is (are) to be sampled and converted. For AD0, bit 0 selects Pin AD0.0, and bit 7 selects pin AD0.7. In software-controlled mode, only one of these bits should be 1. In hardware scan mode, any value containing 1 to 8 ones is allowed. All zeroes is equivalent to 0x01	0x01
15:8	CLKDIV		The APB clock (PCLK_ADC0) is divided by (this value plus one) to produce the clock for the A/D converter, which should be less than or equal to 13 MHz. Typically, software should program the smallest value in this field that yields a clock of 13 MHz or slightly less, but in certain cases (such as a high-impedance analog source) a slower clock may be desirable	0
16	BURST	1	The AD converter does repeated conversions at up to 200 kHz, scanning (if necessary) through the pins selected by bits set to ones in the SEL field. The first conversion after the start corresponds to the least significant 1 in the SEL field, then higher numbered 1-bits (pins) if applicable. Repeated conversions can be terminated by clearing this bit, but the conversion that's in progress when this bit is cleared will be completed **Remark:** START bits must be 000 when BURST = 1 or conversions will not start	0
		0	Conversions are software controlled and require 65 clocks	
20:17	–		Reserved, user software should not write ones to reserved bits. The value read from a reserved bit is not defined	NA
21	PDN	1	The A/D converter is operational	0
		0	The A/D converter is in power-down mode	
23:22	–		Reserved, user software should not write ones to reserved bits. The value read from a reserved bit is not defined	NA
26:24	START		When the BURST bit is 0, these bits control whether and when an A/D conversion is started:	0
		000	No start (this value should be used when clearing PDN to 0)	
		001	Start conversion now	
Further more advanced options for START control are available, which also then apply EDGE				

Redrawn from NXP Semiconductors UM10360 LPC17xx User Manual, Table 531: A/D Control Register (AD0CR – address 0x4003 4000) bit description, with permission of NXP.

switched off. Therefore, to enable it we will have to set bit 12 in the **PCONP** register, seen in Table 14.4.

14.6.2 An ADC Application

Program Example 14.5 provides a good opportunity to see many of the control registers in action. Channel 1 of the ADC is applied, which connects to pin 16 of the mbed.

The configuration of the ADC must be done with care, so let's read the program with diligence, starting from the comment 'initialize the ADC'. The ADC channel we want is multiplexed with bit 24 of Port 0, so first we must allocate this pin to the ADC. This is done through bits 17 and 16 of the **PINSEL1** register (Table 14.2). We then enable the ADC, through the relevant bit in the **PCONP** register (Table 14.4). There follows quite a complex process of configuring the ADC, through the **AD0CR** register. We could set this register by transferring a single word. Instead, we shift in the relevant bits in turn. Check each with Table 14.8.

The data conversion then starts in the **while** loop which follows. The comments contained in the program listing should give you a good picture of each stage.

```
/* Program Example 14.5: A bargraph meter for ADC input, using control registers to set
up ADC and digital I/O
                                                                        */
// variable declarations
char ADC_channel=1;  // ADC channel 1
int ADCdata;         //this will hold the result of the conversion
int DigOutData=0;    //a buffer for the output display pattern

// function prototype
void delay(void);

//define addresses of control registers, as pointers to volatile data
//(i.e. the memory contents)
#define PINSEL1      (*(volatile unsigned long *)(0x4002C004))
#define PCONP        (*(volatile unsigned long *)(0x400FC0C4))
#define AD0CR        (*(volatile unsigned long *)(0x40034000))
#define AD0GDR       (*(volatile unsigned long *)(0x40034004))
#define FIO2DIR0     (*(volatile unsigned char *)( 0x2009C040))
#define FIO2PIN0     (*(volatile unsigned char *)( 0x2009C054))

int main() {
 FIO2DIR0=0xFF;// set lower byte of Port 2 to output, this drives bargraph

//initialize the ADC
 PINSEL1=0x00010000; //set bits 17-16 to 01 to enable AD0.1 (mbed pin 16)
 PCONP |=  (1 << 12);           // enable ADC clock
 AD0CR =   (1 << ADC_channel)   // select channel 1
         | (4 << 8)             // Divide incoming clock by (4+1), giving 4.8MHz
         | (0 << 16)            // BURST = 0, conversions under software control
```

```
        | (1 << 21)                 // PDN = 1, enables power
        | (1 << 24);                // START = 1, start A/D conversion now

 while(1) {                         // infinite loop
  AD0CR = AD0CR | 0x01000000;       //start conversion by setting bit 24 to 1,
                                    //by ORing
   // wait for it to finish by polling the ADC DONE bit
     while ((AD0GDR & 0x80000000) == 0) { //test DONE bit, wait till it's 1
    }
    ADCdata = AD0GDR;           // get the data from AD0GDR
    AD0CR &= 0xF8FFFFFF;        //stop ADC by setting START bits to zero
    // Shift data 4 bits to right justify, and 2 more to give 10-bit ADC
  // value - this gives convenient range of just over one thousand.
    ADCdata=(ADCdata>>6)&0x03FF;   //and mask
    DigOutData=0x00;               //clear the output buffer
  //display the data
    if (ADCdata>200)
     DigOutData=(DigOutData|0x01); //set the lsb by ORing with 1
    if (ADCdata>400)
     DigOutData=(DigOutData|0x02); //set the next lsb by ORing with 1
    if (ADCdata>600)
     DigOutData=(DigOutData|0x04);
    if (ADCdata>800)
     DigOutData=(DigOutData|0x08);
    if (ADCdata>1000)
     DigOutData=(DigOutData|0x10);

    FIO2PIN0 = DigOutData;         // set port 2 to Digoutdata
    delay();    // pause
    }
  }
    void delay(void){              //delay function
//program continues
```

Program Example 14.5 Applying the ADC as a bargraph

Connect an mbed with a potentiometer between 0 and 3.3 V with the wiper connected to mbed pin 16. Connect five LEDs between pin and ground, from pin 22 to pin 26 inclusive. Compile and download your code to the mbed, and press reset. Moving the potentiometer should alter the number of LEDs lit, from none at all to all five.

Exercise 14.6

Add a digital input switch as in the previous example to reverse the operation of the analog input. With the digital switch in one position the LEDs will light from right to left, but with the switch in the alternate position the **DigOutData** variable can be inverted to light LEDs in the opposite direction, from left to right.

Table 14.9: The AD0GDR register

Bit	Symbol	Description	Reset value
3:0	–	Reserved, user software should not write ones to reserved bits. The value read from a reserved bit is not defined	NA
15:4	RESULT	When DONE is 1, this field contains a binary fraction representing the voltage on the AD0[n] pin selected by the SEL field, as it falls within the range of V_{REFP} to V_{REFN}. Zero in the field indicates that the voltage on the input pin was less than, equal to, or close to that on V_{REFN}, while 0x3FF indicates that the voltage on the input was close to, equal to, or greater than that on V_{REFP}	NA
23:16	–	Reserved, user software should not write ones to reserved bits. The value read from a reserved bit is not defined	NA
26:24	CHN	These bits contain the channel from which the RESULT bits were converted (e.g. 000 identifies channel 0, 001 channel 1 ...)	NA
29:27	–	Reserved, user software should not write ones to reserved bits. The value read from a reserved bit is not defined	NA
30	OVERRUN	This bit is 1 in burst mode if the results of one or more conversions was (were) lost and overwritten before the conversion that produced the result in the RESULT bits. This bit is cleared by reading this register	0
31	DONE	This bit is set to 1 when an A/D conversion completes. It is cleared when this register is read and when the ADCR is written. If the ADCR is written while a conversion is still in progress, this bit is set and a new conversion is started	0

Redrawn from NXP Semiconductors UM10360 LPC17xx User Manual, Table 532: A/D Global Data Register (AD0GDR – address 0x4003 4004) bit description, with permission of NXP.

Exercise 14.7

Extend the bargraph so that it has eight or 10 LEDs.

14.6.3 Changing ADC Conversion Speed

One of the limitations of the mbed ADC library is the comparatively slow speed of conversion. This was explored in Exercise 5.6. Let's try now to vary this conversion speed by adjusting the ADC clock speed.

Table 14.8 tells us that the ADC clock frequency should have a maximum value of 13 MHz, and that it takes 65 cycles of the ADC clock to complete a conversion. A quick calculation shows that the minimum conversion time possible is therefore 5 μs. It takes a very careful reading of the LPC1768 user manual (Reference 2.4) to gain a full picture of how the ADC clock frequency is controlled. The ADC clock is derived from the main microcontroller clock; there are several stages of division that the user can control in order to set up a frequency as close to 13 MHz as possible. The first is through register **PCLKSEL0**, detailed in Tables 14.5 and 14.6. Bits 25 and 24 of **PCLKSEL0** control the ADC clock division. We have seen that for most peripherals, including the ADC, the clock can be divided by 1, 2, 4 or 8. On power-up the selection defaults to divide by 4. The clock may be further divided through bits 15–8 of the **AD0CR** register, seen in Table 14.8.

Program Example 14.6 replicates Program Example 5.5, with some interesting results. It is also a useful example, as it combines ADC, DAC and digital I/O, therefore illustrating how these can be used together. It is made up of elements from programs earlier in this chapter, sometimes with adjustments; it should be possible to follow it through without too much difficulty. As sections of the program repeat from earlier examples, only the newer parts are reproduced here. The full program listing can be downloaded from the book website.

```
/* Program Example 14.6: Explore ADC conversion times, programming control registers
directly. ADC value is transferred to DAC, while an output pin is strobed to indicate
conversion duration. Observe on oscilloscope
*/
....
....
int main() {
 FIO2DIR0=0xFF;                   // set lower bits port 2 to output
 PINSEL1=0x00210000; //set bits 21-20 to 10 for analog output (mbed p18)
     //and bits 17-16 to 01 to enable ADC channel 1 (AD0.1, mbed pin 16)

//initialize the ADC.
....
....

 while(1){     // infinite loop
  // start A/D conversion by modifying bits in the AD0CR register
    AD0CR &= (AD0CR & 0xFFFFFF00);
    FIO2PIN0 |= 0x01;      // OR bit 0 with 1 to set pin high
    AD0CR |= (1 << ADC_channel) | (1 << 24);
  // wait for it to finish by polling the ADC DONE bit
    while((AD0GDR & 0x80000000) == 0) {
    }
    FIO2PIN0 &= ~0x01;              // AND bit 0 with 0 to set pin low

    ADCdata = AD0GDR;               // get the data from AD0GDR
    AD0CR &= 0xF8FFFFFF;            //stop ADC by setting START bits to zero
  // shift data 4 bits to right justify, and 2 more to give 10-bit ADC value
    ADCdata=(ADCdata>>6)&0x03FF;  //and mask
```

```
    DACR=(ADCdata<<6);              //could be merged with previous line,
                                   // but separated for clarity
  //delay();                       //insert delay if wished
   }
  }
  ....
```

Program Example 14.6 Applying ADC, DAC and digital output, to measure conversion duration

You can use the same mbed configuration for this as you did for Program Example 14.5, although only the LED on pin 26 is necessary. Compile and run the program. First put an oscilloscope probe on pin 18, the DAC output. The voltage on this pin should change as the potentiometer is used. This confirms that the program is running. Now move the probe to pin 26; you will see the pin pulsing high for the duration of the ADC conversion. If you measure this, you should find it is 14 μs, or just under.

To calculate the ADC clock frequency, and hence conversion time, remember that the mbed CCLK frequency is 96 MHz. We have not touched the **PCLKSEL0** register, so the clock setting for the ADC will be 96 MHz divided by the reset value of 4, or 24 MHz. This is further divided by 5 in the **ADC0CR** setting seen in the program example, leading to an ADC clock frequency of 4.8 MHz, or a period of 0.21 μs. Sixty-five cycles of 0.21 μs leads to the measurement duration mentioned in the previous paragraph.

Exercise 14.8

1. Adjust the setting of the CLKDIV bits in **AD0CR** in Program Example 14.7 to give the fastest permissible conversion time. Run the program, and check that your measured value agrees with the predicted value.
2. Can you now account for the value of ADC conversion time you measured in Exercise 5.6?

14.7 A Conclusion on using the Control Registers

This chapter has explored the use of the LPC1768 control registers, in connection with use of the digital I/O, ADC and DAC peripherals. This discussion has demonstrated how these registers allow the peripherals to be controlled directly, without using the mbed libraries. This has allowed greater flexibility of use of the peripherals, at the cost of getting into the tiny detail of the registers and programming at the level of the bits that make them up. Ordinarily, we probably would not want to program like this; it is time consuming, inconvenient and error prone. However, if we need a configuration or setting not offered by the mbed libraries, this approach can be a way forward. While we have only worked in this way in connection with three of the peripherals, it is possible to do it with any of them. It is worth mentioning that these three peripherals are some of the simpler ones; some of the others require even more attention to detail.

Chapter Review

- This chapter introduced a different way of controlling the mbed peripherals. It demands a much deeper understanding of the mbed microcontroller, but allows for much greater flexibility.
- Some registers relate to just one peripheral and others relate to microcontroller performance as a whole.
- We have begun to implement features that are not currently available in the mbed library, for example in the change of the ADC conversion speed.
- The chapter only introduces a small range of the control registers that are used by the LPC1768. However, it should have given you the confidence to look up and begin to apply any that you need.

Quiz

1. The initialization section of a certain program reads:
   ```
   FIO0DIR0=0xF0;
   PINMODE0=0x0F;
   PINSEL1=0x00204000;
   ```
   Explain the settings that have been made.
2. An LPC1768 is connected with a 3.0 V reference voltage. Its DAC output reads 0.375 V. What is its input digital value?
3. On an LPC1768, the ADC clock is set at 4 MHz. How long does one conversion take?
4. A user wants to sample an incoming signal with an mbed ADC at 44 kHz or greater. What is the minimum ADC clock value?
5. Describe how the ADC clock frequency calculated in Question 4 can be set up.

Reference

14.1. ARM Technical Support Knowledge Articles. Placing C variables at specific addresses to access memory-mapped peripherals. http://infocenter.arm.com/help/index.jsp?topic=/com.arm.doc.faqs/ka3750.html. Accessed January 2012.

CHAPTER 15

Extension Projects

Chapter Outline

15.1 Where do We Go from Here?

A vast number of peripheral devices and applications can be used with mbed projects. In the scope of this book alone it is not possible to explore every technology or peripheral device, but the design and programming techniques covered will allow you to investigate wider concepts to develop advanced projects using the mbed.

This chapter discusses a number of extension topics which suggest other areas for investigating and enjoying, as well as helping to bridge the gap between laboratory prototyping and the world of mass manufacturing. It is not possible to cover each of these aspects in detail, but an overview is provided to allow you to be creative in your approach to mbed projects and to give a springboard to more advanced and innovative systems.

15.2 Pololu Robot for mbed

Pololu manufactures a number of robotics and electronics kits for developing embedded systems. In particular, the Pololu m3pi robot, which features an mbed control board and a host of sensors and actuators, is shown in Figure 15.1. The robot has two wheels and is able to turn on its central point by rotating the wheels in opposite directions. For full details, see Reference 15.1.

Fast and Effective Embedded Systems Design. DOI: 10.1016/B978-0-08-097768-3.00015-5

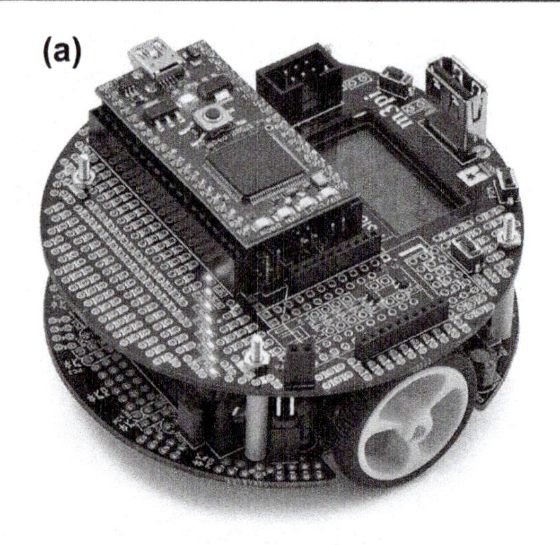

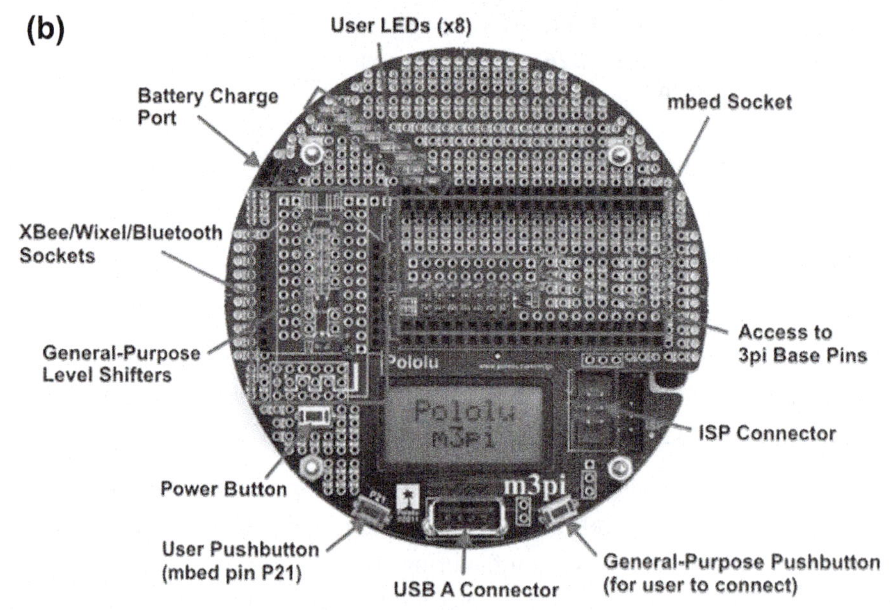

Figure 15.1:
(a) Pololu m3pi robot with (b) mbed control board (Reference 15.1). *(Image reproduced with permission of Pololu Corporation)*

The mp3i has an 8x2 character liquid crystal display (LCD) screen built in, as well as a number of buzzers and switches for input and output control. There are also several reflective sensors on the underside to allow development of line-tracking and maze-solving systems. Beyond this, there is a universal serial bus (USB) port and an XBee socket allowing wireless connectivity. Indeed, various projects have been prototyped by ARM to show the capabilities of the Pololu m3pi, in particular a system that allows control of the robot via a Nintendo Wii controller (see Reference 15.2).

15.3 Advanced Audio Projects

So far, the audio projects we have discussed have used very simple implementations of the analog-to-digital converter (ADC), the digital-to-analog converter (DAC) or the pulse width modulation ports. In addition, some very simple analog circuitry has been used when implementing simple digital audio sampling and analog reconstruction systems. However, the LPC1768 ADC is only 12-bit and the DAC is only 10-bit resolution, whereas high-quality audio generally requires 16-bit or even 24-bit resolution. Furthermore, having only a single DAC means that the mbed cannot directly output stereo (two-channel) audio. If high-quality stereo audio projects are to be developed it is possible to use a specialist ADC/DAC integrated circuit, such as the Texas Instruments TLV320AIC23b (Reference 15.3), which boasts stereo 24-bit resolution at sampling frequencies up to 96 kHz.

RS Components has announced the release of an mbed demo board (Reference 15.4) which includes a TLV320AIC23b ADC/DAC, and ARM has developed a number of libraries and support pages to interface the device (Reference 15.5). The mbed communicates with the TLV320 through an inter-integrated circuit sound (I^2S) port. This is a serial interface specifically designed for connecting digital audio devices together. We have not used it in this book, as it is not part of the 'official' mbed interconnection of Figure 2.1. However, it does appear in the LPC1768 block diagram of Figure 2.3, and there are means of routing it externally.

15.4 The Internet of Things

The Internet of Things is a phrase that has evolved to refer to a system of everyday objects and sensors being connected to the Internet, allowing remote access to information that can be used to control and enhance everyday activities, while interacting with the mobile Internet devices that people are now carrying (for example smartphones). Initially, the Internet of Things referred to global logistical systems that allow advanced stock control and real-time tracking of delivery items. For example, when an item is purchased from an online shop, the shop will have a detailed inventory of all the items held in stock. In general, when a shop receives items to its warehouse, the items are scanned in by an electronic system which evaluates some barcode data and automatically maintains an accurate stock register. When an item is dispatched to a customer, the item is scanned out of the warehouse and is subsequently scanned at various checkpoints along the delivery route, allowing the customer to log into an online portal and track the delivery of their purchase. On delivery, the customer will write an electronic signature into a portable device which will update the online delivery system to identify the item as delivered, and perhaps generate an invoice to be sent to the customer. The whole process happens automatically, allowing the item to have an online status and the customer access to the status information. The technology that allows this to happen is in the barcodes, the barcode scanners, handheld scanning and signature systems, as well as the Internet server hardware and software that manage the tracking information.

Technical advances in scanning and tracking processes are expected to continue, particularly with the development of cheaper and more advanced barcode scanners and the introduction of Quick Response (QR) codes which allow more data than a standard barcode to be encoded into an image. Figure 15.2 shows a QR code which, when scanned, will open the ARM mbed website. One advantage of using QR codes is that they effectively allow a one-way communication protocol with no wires and no electronic requirement in the slave system, which is very cost effective. They are also very tolerant to noise and image degradation. Using a QR code printed on an item, a QR scanner can find out a number of details about that item, essentially in a similar way to requesting a status message via any of the serial communication protocols previously discussed. People also have access to very accurate code scanners these days, as most smartphones are enabled with high-resolution cameras which, when equipped with barcode scanning software, such as that from RedLaser (Reference 15.6), allow many people to scan items with little additional hardware cost. The QR code cannot be changed, of course, so there are limitations, but dynamic QR codes may be an innovation for the future!

More recently, the Internet of Things concept has expanded to include everyday objects attached to the Internet; for example, allowing the data from a temperature sensor on one side of the world to be used in a bespoke mobile phone application being accessed on the other side of the world; or allowing people to control their household lighting and heating systems remotely over the Internet or via a smartphone, bringing potential fuel efficiency and home security improvements. In some cases the ideal application for this globally connected network of data and information has yet to be realized, but it is widely thought that this type of

Figure 15.2:
QR code linking to www.mbed.org

data access and control will become commonplace, especially if systems are made open access (i.e. at no cost) to software developers and network users.

ARM has recently developed a new concept for the mbed platform called 'WebSockets' (Reference 15.7). WebSockets allow mbed developers to prototype systems that adhere to the Internet of Things concept by giving webserver access to wireless- and Ethernet-equipped mbed systems. An mbed system can push sensor data to an ARM-managed WebSocket (a remote server) and the real-time data can be accessed globally by any Internet-equipped device. It is also possible to push data to the mbed through a WebSocket, hence allowing remote Internet control of the mbed system's features.

15.5 Introducing the mbed LPC11U24

In early 2012 ARM released a second mbed platform, the mbed LPC11U24, based on the ARM Cortex-M0 microprocessor architecture. This device is aimed predominantly at prototyping low-power, portable applications and specifically those requiring USB functionality. The NXP LPC11U24 microcontroller is a 48 MHz device with 8 kB of RAM and 32 kB of flash memory. It also supports USB, two serial peripheral interface (SPI) buses, inter-integrated circuit (I^2C) and universal asynchronous receiver/transmitter (UART) serial, and has six analog inputs and up to 30 digital input/output pins. The full pinout diagram for the mbed LPC11U24 is shown in Figure 15.3.

The mbed LPC11U24 is designed to have the same form factor and pin layout as the original mbed LPC1768, but with reduced functionality. So, in general, any programs compiled to run on the lower spec mbed LPC11U24 should execute with identical functionality on the mbed LPC1768. The mbed LPC11U24 is also specifically designed to have lower power consumption, so enabling prototyping of portable systems. Libraries for enabling sleep mode and interrupt-driven wake functions have also been implemented by ARM, and recent tests showed the device drawing 16 mA while awake and less than 2 mA while asleep. However, it must be remembered that the device itself is for prototyping and any resultant production design based on the LPC11U24 could readily do away with a number of mbed hardware features and potentially reduce power consumption to lower levels still.

A further application of the mbed LPC11U24 is as a key enabler for the Internet of Things and WebSockets concepts discussed earlier. The Internet of Things relies heavily on portable sensing devices and mass commercial take-up, so mobile communications, portability, low cost and low power prototyping are essential for these technologies to evolve. In addition to the mbed LPC11U24, ARM has developed a number of WebSockets libraries to allow mobile connectivity to the Internet, using the Roving Networks RN-131C WiFly module, which communicates with the mbed over standard UART serial (see Reference 15.8).

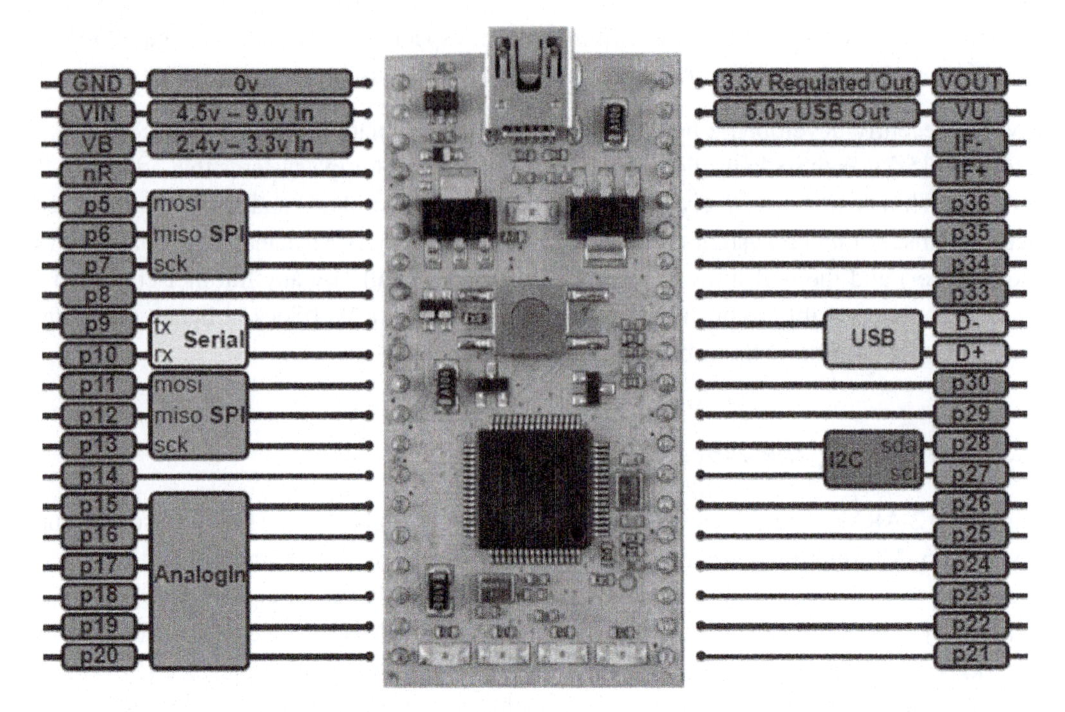

Figure 15.3:
mbed LPC11U24 pinout diagram. *(Image reproduced with permission of ARM Holdings)*

15.6 From mbed to Manufacture

The main thrust of this book has been focused towards rapid prototyping of embedded systems, using the ARM mbed. Rapid prototyping is an essential part of the design process, as it is the point where engineers need to see whether a design concept will work and how it should be implemented in hardware and software. The speed of this process is important, because companies need to know where to invest their research and development resources and which products to develop into mass-market systems. The mbed is an excellent device for accelerating this process as its ease of use and high-level libraries allow continuous testing during the prototyping and proof-of-concept cycle. The easier it is to test a system, the easier it is to develop accurate solutions in a short space of time.

There comes a time, however, when, if a prototype proves successful, a developer will wish to consider how the system could be engineered for mass manufacture and consumer use. At this stage a number of issues should be considered, in particular size, cost, power consumption, manufacturing methods, component availability and reliability, and quality control. So, the move from a working prototype to a commercially ready product is not an insignificant one, and the sooner these considerations can be incorporated into the design the lower the risk of drastic feature changes being required at a later stage. Figure 15.4 summarizes a product design cycle, from initial idea and concept development, through prototyping to final commercialization. The role of the mbed in proof of concept can be very clearly seen.

Although the mbed is designed as a rapid prototyping device, it is intended to allow a simple development path from prototyping to commercialization. In large companies, research and development engineers will prototype and evaluate design concepts and a different set of engineers may be required to take successful prototypes to commercialization. However, in small industries it is not uncommon for a single engineer to see a product through the entire development cycle, so engineers must also have a working knowledge of the manufacturing constraints when working on prototyping and product development.

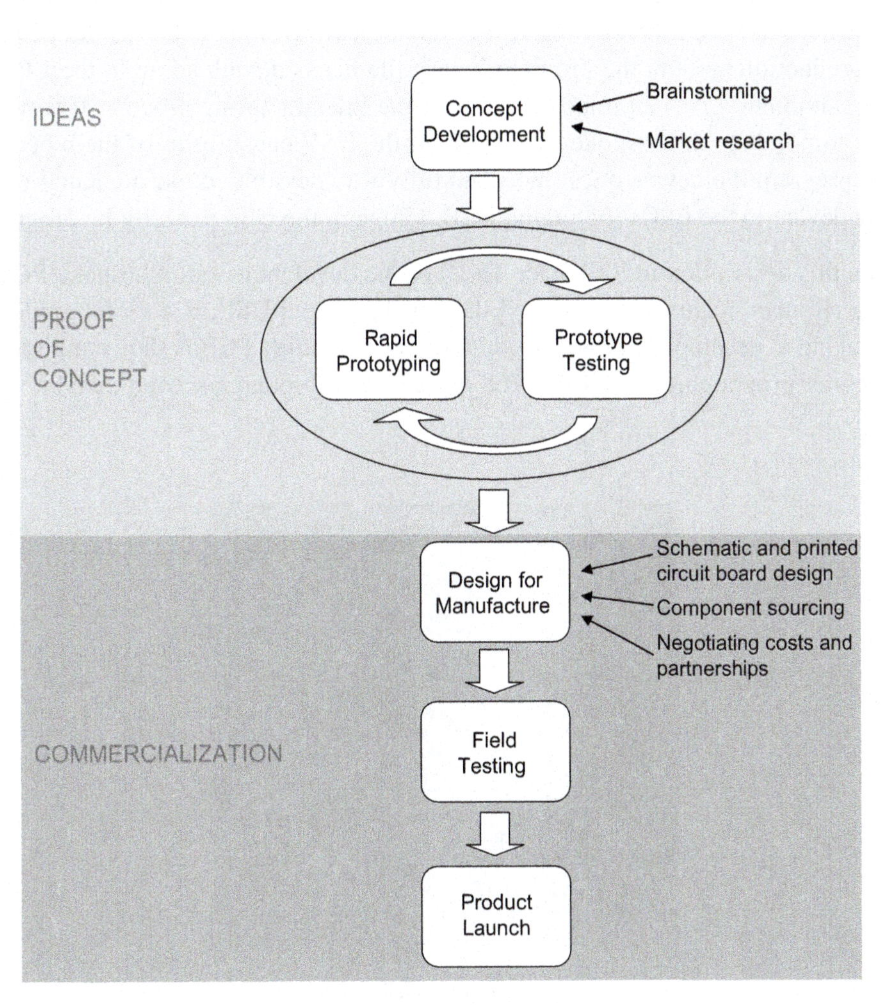

Figure 15.4:
A simplified product development cycle

If only a few commercial units are to be built, then it is quite feasible to use the mbed as it stands as the core controller of the system. However, where thousands of units are to be developed and sold, cost and size become very important factors. It is therefore desirable to implement the mbed hardware components on a custom printed circuit board (PCB) design, in order to minimize size and cost. Indeed, a number of hardware features on the mbed can be removed for certain hardware implementations, if developing a bespoke PCB, hence reducing cost and size further still.

Guidance on the process involved in taking an mbed prototype to the manufacture stage can be found in Reference 15.9, while Reference 15.10 gives a good example. Essentially, the USB bootloader hardware features on the mbed (those which allow drag-and-drop downloading of the .bin file) are only required for prototyping. Furthermore, if the application does not use Ethernet, then the Ethernet features can be left off the PCB design too. In a production system, the program binary file needs downloading to the LPC1768 only once, although this is an important part of the manufacturing process. It would be inefficient to populate the bespoke PCB with all the USB capabilities of the mbed simply to program the device once, but thankfully it is possible to use an actual mbed as a gateway device to the LPC1768 on the PCB, allowing the chip to easily be programmed.

Martin Smith's description in Reference 15.10 of the development of a bespoke PCB containing all mbed features, except for Ethernet, JTAG and USB, is a useful guide for anyone looking to develop their own mbed PCBs. The resulting PCB is shown in Figure 15.5. The process of programming the LPC1768 on this PCB through a second mbed is also described.

Figure 15.5:
Bespoke PCB based on the mbed, showing the LPC1768 in place. *(Image reproduced with permission of Martin Smith)*

15.7 Closing Thoughts

The mbed is shown to be a flexible and powerful device for the rapid prototyping of a number of electronic systems from electromechanical control to advanced Internet communications. The possibilities for projects involving the mbed are perhaps endless, and these latter chapters are intended to widen the developer's creative perspective and give some examples of how the technology described can be applied to bigger and more advanced projects.

The mbed itself is an innovative design, using a novel online compiler and a simple drag-and-drop USB interface. The aim is to make working with embedded systems simple, as often the learning curve is so steep that non-engineers rarely overcome the initial hurdles. The mbed is an excellent device to use for prototyping, education (both at early learning and higher education levels) and engagement between artistic and technological fields, where developers may not have a conventional engineering background. The idea is to make embedded systems design fun, accessible, yet productive, which the mbed does with ease.

References

15.1. Pololu m3pi robot with mbed socket. http://www.pololu.com/catalog/product/2151
15.2. mbed Robot Racing Wii. http://mbed.org/cookbook/mbed-Robot-Racing-Wii
15.3. TLV320AIC23B low-power stereo CODEC with HP amplifier. http://www.ti.com/product/tlv320aic23b
15.4. mbed now does audio. http://www.designspark.com/content/mbed-now-does-audio
15.5. TLV320AIC23B. http://mbed.org/cookbook/TLV320AIC23B
15.6. RedLaser is a scanning application for iPhone. http://redlaser.com/
15.7. The Internet of Things. http://mbed.org/cookbook/IOT
15.8. Roving Networks RN-131 Wi-Fi module http://www.rovingnetworks.com/products/RN_131
15.9. Prototype to hardware. http://mbed.org/users/chris/notebook/prototype-to-hardware/
15.10. Turning an mbed into a custom PCB. http://mbed.org/users/ms523/notebook/turning-an-mbed-circuit-into-a-custom-pcb/

APPENDIX A

Some Number Systems

A.1 Binary, Decimal and Hexadecimal

The number system we are most familiar with, decimal, makes use of 10 different symbols to represent numbers, i.e. 0, 1, 2, 3, 4, 5, 6, 7, 8 and 9. Each of these symbols represents a number, and we make larger numbers by using groups of symbols. In this case the digit most to the right represents units, the next represents tens, the next hundreds, and so on. For example, the number 249, shown in Figure A.1, is evaluated by adding the values in each position:

$$2\ hundreds + 4\ tens + 9\ units = 249$$

or

$$2 \times 10^2 + 4 \times 10^1 + 9 \times 10^0 = 249$$

The *base* or *radix* of the decimal system, just described, is 10. We almost certainly count in the decimal system owing to the accident of having 10 fingers and thumbs on our hands. There is nothing intrinsically correct or superior about it. It is quite possible to count in other bases, and the world of digital computing almost forces us to do this.

The binary counting system has a base or radix of 2. It therefore uses just two symbols, normally 0 and 1. These are called binary digits, or bits. Numbers are made up of groups of digits. The value each digit represents again depends upon its position in the number. Therefore, the 4-bit number 1101, shown in Figure A.2, is interpreted as:

$$1 \times 8 + 1 \times 4 + 0 \times 2 + 1 \times 1 = 8 + 4 + 1 = 13$$

Similarly, the value 0110 binary = 6 decimal. Note that we refer to the units digit as 'bit 0', or the least significant bit (LSB). The twos digit is called 'bit 1' and so on, up to the most significant bit (MSB).

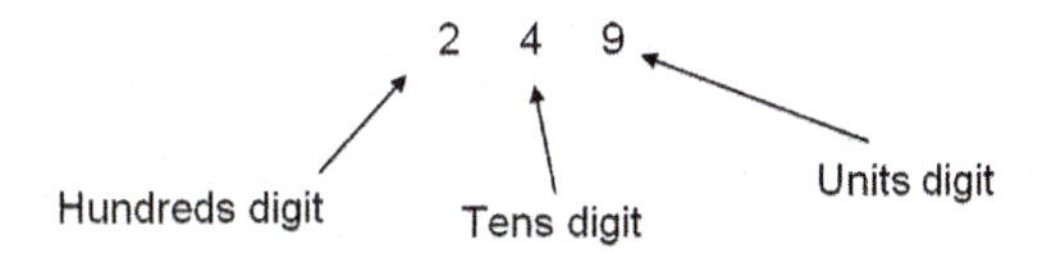

Figure A.1:
The decimal number 249

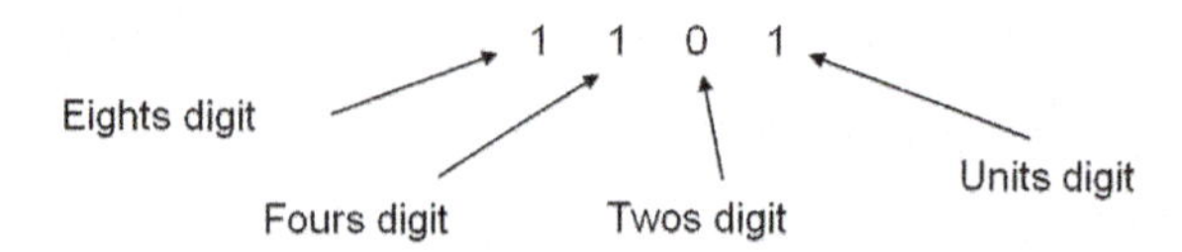

Figure A.2:
The binary number 1101

We are obviously interested in binary representation of numbers, because that is how digital machines perform mathematical operations. But sometimes there are just too many 0s and 1s to keep track of. We often group bits in 4s, as we often look at 4-bit numbers, 8-bit numbers, 12, 16, 20, 24, 28 and 32-bit numbers.

A single *byte* is made up of 8 bits. With one byte we can count up to 255 in decimal; for example:

0000 0000 binary = 0 decimal
1111 1111 binary = 255 decimal
1001 1100 binary = 128 + 16 + 8 + 4 = 156

The range of 0 to 255, offered by a single byte, is not very great, however. To perform mathematical calculations to a high accuracy we need to work with 16, 24 or 32-bit systems. With 16 bits we can count or resolve up to 65 535; for example:

1111 1111 1111 1111 binary = 65 535 decimal
0111 0111 0111 0110 binary = 30 582 decimal

Working with large binary numbers is something that is just not easy for a human; we just cannot absorb all those 1s and 0s. A very convenient alternative is to use hexadecimal, which works to base 16. This means that we can now count up to 15 (decimal) before there is an overflow, and each new overflow column is an increased power of 16. Consider the hexadecimal number 371 as shown in Figure A.3. Here, we see that 371 in hexadecimal is $(3 \times 256) + (7 \times 16) + 1 = 881$ in decimal.

Now we need to represent 16 numbers with just a single digit. We use the decimal digits for numbers 0 to 9, but to represent numbers 10 to 15 we use letters: 'A' or 'a' is adopted to represent decimal 10. Similarly, B or b = 11, C = 12, D = 13, E = 14 and F = 15. As convention, we put '0x' before a hexadecimal number; this means that all the bits before the

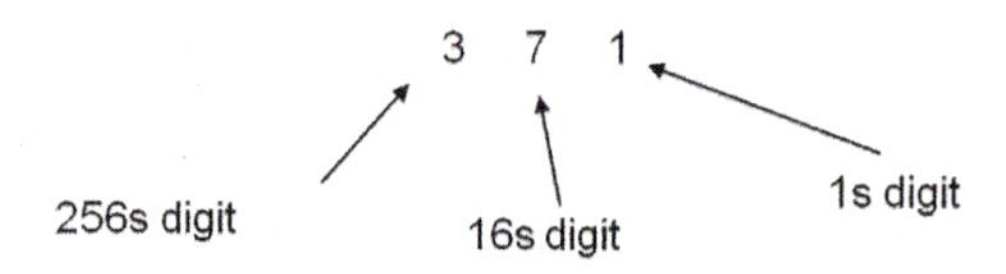

Figure A.3:
Hexadecimal number system

Table A.1: 4-bit values in binary, hexadecimal and decimal

4-bit binary number				Hexadecimal equivalent	Decimal equivalent
0	0	0	0	0x0	0
0	0	0	1	0x1	1
0	0	1	0	0x2	2
0	0	1	1	0x3	3
0	1	0	0	0x4	4
0	1	0	1	0x5	5
0	1	1	0	0x6	6
0	1	1	1	0x7	7
1	0	0	0	0x8	8
1	0	0	1	0x9	9
1	0	1	0	0xA	10
1	0	1	1	0xB	11
1	1	0	0	0xC	12
1	1	0	1	0xD	13
1	1	1	0	0xE	14
1	1	1	1	0xF	15

number are zero or clear. For example, $0xE5 = 14 \times 16 + 5 = 229$. Table A.1 shows the equivalent 4-bit values in binary, hexadecimal and decimal.

The great advantage of hexadecimal numbers is that we can represent 4-bit numbers with a single digit. An 8-bit number is represented with 2 hexadecimal digits, and 16-bit numbers with 4 digits. You can then see that by individually looking at groups of four bits, we can easily generate the hexadecimal equivalent, as with the following examples:

255 decimal = 1111 1111 = 0xFF
156 decimal = 1001 1100 = 0x9C
65 535 decimal = 1111 1111 1111 1111 = 0xFFFF
30 582 decimal = 0111 0111 0111 0110 = 0x7776

While we correctly use the two values of 0 and 1 in all binary numbers above, it is worth noting that different terminology is sometimes used when we apply electronic circuits to represent these numbers. This is done particularly by those who are thinking more in terms of the circuit than of the numbers. This terminology is shown in Table A.2.

Table A.2: Some terminology for logic levels

Logic 0	Logic 1
0	1
Off	On
Low	High
Clear	Set
Open	Closed

A.2 Representation of Negative Numbers: Two's Complement

Simple binary numbers allow only the representation of unsigned numbers, which under normal circumstances are considered to be positive. Yet we must have a way of representing negative numbers as well. A simple way of doing this is by offsetting the available range of numbers. We do this by coding the largest anticipated negative number as zero and counting up from there. In the 8-bit range, with symmetrical offset, −128 can be represented as 00000000, 1000000 then represents zero, and 11111111 represents +127. This method of coding is called *offset binary*, and is illustrated in Table A.3. It is used on occasions (e.g. in

Table A.3: Two's complement and offset binary

Two's complement	Decimal	Offset binary
0111 1111	+127	1111 1111
0111 1110	+126	1111 1110
	:	
	:	
0000 0001	+1	1000 0001
0000 0000	0	1000 0000
1111 1111	−1	0111 1111
1111 1110	−2	0111 1110
	:	
	:	
1000 0010	−126	0000 0010
1000 0001	−127	0000 0001
1000 0000	−128	0000 0000

analog-to-digital converter outputs), but its usefulness is limited, as it is not easy to do arithmetic with.

Let us consider an alternative approach. Suppose we took an 8-bit binary down counter, and clocked it from any value down to, and then below, zero. We would get this sequence of numbers:

Binary	Decimal
0000 0101	5
0000 0100	4
0000 0011	3
0000 0010	2
0000 0001	1
0000 0000	0
1111 1111	−1?
1111 1110	−2?
1111 1101	−3?
1111 1100	−4?
1111 1011	−5?

This gives a possible means of representing negative numbers: effectively, we subtract the magnitude of the negative number from zero, within the limits of the 8-bit number, or whatever other size is in use. This representation is called *two's complement*. It can be shown that using two's complement leads to correct results when simple binary addition and subtraction is applied. Two's complement notation can be applied to binary words of any size.

The two's complement of an n-bit number is found by subtracting it from 2^n; that is where the terminology comes from. Rather than doing this error-prone subtraction, an easier way of reaching the same result is to complement (i.e. change 1 to 0, and 0 to 1) all the bits of the positive number, and then add 1. Hence, to find −5 we follow the procedure:

original number		complement all		add one
0000 0101 (+5)	→	1111 1010	→	1111 1011 (−5 in two's complement)

To convert back, simply subtract 1 and complement again. Note that the most significant bit of a two's complement number acts as a 'sign bit', 1 for negative, 0 for positive. The 8-bit binary range, shown for both two's complement and offset binary, appears in Table A.3.

A.2.1 Range of Two's Complement

In general, the range of an n-bit two's complement number is from $-2^{(n-1)}$ to $+\{2^{(n-1)} - 1\}$. Table A.4 summarizes the ranges available for some commonly used values of n.

A.3 Floating Point Number Representation

The numbers described so far all appear to be integers; we have not been able to represent any sort of fractional number, and their range is limited by the size of the binary word representing them. Suppose we need to represent really large or small numbers? Another way of expressing a number, which greatly widens the range, is called *floating point* representation.

In general, a number can be represented as $a \times r^e$, where a is the *mantissa*, r is the radix, and e is the exponent. This is sometimes called scientific notation. For example, the decimal number 12.3 can be represented as:

1.23×10^1 *or*
$.123 \times 10^2$ *or*
12.3×10^0 *or*
123×10^{-1} *or*
1230×10^{-2}

Floating point notation adapts and applies scientific notation to the computer world. The name is derived from the way the binary point can be allowed to float, by adjusting the value of the exponent, to make best use of the bits available in the mantissa. Standard formats exist for representing numbers by their sign, mantissa and exponent, and a host of hardware and software techniques exists to process numbers represented in this way. Their disadvantage lies in their greater complexity, and hence usually slower processing speed and higher cost. For flexible use of numbers in the computing world they are, however, essential.

The most widely recognized and used format is the *IEEE Standard for Floating-Point Arithmetic* (known as IEEE 754). In single precision form this makes use of 32-bit

Table A.4: Number ranges for differing word sizes

Number of bits	Unsigned binary	Two's complement
8	0 to 255	−128 to +127
12	0 to 4095	−2048 to +2047
16	0 to 65 535	−32 768 to +32 767
24	0 to 16 777 215	−8 388 608 to +8 388 607
32	0 to 4 294 967 295	−2 147 483 648 to +2 147 483 647

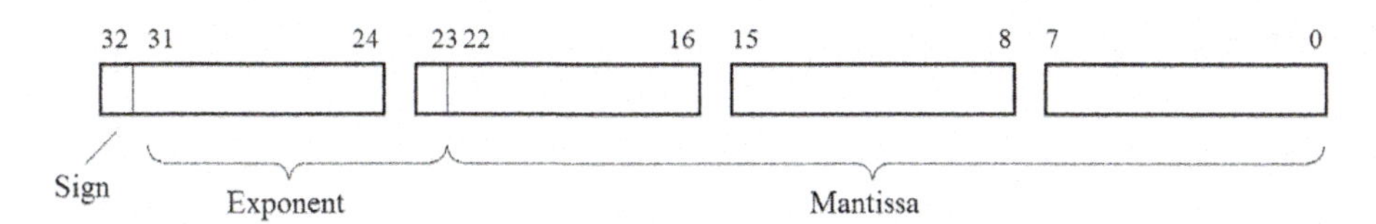

Figure A.4:
IEEE 754 32-bit floating point format

representation for a number, with 23 bits for the mantissa, 8 bits for the exponent and a sign bit, as seen in Figure A.4. The binary point is assumed to be just to the left of the MSB of the mantissa. A further bit, always 1 for a non-zero number, is added to the mantissa, making it effectively a 24-bit number. Zero is represented by 4 zero bytes. The number 127 is subtracted from the exponent, leading to an effective range of exponents from −126 to +127. Exponent 255 (leading to 128, when 127 is subtracted) is reserved to represent infinity. The value of a number represented in this format is then:

$$(-1)^{\text{sign}} \times 2^{(\text{exponent} \; -127)} \times 1.\text{mantissa}$$

This allows number representation in the range:

$$\pm 1.175494 \times 10^{-38} \text{to} \pm 3.402823 \times 10^{+38}$$

APPENDIX B

Some C Essentials

B.1 A Word about C

This appendix aims to summarize all features of the C language used in this book, though not of the language as a whole. It is intended to be just adequate for the purpose of supporting the book, and should not be viewed as complete. With care and experimentation it can be used as an adequate introduction to the main features of the language. If you are a C novice, however, it is well worth having another reference source available, particularly as you consider the more advanced features. References B.1 and B.2 are both good.

As you progress through the book, you will find yourself jumping around within the material of this appendix. Do not feel you need to read it sequentially. Instead, read the different sections as they are referenced from the book.

B.2 Elements of a C Program

B.2.1 Keywords

C has a number of keywords whose use is defined. A programmer cannot use a keyword for any other purpose, for example as a data name. Keywords are summarized in Tables B.1–B.3.

B.2.2 Program Features and Layout

Simply speaking, a C program is made up of:

Declarations

All variables in C must be declared before they can be applied, giving as a minimum the variable name and its data type. A declaration is terminated with a semicolon. In simple programs, declarations appear as one of the first things in the program. They can also occur within the program, with significance attached to the location of the declaration.

For example:

```
float exchange_rate;
int new_value;
```

Table B.1: C keywords associated with data type and structure definition

Word	Summary meaning	Word	Summary meaning
char	A single character, usually 8-bit	**signed**	A qualifier applied to **char** or **int** (default for **char** and **int** is signed)
const	Data that will not be modified	**sizeof**	Returns the size in bytes of a specified item, which may be variable, expression or array
double	A 'double precision' floating-point number	**struct**	Allows definition of a data structure
enum	Defines variables that can only take certain integer values	**typedef**	Creates new name for existing data type
float	A 'single precision' floating-point number	**union**	A memory block shared by two or more variables, of any data type
int	An integer value	**unsigned**	A qualifier applied to **char** or **int** (default for **char** and **int** is signed)
long	An extended integer value; if used alone, integer is implied	**void**	No value or type
short	A short integer value; if used alone, integer is implied	**volatile**	A variable which can be changed by factors other than the program code

Table B.2: C keywords associated with program flow

Word	Summary meaning	Word	Summary meaning
break	Causes exit from a loop	**for**	Defines a repeated loop – loop is executed as long as condition associated with **for** remains true
case	Identifies options for selection within a **switch** expression	**goto**	Program execution moves to labeled statement
continue	Allows a program to skip to the end of a **for**, **while** or **do** statement	**if**	Starts conditional statement; if condition is true, associated statement or code block is executed
default	Identifies default option in a **switch** expression, if no matches found	**return**	Returns program execution to calling routine, causing also return of any data value specified by function
do	Used with **while** to create loop, in which statement or code block following **do** is repeated as long as **while** condition is true	**switch**	Used with **case** to allow selection of a number of alternatives; **switch** has an associated expression which is tested against a number of **case** options
else	Used with **if**, and precedes alternative statement or code block to be executed if **if** condition is not true	**while**	Defines a repeated loop – loop is executed as long as condition associated with **while** remains true

Table B.3: C keywords associated with data storage class

Word	Summary meaning	Word	Summary meaning
auto	Variable exists only within block within which it is defined. This is the default class	**register**	Variable to be stored in a CPU register; thus, address operator (&) has no effect
extern	Declares data defined elsewhere	**static**	Declares variable which exists throughout program execution; the location of its declaration affects in what part of the program it can be referenced

declare a variable called **exchange_rate** as a floating point number, and another variable called **new_value** as an integer. The data types are keywords seen in Tables B.1–B.3.

Statements

Statements are where the action of the program takes place. They perform mathematical or logical operations, and establish program flow. Every statement that is not a block (see below) ends with a semicolon. Statements are executed in the sequence they appear in the program, except where program branches take place.

For example, this line is a statement:

```
counter = counter + 1;
```

Space and Layout

There is no strict layout format to which C programs must adhere. The way the program is laid out and the use of space are both used to enhance clarity. Blank lines and indents in lines, for example, are ignored by the compiler, but used by the programmer to optimize the program layout.

As an example, the program that the mbed compiler always starts up with, shown as Program Example 2.1, could be written as shown here. It would not be easy to read, however. It is the semi-colons at the end of each statement, and the brackets, which in reality define much of the program structure.

```
#include "mbed.h"
DigitalOut myled(LED1); int main() {while(1) {myled = 1; wait(0.2); myled = 0; wait(0.2);}}
```

Comments

Two ways of commenting are used. One is to place the comment between the markers /* and */. This is useful for a block of text information running over several lines. Alternatively, when two forward slash symbols (//) are used, the compiler ignores any text that follows on that line only, which can then be used for comment.

For example:

```
/*A program which flashes mbed LED1 on and off,
Demonstrating use of digital output and wait functions. */
#include "mbed.h"  //include the med header file as part of this program
```

Code Blocks

Declarations and statements can be grouped together into *blocks*. A block is contained within braces, i.e. { and }. Blocks can and are written within other blocks, each within its own pair of braces. Keeping track of these pairs of braces is an important pastime in C programming, as in a complex piece of software there can be numerous ones nested within each other.

B.2.3 Compiler Directives

Compiler directives are messages to the compiler, and do not directly lead to program code. Compiler directives all start with a hash, #. Two examples follow.

#include

The **#include** directive directly inserts another file into the file that invokes the directive. This provides a feature for combining a number of files as if they were one large file. Angled brackets (< >) are used to enclose files held in a directory different from the current working directory, hence often for library files not written by the current author. Quotation marks are used to contain a file located within the current working directory, hence often user defined.

For example:

```
#include "mbed.h"
```

#define

The **#define** directive allows use of names for specific constants. For example, to use the number $\pi = 3.141592$ in the program, we could create a #define for the name 'PI' and assign that number to it, as shown:

```
#define PI   3.141592
```

The name 'PI' is then used in the code whenever the number is needed. When compiling, the compiler replaces the name in the **#define** with the value that has been specified.

B.3 Variables and Data

B.3.1 Declaring, Naming and Initializing

Variables must be named, and their data type defined, before they can be used in a program. Keywords from Table B.1 are used for this. For example,

```
int MyVariable;
```

defines 'MyVariable' as a data type int (integer).

It is possible to initialize the variable at the same time as declaration; for example,

```
int MyVariable=25;
```

initializes **MyVariable**, and sets it to an initial value of 25.

It is possible to give variables meaningful names, while still avoiding excessive length, for example 'Height', 'InputFile', 'Area'. Variable names must start with a letter or an underscore; no other punctuation marks are allowed. Variable names are case sensitive.

B.3.2 Data Types

When a data declaration is made, the compiler reserves for it a section of memory, whose size depends on the type invoked. Examples of the link between data type, number range and memory size are shown in Table B.4. It is interesting to compare these with information on number types given in Appendix A. Note that the actual memory size applied to data types can vary between compilers. A full listing for the mbed compiler can be found in Reference B.3.

B.3.3 Working with Data

In C we can work with numbers in binary, fixed or floating point decimal, or hexadecimal format, depending on what is most convenient, and what number type and

Table B.4: Example C data types, as implemented by the mbed compiler

Data type	Description	Length (bytes)	Range
char	Character	1	0 to 255
signed char	Character	1	−128 to +127
unsigned char	Character	1	0 to 255
short	Integer	2	−32768 to +32767
unsigned short	Integer	2	0 to 65 535
int	Integer	4	−2147483648 to +2147483647
long	Integer	4	−2147483648 to +2147483647
unsigned long	Integer	4	0 to 4294967295
float	Floating point		$1.17549435 \times 10^{-38}$ to $3.40282347 \times 10^{+38}$
double	Floating point, double precision		$2.22507385850720138 \times 10^{-308}$ to $1.79769313486231571 \times 10^{+308}$

range is required. For time-critical applications it is important to remember that floating point calculations can take much longer than fixed point. In general, it is easiest for us to work in decimal, but if a variable represents a register bit field or a port address, then it is usually more appropriate to manipulate the data in hexadecimal. When writing numbers in a program, the default radix for integers is decimal, with no leading 0 (zero). Octal numbers are identified with a leading 0. Hexadecimal numbers are prefixed with 0x.

For example, if a variable **MyVariable** is of type **char** we can perform the following examples to assign a number to that variable:

```
MyVariable = 15;  //a decimal example
MyVariable = 0x0E; //a hexadecimal example
```

The value for both is the same.

B.4 Functions

A function is a section of code that can be called from another part of the program. So, if a particular piece of code is to be used or duplicated many times, we can write it once as a function, and then call that function whenever the specific operation is required. Using functions saves coding time and improves readability by making the code neater.

Data can be passed to functions, and returned from them. Such data elements, called *arguments*, must be of a type that is declared in advance. Only one return variable is allowed, whose type must also be declared. The data passed to the variable is a *copy* of the original. Therefore, the function does not itself modify the value of the variable named. The impact of the function should thus be predictable and controlled.

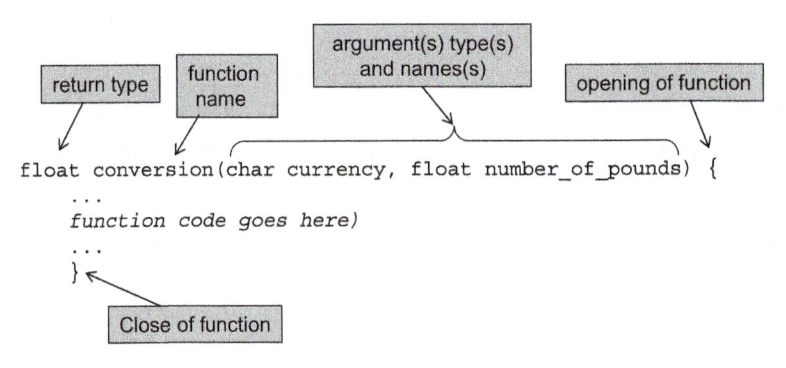

Figure B.1:
Function example

A function is defined in a program by a block of code having particular characteristics. Its first line forms the function header, with the format:

```
Output_type function_name (variable_type_1 variable_name_1, variable_type_2
variable_name_2,...)
```

An example is shown in Figure B.1. The return type is given first. In this example, the keyword **float** is used. After the function name, in brackets, one or more data types may be listed, which identify the arguments which must be passed *to* the function. In this case two arguments are sent, one of type **char** and one of type **float**. Following the function header, a pair of braces encloses the code which makes up the function itself. This could be anything from a single line to many pages. The final statement of the function may be a **return**, which will specify the value returned to the calling program. This is not essential if no return value is required.

B.4.1 The main Function

The core code of any C program is contained within its 'main' function. Other functions may be written outside **main( )**, and called from within it. Program execution starts at the beginning of **main( )**. It must follow the structure just described. However, as **main( )** contains the central program, one does not expect to send anything to it or to receive anything from it. Therefore, usual patterns for **main( )** are:

```
void main (void){

void main (){

int main (){
```

The keyword **void** indicates that no data is specified. The mbed **main( )** function applies the third option, as in C++ **int** is the return type for **main( )** normally used.

B.4.2 Function Prototypes

Just like variables, functions must be declared at the start of a program, before the main function. The declaration statements for functions are called prototypes. Each function in the code must have an associated prototype for it to run. The format is the same as for the function header.

For example, the following function prototype applies to the function header seen above:

```
float conversion(char currency, float number_of_pounds)
```

This describes a function that takes inputs of a character value for the selected currency and a floating point (decimal) value for the number of pounds to be converted. The function returns the decimal monetary value in the specified currency.

B.4.3 Function Definitions

The actual function code is called the *function definition*. For example:

```
float conversion(char currency, float number_of_pounds) {
 float exchange_rate;
 switch(currency) {
    case 'U': exchange_rate = 1.50;       // US dollars
       break;
    case 'E': exchange_rate = 1.12 );    // Euros
       break;
     case 'Y': exchange_rate = 135.4);    // Japan yen
        break;
     default: exchange_rate = 1);
 }
 exchange_value=number_of_pounds*exchange_rate;
 return(exchange_value);
}
```

This function can be called any number of times from within the main C program, or from another function, for example in this statement:

```
ten_pounds_in_yen=conversion('Y',10.00);
```

B.4.4 Using the static Storage Class with Functions

The static data type is useful for defining variables within functions, where the data inside the function must be remembered between function calls. For example, if a function within a real-time system is used to calculate a digital filter output, the function should always remember its previous data values. In this case, data values inside the function should be defined as static; for example:

```
float movingaveragefilter(float data_in) {
   static float data_array[10];        // define static float data array
   for (int i=8;i>=0;i--) {
     data_array[i+1]=data_array[i]; // shift each data value along
   }                                    // (the oldest data value is discarded)
   data_array[0]=data_in;              // place new data at index 0
   float sum=0;
   for (int i=0;i<=9;i++) {
     sum=sum+data_array[i];          // calculate sum of data array
   }
      return sum/10;                      // return average value of array
}
```

B.5 Operators

C has a wide set of operators, shown in Table B.5. The symbols used are familiar, but their application is *not* always the same as in conventional algebra. For example, a single 'equals' symbol, =, is used to assign a value to a variable. A double equals sign, = =, is used to represent the conventional 'equal to'.

Table B.5: C operators

<table>
<tr><th>Precedence and order</th><th>Operation</th><th>Symbol</th><th>Precedence and order</th><th>Operation</th><th>Symbol</th></tr>
<tr><td colspan="6">Parentheses and array access operators</td></tr>
<tr><td>1, L to R</td><td>Function calls</td><td>( )</td><td>1, L to R</td><td>Point at member</td><td>X->Y</td></tr>
<tr><td>1, L to R</td><td>Subscript</td><td>[ ]</td><td>1, L to R</td><td>Select member</td><td>X.Y</td></tr>
<tr><td colspan="6">Arithmetic operators</td></tr>
<tr><td>4, L to R</td><td>Add</td><td>X+Y</td><td>3, L to R</td><td>Multiply</td><td>X*Y</td></tr>
<tr><td>4, L to R</td><td>Subtract</td><td>X–Y</td><td>3, L to R</td><td>Divide</td><td>X/Y</td></tr>
<tr><td>2, R to L</td><td>Unary plus</td><td>+X</td><td>3, L to R</td><td>Modulus</td><td>%</td></tr>
<tr><td>2, R to L</td><td>Unary minus</td><td>–X</td><td></td><td></td><td></td></tr>
<tr><td colspan="6">Relational operators</td></tr>
<tr><td>6, L to R</td><td>Greater than</td><td>X>Y</td><td>6, L to R</td><td>Less than or equal to</td><td>X<=Y</td></tr>
<tr><td>6, L to R</td><td>Greater than or equal to</td><td>X>=Y</td><td>7, L to R</td><td>Equal to</td><td>X= =Y</td></tr>
<tr><td>6, L to R</td><td>Less than</td><td>X<Y</td><td>7, L to R</td><td>Not equal to</td><td>X!=Y</td></tr>
<tr><td colspan="6">Logical operators</td></tr>
<tr><td>11, L to R</td><td>AND (1 if both X and Y are not 0)</td><td>X&&Y</td><td>2, R to L</td><td>NOT (1 if X=0)</td><td>!X</td></tr>
<tr><td>12, L to R</td><td>OR (1 if either X or Y are not 0)</td><td>X||Y</td><td></td><td></td><td></td></tr>
<tr><td colspan="6">Bitwise operators</td></tr>
<tr><td>8, L to R</td><td>Bitwise AND</td><td>X&Y</td><td>2, L to R</td><td>Ones complement (bitwise NOT)</td><td>≈X</td></tr>
<tr><td>10, L to R</td><td>Bitwise OR</td><td>X|Y</td><td>5, L to R</td><td>Right shift. X is shifted right Y times</td><td>X≫Y</td></tr>
<tr><td>9, L to R</td><td>Bitwise XOR</td><td>X^Y</td><td>5, L to R</td><td>Left shift. X is shifted left Y times</td><td>X≫Y</td></tr>
<tr><td colspan="6">Assignment operators</td></tr>
<tr><td>14, R to L</td><td>Assignment</td><td>X=Y</td><td>14, R to L</td><td>Bitwise AND assign</td><td>X&=Y</td></tr>
<tr><td>14, R to L</td><td>Add assign</td><td>X+=Y</td><td>14, R to L</td><td>Bitwise inclusive OR assign</td><td>X|=Y</td></tr>
<tr><td>14, R to L</td><td>Subtract assign</td><td>X–=Y</td><td>14, R to L</td><td>Bitwise exclusive OR assign</td><td>X^=Y</td></tr>
<tr><td>14, R to L</td><td>Multiply assign</td><td>X * = Y</td><td>14, R to L</td><td>Right shift assign</td><td>X ≫=Y</td></tr>
</table>

Continued

Table B.5: C operators—cont'd

Precedence and order	Operation	Symbol	Precedence and order	Operation	Symbol
14, R to L	Divide assign	X/=Y	14, R to L	Left shift assign	X ≪=Y
14, R to L	Remainder assign	X%=Y			
Increment and decrement operators					
2, R to L	Preincrement	++X	2, R to L	Postincrement	X++
2, R to L	Predecrement	−−X	2, R to L	Postdecrement	X−−
Conditional operators					
13, R to L	Evaluate either X (if Z≠0) or Y (if Z=0)	Z?X:Y	15, L to R	Evaluate X first, followed by Y	X,Y
"Data interpretation" operators					
2, R to L	The object or function pointed to by X	*X	2, R to L	The address of X	&X
2, R to L	Cast – the value of X, with (scalar) type specified	(type) X	2, R to L	The size of X, in bytes	Sizeof X

Operators have a certain order of precedence, shown in Table B.5. The compiler applies this order when it evaluates a statement. If more than one operator at the same level of precedence occurs in a statement, then those operators are evaluated in turn, either left to right or right to left, as shown in the table. For example, the line

```
counter = counter + 1;
```

contains two operators. Table B.5 shows that the addition operator has precedence level 4, while all assign operators have precedence 14. The addition is therefore evaluated first, followed by the assign. The outcome is that the variable **counter** is incremented by 1.

B.6 Flow Control: Conditional Branching

Flow control covers the different forms of branching and looping available in C. As branching and looping can lead to programming errors, C provides clear structures to improve programming reliability.

B.6.1 If and else

If statements always start with use of the **if** keyword, followed by a logical condition. If the condition is satisfied, then the code block that follows is executed. If the condition is not satisfied, then the code is not executed. There may or may not also be following **else** or **else if** statements.

Syntax:

```
if (Condition1){
   ... C statements here
}else if (Condition2){
   ... C statements here
}else if (Condition3){
   ... C statements here
}else{
   ... C statements here
}
```

The **if** and **else** statements are evaluated in sequence, i.e.:

- **else if** statements are only evaluated if previous **if** or **else if** conditions have failed
- **else** statements are only executed if all previous conditions have failed.

For example, in the above example, the `else if (Condition2)` will only be executed if Condition1 has failed.

Example:

```
if (data > 10){
  data += 5;              //If we reach this point, data must be > 10
}else if(data > 5){       //If we reach this point, data must be <= 10
  data -= 3;
}else{                    //If we reach this point, data must be <= 5
nVal = 0;
}
```

B.6.2 Switch Statements, and using break

The **switch** statement allows a selection to be made of one out of several actions, based on the value of a variable or an expression given in the statement. An example of this structure has already appeared, in the example function in Section B.4.3. The structure uses no fewer than four C keywords. Selection is made from a list of **case** statements, each with an associated label – note that a colon following a text word defines it as a label. If the label equals the **switch** expression then the action associated with that **case** is executed. The **default** action (which is optional) occurs if none of the **case** statements is satisfied. The **break** keyword, which terminates each **case** condition, can be used to exit from any loop. It causes program execution to continue after the **switch** code block.

B.7 Flow Control: Program Loops

B.7.1 while Loops

A **while** loop is a simple mechanism for repeating a section of code, until a certain condition is satisfied. The condition is stated in brackets after the word **while**, with the conditional code block following. For example:

```
i=1
while (i<10) {
   ... C statements here
   i++    //increment i
}
```

Here, the value of **i** is defined outside the loop; it is then updated within the loop. Eventually **i** increments to 10, at which point the loop will terminate. The condition associated with the while statement is evaluated at the start of each loop iteration; the loop then only runs if the condition is found to be true.

B.7.2 for Loops

The **for** loop allows a different form of looping, in that the dependent variable is updated automatically every time the loop is repeated. It defines an initialized variable, a condition for looping and an update statement. Note that the update takes place at the end of each loop iteration. If the updated variable is no longer true for the loop condition, the loop stops and program flow continues. For example:

```
for(j=0; j<10; j++) {
   ... C statements here
}
```

Here, the initial condition is `j=0` and the update value is `j++`, i.e. **j** is incremented. This means that **j** increments with each loop. When **j** becomes 10 (i.e. after 10 loops), the condition `j<10` is no longer satisfied, so the loop does not continue any further.

B.7.3 Infinite Loops

We often require a program to loop forever, particularly in a super-loop program structure. An infinite loop can be implemented by either of the following loops:

```
while(1) {
   ... continuously called C statements here
}
```

or

```
for(;;) {
   ... continuously called C statements here
}
```

B.7.4 Exiting Loops with break

The **break** keyword can also be used to exit from a **for** or **while** loop, at any time within the loop. For example:

```
while(i>5) {
     ... C statements here
     if (fred == 1)
        break;
```

```
    ... C statements here
}                              //end of while
//execution continues here loop completion, or on break
```

B.8 Derived Data Types

In addition to the fundamental data types, there are further data types that can be derived from them. Example types that we use are described in this section.

B.8.1 Arrays and Strings

An array is a set of data elements, each of which has the same type. Any data type can be used. Array elements are stored in consecutive memory locations. An array is declared with its name and the data type of its elements; it is recognized by the use of the square brackets following the name. The number of elements and their value can also be specified. For example, the declaration

```
unsigned char message1[8];
```

defines an array called **message1**, containing eight characters. Alternatively, it can be left to the compiler to deduce the array length, as seen in the two examples here:

```
char item1[] = "Apple";
int nTemp[] = {5,15,20,25};
```

In each of these the array is initialized as it is declared.

Elements within an array can be accessed with an index, starting with value 0. Therefore, for the first example above, **message1[0]** selects the first element and **message1[7]** the last. An access to **message [8]** would be outside the boundary of the array, and would give invalid data. The index can be replaced by any variable that represents the required value.

Importantly, the name of an array is set equal to the address of the initial element. Therefore, when an array name is passed in a function, what is passed is this address.

A string is a special array of type **char** that is ended by the NULL (\0) character. The null character allows code to search for the end of a string. The size of the string array must therefore be one byte greater than the string itself, to contain this character. For example, a 20-character string could be declared:

```
char MyString[21];   // 20 characters plus null
```

B.8.2 Pointers

Instead of specifying a variable by name, we can specify its address. In C terminology such an address is called a *pointer.* A pointer can be loaded with the address of a variable by using the unary operator '&', like this:

```
my_pointer = &fred;
```

This loads the variable **my_pointer** with the *address* of the variable **fred**; **my_pointer** is then said to *point* to **fred**.

Doing things the other way round, the value of the variable pointed to by a pointer can be specified by prefixing the pointer with the '*' operator. For example, ***my_pointer** can be read as 'the value pointed to by **my_pointer**'. The * operator, used in this way, is sometimes called the *dereferencing* or *indirection* operator. The indirect value of a pointer, for example ***my_pointer**, can be used in an expression just like any other variable.

A pointer is declared by the data type it points to. Thus,

```
int *my_pointer;
```

indicates that **my_pointer** points to a variable of type **int**.

We can also use pointers with arrays, because an array is really just a number of data values stored at consecutive memory locations. So, if the following is defined:

```
int dataarray[]={3,4,6,2,8,9,1,4,6};    // define an array of arbitrary values
int *ptr;                               // define a pointer
ptr = &dataarray[0];                    // assign pointer to the address of
                                        // the first element of the data array
```

given the previous declarations, the following statements will therefore be true:

```
*ptr == 3;        // the first element of the array pointed to
*(ptr+1) == 4;    // the second element of the array pointed to
*(ptr+2) == 6;    // the third element of the array pointed to
```

Thus, array searching can be done by moving the pointer value to the correct array offset. Pointers are required for a number of reasons, but one simple reason is because the C standard does not allow us to pass arrays of data to and from functions, so we must use pointers instead to get around this.

B.8.3 Structures and Unions

Structures and *unions* are both sets of related variables, defined through the C keywords **struct** and **union**. In a way they are like arrays, but in both cases they can be of data elements of *different* types.

Structure elements, called *members*, are arranged sequentially, with the members occupying successive locations in memory. A structure is declared by invoking the **struct** keyword, followed by an optional name (called the structure *tag*), followed by a list of the structure members, each of these itself forming a declaration. For example:

```
struct resistor {int val; char pow; char tol;};
```

declares a structure with tag **resistor**, which holds the value (**val**), power rating (**pow**) and tolerance (**tol**) of a resistor. The tag may come before or after the braces holding the list of structure members.

Structure elements are identified by specifying the name of the variable and the name of the member, separated by a full stop (period). Therefore, **resistor.val** identifies the first member of the example structure above.

Like a structure, a union can hold different types of data. Unlike the structure, union elements all begin at the same address. Hence, the union can represent only one of its members at any one time, and the size of the union is the size of the largest element. It is up to the programmer to track which type is currently stored. Unions are declared in a format similar to that of structures.

Unions, structures and arrays can occur within each other.

B.9 C Libraries and Standard Functions

B.9.1 Header Files

All but the simplest of C programs are made up of more than one file. In general, many files are combined together in the process of compiling, for example original source files combining with standard library files. To aid this process, a key section of any library file is detached and created as a separate *header* file.

Header file names end in **.h**; the file typically includes declarations of constants and function prototypes, and links on to other library files. The function definitions themselves stay in the associated **.c** or **.cpp** files. In order to use the features of the header file and the file(s) it invokes, it must be included within any program accessing it, using **#include**. We see **mbed.h** being included in every program in the book. Note also that the **.c** file where the function declarations appear must also include the header file.

B.9.2 Libraries, and the C Standard Library

Because C is a simple language, much of its functionality derives from standard functions and macros that are available in the libraries accompanying any compiler. A C library is a set of precompiled functions that can be linked in to the application. These may be supplied with a compiler, available in-company, or may be public domain. Notably, there is a *Standard Library,* defined in the C ANSI standard. There are a number of standard header files, used for different groups of functions within the standard library. For example, the <math.h> header file is used for a range of mathematical functions (including all trigonometric functions), while <stdio.h> contains the standard input and output functions, including the **printf** function.

B.9.3 Using printf

This versatile function provides formatted output, typically for sending display data to a PC screen. Text, data, formatting and control formatting can be specified. Only summary information is provided here; a full statement can be found in Reference B.1. Examples below are taken from book chapters. In each case the function appears in the form **pc.printf( )**, indicating that **printf( )** is being used as a member function of a C++ class **pc** created in the example program. This does not affect the format applied.

Simple Text Messages

```
pc.printf("ADC Data Values...\n\r"); \\send an opening text message
```

This prints the text string 'ADC Data Values...' to screen, and uses control characters \n and \r to force a new line and carriage return, respectively.

Data Messages

```
pc.printf("%1.3f",ADCdata);
```

This prints the value of the **float** variable **ADCdata**. A *conversion specifier*, initiated by the % character, defines the format. Within this, the 'f' specifies floating point, and the .3 causes output to three decimal places.

```
pc.printf("%1.3f \n\r",ADCdata); \\send the data to the terminal
```

As above, but includes \n and \r to force a new line and carriage return.

Combination of Text and Data

```
pc.printf("random number is %i\n\r", r_delay);
```

This prints a text message, followed by the value of **int** variable (indicated by the 'i' specifier) **r_delay**.

```
pc.printf("Time taken was %f seconds\n", t.read()); //print timed value to pc
```

This prints a text message, followed by the return value of function **t.read( )**, which is of type **float**.

B.10 File Access Operations

B.10.1 Overview

In C we can open files, read and write data and also scan through files to specific locations, even searching for particular types of data. The commands for input and output operations are all defined by the C stdio library, already mentioned in connection with **printf( )**.

The stdio library uses the concept of *streams* to manage the data flow. All streams have similar properties, even though the actual application of the data flow may be very varied. Streams are represented in the stdio library as pointers to FILE objects, which uniquely identify the stream. We can store data in files (as chars) or we can store words and strings (as character arrays). A summary of useful stdio file access library functions is given in Table B.6.

B.10.2 Opening and Closing Files

A file can be opened with the following command:

```
FILE* pFile = fopen("datafile.txt","w");
```

This assigns a pointer with name **pFile** to the file at the specific location given in the **fopen** statement. In this example the *access mode* is specified with a 'w', and a number of other file open access modes, and their specific meanings, are shown in Table B.7. When 'w' is the access mode (meaning *write access*), if the file does not already exist then the **fopen** command will automatically create it in the specified location.

When you have finished using a file for reading or writing it is good practice to close it, for example with

```
fclose(pFile);
```

Table B.6: Useful stdio library functions

Function	Format	Summary action
fclose	`int fclose ( FILE * stream );`	Closes a file
fgetc	`int fgetc ( FILE * stream );`	Gets a character from a stream
fgets	`char * fgets ( char * str, int num, FILE * stream );`	Gets a string from a stream
fopen	`FILE * fopen ( const char * filename, const char * mode);`	Opens the file of type file and name filename
fprintf	`int fprintf ( FILE * stream, const char * format, ... );`	Writes formatted data to a file
fputc	`int fputc ( int character, FILE * stream );`	Writes a character to a stream
fputs	`int fputs ( const char * str, FILE * stream );`	Writes a string to a stream
fseek	`int fseek ( FILE * stream, long int offset, int origin );`	Moves the file pointer to the specified location

str: an array containing the null-terminated sequence of characters to be written; stream: pointer to a FILE object that identifies the stream where the string is to be written.

Table B.7: Access modes for fopen

Access mode	Action
'r'	Open an existing file for reading
'w'	Create a new empty file for writing. If a file of the same name already exists it will be deleted and replaced with a blank file
'a'	Append to a file. Write operations result in data being appended to the end of the file. If the file does not exist a new blank file will be created
'r+'	Open an existing file for both reading and writing
'w+'	Create a new empty file for both reading and writing. If a file of the same name already exists it will be deleted and replaced with a blank file
'a+'	Open a file for appending or reading. Write operations result in data being appended to the end of the file. If the file does not exist a new blank file will be created

B.10.3 Writing and Reading File Data

If the intention is to store numerical data, this can be done in a simple way by storing individual 8-bit data values. The **fputc** command allows this, as follows:

```
char write_var=0x0F;
fputc(write_var, pFile);
```

This stores the 8-bit variable **write_var** to the data file. The data can also be read from a file to a variable as follows:

```
read_var = fgetc(pFile);
```

Using the stdio.h commands, it is also possible to read and write words and strings with **fgets( )** and **fputs( ),** and to write formatted data with **fprintf( )**.

It is also possible to search or move through files looking for particular data elements. When reading data from a file, the file pointer can be moved with the fseek command. For example, the following command will reposition the file pointer to the 8th byte in the text file:

```
fseek (pFile , 8 , SEEK_SET ); // move file pointer to 8 bytes from the start
```

The first term of the **fseek( )** function is the stream (the file pointer name), the second term is the pointer offset value and the third term is the origin for where the offset should be applied. There are three possible values for the origin as shown in Table B.8. The origin values have predefined names as shown.

Further details on the stdio commands and syntax can also be found in References B.1 and B.2.

Table B.8: fseek origin values

Origin value	Description
SEEK_SET	Beginning of file
SEEK_CUR	Current position of the file pointer
SEEK_END	End of file

B.11 Towards Professional Practice

The readability of a C program is much enhanced by good layout on the page or screen. This helps to produce good and error-free code, and enhances the ability to understand, maintain and upgrade it. Many companies impose a 'house style' on C code written for them; such guides can be found online.

The very simple style guide adopted in this book should be exemplified in any of the programs reproduced. It includes:

- Courier New font applied
- opening header text block, giving overview of program action
- blank lines used to separate major code sections
- extensive commenting
- opening brace of any code block placed on the line which initiates it
- code within any code block indented 2 spaces compared with code immediately outside the block
- closing brace stands alone, indented to align with line initiating code block.

Version control is essential practice in any professional environment. For clarity, we have not displayed version control information within the programs as reproduced in the book, although this was done in the development process. Minimal version control within the source code (spaced within the header text block) would include the date of origin, name of the original author, name of the person making the most recent revision, and date.

References

B.1. Prinz, P. and Kirch-Prinz, U. (2003). C Pocket Reference. O'Reilly.

B.2. The C++ Resources Network. www.cplusplus.com

B.3. ARM. RealView® Compilation Tools. Version 4.0. Compiler Reference Guide. December 2010. DUI 0348C (ID101213).

APPENDIX C

mbed Technical Data

C.1 Summarizing Technical Details of the mbed

The mbed is built around an LPC1768 microcontroller, made by NXP Semiconductors. The LPC1768 is, in turn, designed around an ARM Cortex-M3 core, with 512 kB flash memory, 64 kB RAM, and a wide range of interfaces, including Ethernet, universal serial bus (USB) device, Controller Area Network (CAN), serial peripheral interface (SPI), inter-integrated circuit (I^2C) and other input/output (I/O). It runs at 96 MHz.

Full technical details of the mbed and LPC1768 can be found in References 2.1 to 2.4. Summary mbed operating conditions are given directly below.

Package:

- 40-pin DIP package, with 0.1 inch spacing between pins, and 0.9 inches between the two rows.
- Overall size: 2 inches × 1 inch, 53 mm × 26 mm.

Power:

- Powered through the USB connection, or 4.5−9.0 V applied to VIN (see Figure 2.1).
- Power consumption < 200 mA (100 mA approx. with Ethernet disabled).
- Real-time clock battery backup input VB; 1.8−3.3 V at this input keeps the Real Time Clock running. This requires 27 μA, which can be supplied by a coin cell.
- 3.3 V regulated output on VOUT available to power peripherals.
- 5.0 V from USB available on VU, available only when USB is connected.
- Total supply is current limited to 500 mA.
- Digital I/O pins are 3.3 V, 40 mA each, 400 mA maximum total.

Reset:

- nR − Active-low reset pin with identical action to the reset button. Pull-up resistor is on the board.

C.2 LPC1768 Electrical Characteristics

To interface successfully to any microcontroller, certain electrical conditions must be satisfied. Many digital components are designed to be compatible with each other, so in many situations we do not need to worry about this. Once it comes to applying non-standard devices, however, it is very important to gain an understanding of interfacing requirements. Input signals have to lie within certain thresholds in order to be correctly interpreted as logic levels. We need also to understand the ability of an output to drive external loads that may be connected to it.

These operating conditions for the mbed are referred to in Chapter 3 and from time to time onwards. They are specified precisely, and in very great detail, by the manufacturer of any electronic component or integrated circuit, in the relevant datasheet. It is part of the skill of a professional design engineer to know where to access these details, to know how to interpret them and to be able to design to meet the criteria specified. The hobbyist can engage creatively or intuitively in some trial and error, and hope for the best. The professional needs to be able to predict analytically the performance of a design.

Inside the mbed is the LPC1768 microcontroller; when interfacing with the mbed we are actually directly interfacing with an LPC1768. We therefore turn to the LPC1768 datasheet (Reference 2.3). For a novice reader this is a very complex document. We have therefore extracted a very limited amount of data which contains some of the main details required. Some of these lead to the mbed figures quoted above; others indicate operating conditions within which the mbed figures apply.

C.2.1 Port Pin Interface Characteristics

Table C.1 is drawn mainly from Table 7 of Reference 2.3, and defines port pin operating characteristics. The format of Reference 2.3 is retained as far as possible, but for simplicity a few of the finer details of operation are removed. Each parameter that appears in the table is described here.

- **Supply voltage (3.3 V), $V_{DD(3V3)}$** – This is the power supply voltage connected to the $V_{DD(3V3)}$ pin of the microcontroller. Although nominally 3.3 V, the minimum acceptable supply voltage for this pin is seen to be 2.4 V, and the maximum 3.6 V. This pin is one of several supply inputs, and supplies the port pins.
- **LOW-level input current, I_{IL}** – This is the current flowing into a port pin, with pull-up resistor disabled, when the input voltage is at 0 V. This very low current implies a very high input impedance.
- **HIGH-level input current, I_{IH}** – This is the current flowing into a port pin, with pull-down resistor disabled, when the input voltage is equal to $V_{DD(3V3)}$. This very low current implies a very high input impedance.
- **Input voltage, V_I** – This indicates the legal range of input voltages to any port pin. Unsurprisingly, the minimum is 0 V. Interestingly, the maximum is 5 V, showing that the pin input can actually exceed the supply voltage. This is a very useful feature, and allows interfacing to a system supplied from 5 V.

Table C.1: Selected port pin characteristics

Symbol	Parameter	Conditions	Minimum	Typical	Maximum	Unit
$V_{DD(3V3)}$	Supply voltage (3.3 V)	External rail	2.4	3.3	3.6	V
I_{IL}	LOW-level input current	$V_I = 0$ V; on-chip pull-up resistor disabled	–	0.5	10	nA
I_{IH}	HIGH-level input current	$V_I = V_{DD(3V3)}$; on-chip pull-down resistor disabled	–	0.5	10	nA
V_I	Input voltage	pin configured to provide a digital function	0	–	5	V
V_O	Output voltage	output active	0	–	$V_{DD(3V3)}$	V
V_{IH}	HIGH-level input voltage		$0.7V_{DD(3V3)}$	–	–	V
V_{IL}	LOW-level input voltage		–	–	$0.3V_{DD(3V3)}$	V
V_{OH}	HIGH-level output voltage	$I_{OH} = -4$ mA	$V_{DD(3V3)} - 0.4$	–	–	V
V_{OL}	LOW-level output voltage	$I_{OL} = 4$ mA	–	–	0.4	V
I_{OH}	HIGH-level output current	$V_{OH} = V_{DD(3V3)} - 0.4$ V	−4	–	–	mA
I_{OL}	LOW-level output current	$V_{OL} = 0.4$ V	4	–	–	mA
I_{OHS}	HIGH-level short-circuit output current	$V_{OH} = 0$ V	–	–	−45	mA
I_{OLS}	LOW-level short-circuit output current	$V_{OL} = V_{DD(3V3)}$	–	–	50	mA
I_{pd}	Pull-down current	$V_I = 5$ V	10	50	150	μA
I_{pu}	Pull-up current	$V_I = 0$ V	−15	−50	−85	μA

- **Output voltage, V_O** – This indicates the range of output voltages that a port pin can source. The limits are 0 V and the supply voltage.
- **HIGH-level input voltage, V_{IH}** – This parameter defines the range of voltages for an input to be recognized as a Logic 1. Any input voltage exceeding ($V_{DD(3V3)}$ – 0.4) V is interpreted as Logic 1. In this context there is no need for a typical or maximum value, although the maximum is contained within the definition for V_I. V_{IH} is illustrated in Figure 3.1.

- **LOW-level input voltage, V_{IL}** – This parameter defines the range of voltages for an input to be recognized as a Logic 0. Any input voltage less than 0.4 V is interpreted as Logic 0. In this context there is no need for a typical or minimum value, although the minimum is contained within the definition for V_I. V_{IL} is illustrated in Figure 3.1.
- **HIGH-level output voltage, V_{OH}** – This indicates the output voltage, for a Logic 1 output. The value is defined for a load current of 0.4 mA, where the convention is applied that a current flowing out of a logic gate terminal is negative. Consider that a circuit such as Figure 3.3b applies. For this output current the minimum output voltage is ($V_{DD(3V3)}$ – 0.4) V. With no output current the output voltage will be equal to $V_{DD(3V3)}$. The values quoted imply an approximate output resistance of 100 Ω under these operating conditions.
- **LOW-level output voltage, V_{OL}** – This indicates the output voltage, for a Logic 0 output. The value is defined for a load current of 0.4 mA, where the convention is applied that a current flowing out of a logic gate terminal is positive. Consider that a circuit such as Figure 3.3c applies. For this output current the maximum output voltage is 0.4 V. With no output current the output voltage will be equal to 0 V. The values quoted imply an approximate output resistance of 100 Ω under these operating conditions.
- **HIGH-level output current, I_{OH}** – This gives the same information as the HIGH-level output voltage, V_{OH}.
- **LOW-level output current, I_{OL}** – This gives the same information as the LOW-level output voltage, V_{OL}.
- **HIGH-level short-circuit output current, I_{OHS}** – This gives the maximum output current if the output is at Logic 1, but is short-circuited to ground.
- **LOW-level short-circuit output current, I_{OLS}** – This gives the maximum output current if the output is a Logic 0, but is connected to the supply, $V_{DD(3V3)}$.
- **Pull-down current, I_{pd}** – This is the current which flows due to the internal pull-down resistor, when enabled, when $V_I = 5$ V.
- **Pull-up current, I_{pu}** – This is the current which flows due to the internal pull-up resistor, when enabled, when $V_I = 0$ V.

Table C.2: Selected limiting values

Symbol	Parameter	Conditions	Minimum	Maximum	Unit
$V_{DD(3V3)}$	Supply voltage (3.3 V)	External rail	2.4	3.6	V
V_I	Input voltage	5 V tolerant I/O pins; only valid when the $V_{DD(3V3)}$ supply voltage is present	−0.5	+5.5	V
I_{DD}	Supply current	Per supply pin	–	100	mA
I_{SS}	Ground current	Per ground pin	–	100	mA

C.2.2 Limiting Values

The values in the previous section showed the limits to which operating conditions could be taken, while maintaining normal operation. Limiting values (Table C.2), also called absolute maximum values, define the limits that must be maintained, otherwise device damage occurs. These must, of course, always be observed. For example, although a single port pin can supply up to 45 mA when short-circuited, one must also take note of the limiting value for supply and ground currents.

APPENDIX D

Parts List

To get started with the practical exercises in this book you will of course need an mbed. There is some flexibility in almost any other component or subsystem after that. The items listed below are the ones we used, sourced from the supplier shown; in most cases it is easy to substitute others. As is the way with electronics, some of these parts are likely to become obsolete and you will have to seek an alternative. Although we identify certain suppliers, there are many alternatives.

Table D.1: Components and devices used in the book

Description	Supplier	Supplier code
Chapter 2		
NXP mbed prototyping board LPC1768	www.rs-online.com	703-9238
Prototyping breadboard	www.rapidonline.com	34-0664 OR 34-0550
Jumper wire kit	www.rapidonline.com	34-0495 OR 34-0555
	Chapter 3	
LED, 5 V, red	www.rapidonline.com	56-1500
LED, 5 V, green	www.rapidonline.com	56-1505
Switch PCB mount SPDT	www.rapidonline.com	76-0200
Photointerrupter/Slotted Optoswitch	www.rapidonline.com	58-0944 OR 58-0303
Kingbright 7-segment display	www.rapidonline.com	57-0115
Switching transistor, ZVN4206	www.rapidonline.com	47-0162
Motor, 6 V, DC	www.rapidonline.com	37-0161
Battery box, 4×AA	www.rapidonline.com	18-2913 OR 18-2909
Diode, IN4001	www.rapidonline.com	47-3420
	Chapter 4	
Servo Hitec HS-422	www.active-robots.com	HS-422
Piezo transducer ('buzzer')	www.rapidonline.com	35-0200
	Chapter 5	
10k linear potentiometer	www.rapidonline.com	65-0715

Continued

Table D.1: Components and devices used in the book—cont'd

Description	Supplier	Supplier code
NORPS12 light-dependent resistor	www.rapidonline.com	58-0132
10k resistor	www.rapidonline.com	62-2146
LM35 temperature sensor	www.rapidonline.com	82-0240
	Chapter 6	
(No new items)		
	Chapter 7	
(two mbeds and breadboards required for some builds)		
Triple-axis accelerometer breakout, ADXL345	www.sparkfun.com	SEN-09836
4k7 resistor	www.rapidonline.com	62-2130
Push-button switch, red	www.rapidonline.com	78-0160
Push-button switch, green	www.rapidonline.com	78-0155
Digital temperature sensor breakout, TMP102	www.sparkfun.com	SEN-09418
Ultrasonic range finder, SRF08	www.rapidonline.com	78-1086
	Chapter 8	
PC1602 alphanumeric LCD display	www.rapidonline.com	57-0913
Nokia 6610 display	www.coolcomponents.co.uk	000147
	Chapter 9	
ICL7611 op amp	www.rapidonline.com	82-0782
	Chapter 10	
Breakout board for MicroSD transflash	www.sparkfun.com	BOB-00544
Transcend MicroSD card	www.rapidonline.com	19-9123
	Chapter 11	
1.5k resistor	www.rapidonline.com	62-2106
47k resistor	www.rapidonline.com	62-2178
4.7 nF capacitor	www.rapidonline.com	08-0995
10 μF capacitor	www.rapidonline.com	11-0840
0.1 μF capacitor	www.rapidonline.com	08-1020

Chapter 12		
RN-41 Bluetooth module (2 required)	www.sparkfun.com	WRL-10559
Ethernet RJ45 8-pin connector	www.sparkfun.com	PRT-00643
Ethernet RJ45 breakout board	www.sparkfun.com	BOB-00716
Chapter 13		
HMC6352 digital compass module	www.sparkfun.com	SEN-07915
Servo – continuous rotation	www.oomlout.co.uk	NA
MCP2551 CAN transceiver (2 required)	www.farnell.com	9758569
Chapter 14		
(No new items)		
Chapter 15		
Pololu m3pi robot with mbed socket	www.pololu.com	2151
WiFly GSX breakout RN-131C	www.sparkfun.com	WRL-10050
NXP mbed prototyping board LPC11U24	www.rs-online.com	751-0725

Note: Order codes and suppliers change very rapidly. Check that each code is correct for the part you wish to order, before placing the order.

APPENDIX E

The Tera Term Terminal Emulator

E.1 Introducing Terminal Applications

A terminal emulator allows a host computer to send or receive data from another computer, through a variety of links. Terminal emulators simply take the form of a software package running on the host computer which creates an image on the computer screen, through which settings can be selected. It then provides a context for data transfer, using the host computer keyboard for data input. Such a terminal is particularly useful for displaying messages from the mbed to a computer screen. It can become a very useful tool for user interfacing and software debugging. For example, status or error messages can be embedded into a program running on the mbed, and transferred to the terminal emulator when certain points in the program are reached.

While a number of terminal emulators are available, Tera Term is recommended by the mbed team. This is a free and open source terminal emulator which allows host PC communication with the mbed. In order to use Tera Term with a Windows PC, you will first have to install the Windows Serial Driver; this can be installed from the Handbook section of the mbed website, at time of writing http://mbed.org/handbook/Windows-serial-configuration. Note that you need to run the installer for every mbed, as Windows loads the driver based on the mbed serial number. The Tera Term support page can be viewed at http://logmett.com/

E.2 Setting up and Testing Tera Term

Open the Tera Term application. You will need to perform the following configuration:

- Select File → New Connection (or just press Alt+N).
- Select the *Serial* radio button and select the *mbed Serial Port* from the drop-down menu, as seen in Figure E.1.
- Click *OK*.

If *mbed Serial Port* is not in the drop-down menu, the Windows Serial Driver may not be installed.

Set up New-line format (to print out new line characters correctly) by following:

- Setup → Terminal ...
- Under 'New-line', set Receive to 'LF'.

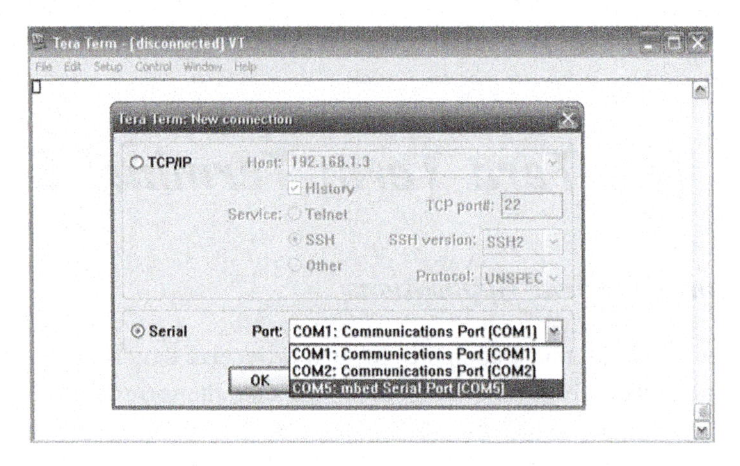

Figure E.1:
Setting up the Tera Term connection

To test Tera Term running with an mbed, create a new project of suitable name in the compiler, and enter the code of Program Example E.1. This code simply greets the world, and then reads your keyboard input and displays it to Tera Term. To show that the character value is being seen by the mbed, the program adds one to the ASCII code, meaning that every letter is echoed back to the terminal as the letter following it in the alphabet: b instead of a, c instead of b, and so on.

```
// Print to the PC, then pass back characters (slightly modified!)
*/
#include "mbed.h"
Serial pc(USBTX, USBRX);      // define transmitter and receiver
int main() {
  pc.printf("Hello World!");
  while(1) {
    pc.putc(pc.getc() + 1); //adds 1 to the ASCII code, and returns it
  }
}
```

Program Example E.1 Transferring keyboard characters to the screen

A more advanced program using terminal access is given in Program Example E.2. This is taken from the mbed site, and uses a pulse width modulated signal to increase and decrease the brightness of a light-emitting diode (LED). The pulse width modulation (PWM) is controlled from the keyboard and displayed on the Tera Term screen. To test the program, connect an LED between pin 21 and ground, and enable Tera Term.

```
/*Connects to the mbed with a Terminal program and uses the 'u' and 'd' keys to make LED1 brighter
or dimmer
*/
#include "mbed.h"
Serial pc(USBTX, USBRX);     // define transmitter and receiver
```

```
PwmOut led(p21);
float brightness=0.0;

int main() {
  pc.printf("Press 'u' to turn LED1 brightness up, 'd' for down\n\r");
  while(1) {
  char c = pc.getc();
  wait(0.001);
 if((c == 'u') && (brightness < 0.1)) {
    brightness += 0.001;
    led = brightness;
 }
 if((c == 'd') && (brightness > 0.0)) {
    brightness -= 0.001;
    led = brightness;
 }
  pc.printf("%c %1.3f \n \r",c,brightness);
  }
}
```

Program Example E.2 Controlling PWM setting from the keyboard

Index

Note: Page numbers with "f" denote figures; "t" tables

CPSIA information can be obtained at www.ICGtesting.com
Printed in the USA
BVOW07s0031260815

415044BV00004B/43/P